“All families and genera”

"All families and genera"

Exploring the Corpus of English Life Sciences Texts

Edited by

Isabel Moskowich
University of A Coruña

Inés Lareo
University of Vigo

Gonzalo Camiña
University of Nantes

John Benjamins Publishing Company

Amsterdam / Philadelphia

The paper used in this publication meets the minimum requirements of the American National Standard for Information Sciences – Permanence of Paper for Printed Library Materials, ANSI Z39.48-1984.

DOI 10.1075/z.237

Cataloging-in-Publication Data available from Library of Congress:
LCCN 2021020976

ISBN 978 90 272 0924 5 (HB)
ISBN 978 90 272 5962 2 (E-BOOK)

Cover image: Anabella Barsaglini-Castro and Inés Lareo.

John Benjamins Publishing Company · https://benjamins.com

Table of contents

About this book

The compilation of the Corpus of English Life Sciences Texts (CELiST) started in 2006 with the search for samples and the reading and researching needed prior to compilation itself. Although CELiST shared the structure of its sister-corpora withing the Coruña Corpus of English Scientific Writing, this still took some time. As with other subcorpora in the collection, the compilation and works on CELiST coexisted with those for the Corpus of History English Texts (CHET) and these, in turn, with the preparation for the open-access publication of the other two, Corpus of English Texts on Astronomy (CETA) and Corpus of English Philosophy Texts (CEPhiT). All of them can now be freely reached at the repository of the University of A Coruña (Spain), https://ruc.udc.es/dspace/handle/2183/21846.

The compilation of CELiST finished in 2016 but, for several reasons (and revisions), it was not published in this repository (https://ruc.udc.es/dspace/handle/2183/25720, DOI: https://doi.org/10.17979/spudc.9788497497848) until 2020. This was so as almost at the end of the whole process, some of the compilers decided to revise the role of women in the development of the Life Sciences in Britain to discover their importance had been greater than initially believed, especially in the field of Botany. Although this implied a revision of the criteria and conditions used for the forty samples in the collection (as well as deciding who to exclude in order to incorporate a woman writer), we will always think the effort of reconstruction of this particular corpus was worthwhile. The compilers counted for this project with the help of several research assistants to whom we would like to express our gratitude as well: Iria Bello Viruega, Paula Lojo Sandino, Luis Puente-Castelo and Estefanía Sánchez-Barreiro.

To complete this project, we received the help of several institutions. From 2005 the University of A Coruña provided the MuStE group with a research room and all the facilities of the institution and, more importantly, of its library and staff. We also received the help of several other institutions such as Instituto de Historia de la Medicina y de la Ciencia López Piñeiro (Valencia – Spain) where some interesting conversations took place. Besides, we would like to thank Berkeley University (Bancroft Library) for their kindness in giving us access to their collections. In Europe, we must express our gratitude to the libraries and staff of Humboldt Universität (Berlin), University College Cork (Ireland), RWTH Aachen Universität,

Ruprecht-Karls-Universität Heidelberg. Our visits to all these centres were possible thanks to the funding received from the Autonomous Government of Galicia and Deputación da Coruña.

Explorations of life sciences writing (1700–1900)

A preface

At the beginning of the last decade of the twentieth century, the ancient discipline of *philology* seemed to have run its course. University departments of philology still existed, as did journals with 'philology' in their titles; chairs in philology, in some of the more ancient foundations, still existed: academic institutions are notoriously slow to change their structures and labels. But the subjects and approaches pursued in these universities and publications, and by the researchers holding those positions, were very different from those envisaged by those who had founded or endowed these departments, journals and chairs. Some researchers focused on literary explorations, increasingly driven by developments in philosophy such as postmodernism, while others, more linguistically-inclined, turned to the austere descriptive formalisms that had emerged from the generative schools so dominant in the 1960s and 1970s.

However, these trends were changing as the century drew to a close, and as the new millennium approached different, exciting trends emerged. A new emphasis on interdisciplinary research could be observed, and the old insistence on disciplinary difference that had been dominant since the nineteenth century began to break down (institutional restructurings, albeit with mixed success, were one attempt to respond to these changes). In the linguistic sciences, an important development in qualitative research was the rise of pragmatics, which saw language as a social tool with social functions and thus increasingly focused on contextual factors in linguistic behaviour. And new tools, resulting from the digital revolution, allowed researchers in the humanities more broadly to handle very large bodies of data, and to adopt quantitative approaches in their enquiries.

However, for linguistic historians in particular it was crucial that the evidential basis of any findings derived from quantitative analysis was continually assessed. As a result, leaders in the field increasingly brought qualitative and quantitative approaches into articulation; as the late Matti Rissanen (Helsinki) used to aver, 'research begins when counting ends', and philology, which had always emphasised the connexion between language and context, began to be re-evaluated, and placed on a new theoretical and methodological footing. Terms such as 'the new philology', 'computational philology', and 'reimagined philology' began to appear in the scholarly literature, all with different but converging agendas.

An important element in recent corpus-studies has been the emphasis on focused datasets, where the significance of metadata is highlighted; simply assembling random bodies of material in an opportunistic fashion, as was commonly the case in the first phase of corpus-construction, is now deemed insufficient. The *Corpus of English Life Sciences Texts* (CELiST) is beautifully designed to allow the breaking of new ground in the historical study of English, and the accompanying volume, herewith presented, presents both the underpinning methodology for CELiST's construction, and examples of the new insights it enables.

There are many things that are really impressive about the current set of papers, but from my perspective I would draw attention to three things. First, the collection begins by setting out very clearly, and in exemplary fashion, the methodological underpinning for CELiST: something that in the first phase of corpus studies was (as I have just flagged) rather neglected. The research team has clearly thought through important issues of metadata, ensuring that any findings derived from the analysis are suitably robust. Secondly, in their choice of texts the team has drawn upon a sound knowledge of the history of science, choosing materials with a clear contemporary impact, but at the same time being aware of the axiom that 'every text has its own history', each with peculiarities derived from the dynamically-shifting social conditions within which authors worked. In this regard, their work can be related to the latest trends in book history. As a result – my third point – it is possible for the investigators to identify and – crucially – explain important features of the material under review, hitherto occluded. I have been impressed, for instance, by how the discourse of female scientists is subjected in several of these papers to delicate analysis, and accounted for in terms of the social structures of the discourse communities for whom they catered.

I am honoured to have been asked to write these short prefatory remarks for this excellent set of papers, and recommend them wholeheartedly for anyone interested in English historical linguistics, in book history, and in the history of science.

Jeremy J. Smith
University of Glasgow
December 2020

List of contributors

Elena Alfaya-Lamas

Elena Alfaya-Lamas is deputy dean for the Faculty of Arts and Documentation of the University of A Coruña. She obtained an MA in Germanic Philology from the University of Santiago de Compostela and a PhD in English Linguistics in 2002 in A Coruña. Her dissertation dealt with the geographical localization of several homonymous Mediaeval English manuscripts of the *Cursor Mundi*. From 1998 to 2000 she was a postgraduate worker and scholarship holder in the Department of Linguistics and English Language of the University of Edinburgh. In November 2001 she became an associate lecturer at CESUGA-University College Dublin. She obtained a position as a post-doctoral researcher at the University of A Coruña where she became a lecturer and researcher in the area of Information and Documentation Science in 2004. Her main research interests are historical linguistics, cognitive linguistics, discourse analysis, gender studies and mind-consciousness studies. She is a member of the Mindfulness Association. She teaches Informational Behaviour, Archivistics, Records Management and Scientific Research Techniques.

Francisco Alonso-Almeida

Francisco Alonso-Almeida is full professor of English Philology in the Departamento de Filología Moderna, Traducción e Interpretación, Universidad de Las Palmas de Gran Canaria. His research interests are (historical) discourse analysis and (historical) pragmatics. He has published on these matters in such international peer-reviewed journals as *Neophilologus, Intercultural Pragmatics, Research in Language, Discourse Studies, Journal of Historical Pragmatics*, and *Folia linguistica historica*, among others. Another focal area of research concerns medieval manuscript studies and editorial work. His publication of a volume on Middle English *materia medica* in the *Middle English Texts* collection (Carl T. Winter) is one example. He is currently working very actively on modality from a diachronic perspective and the notion of evidentiality. He also collaborates on a regular basis with the MuStE research group, University of A Coruña, and with the Department of Applied Linguistics, University Polytechnic Valencia.

Francisco J. Alvarez-Gil

Dr. Francisco J. Álvarez-Gil is a lecturer of English language and linguistics in the Departamento de Filología Moderna, Traducción e Interpretación, Universidad de Las Palmas de Gran Canaria. He has delivered talks on linguistic aspects of, say, Modern English scientific writing in international conferences, e.g. *XVIII Congresso Internazionale dell'Associazione Italiana de Linguistica Applicata, 20th International Conference on English Historical linguistics*, and *Colloque International de Linguistique de Corpus*. He has been a visiting scholar in some reputed Universities with leading departments in English for Specific Purposes (ESP). Some of these are the Université Jean Moulin Lyon III (France) and L'Università di Modena e Reggio Emilia (Italy), among others. His research interests primarily concern the analysis of specialized English from a synchronic and diachronic perspective, corpus linguistics, and language pedagogy at tertiary level, and he has published in this respect in international peer-reviewed journals.

Anabella Barsaglini-Castro

Anabella Barsaglini-Castro graduated in English Philology at the University of A Coruña in 2016, and obtained an MA in English Language and Linguistics in 2017. Since then, she has been collaborating with the MuStE group in the compilation of the *Coruña Corpus of English Scientific Writing* and also participating in some international conferences. In 2019, she stayed as a visiting scholar at the University of Liverpool (UK), where she completed the compilation of her own corpus on *Posthumanism English Texts (PET)*. She is currently working on her PhD dissertation, which deals with the study of persuasion in contemporary fiction and non-fiction texts, and also collaborating with some teaching tasks in the English Department of the aforementioned University.

Magdalena Bator

Magdalena Bator obtained her PhD from the University of Adam Mickiewicz in Poznań and D.Litt. from the University of Silesia in Katowice. She works as a university professor at the WSB University in Poznań, where she teaches linguistics. Her research interests cover the history of English, historical semantics, the structure and typology of culinary recipes, language contact and borrowing. She has published on foreign influences on English (mostly medieval), semantic change, (English, German and Polish) historical recipes, linguistic synaesthesia, and English influence on modern Polish.

Iria Bello Viruega
Iria Bello Viruega was awarded her PhD in English Studies at the University of A Coruña (Spain). Currently she works at the Departamento de Filología Española, Moderna y Clásica at the Universitat de les Illes Balears (Spain). Her field of specialization covers topics related to the history of the English Language and L2 acquisition using several empirical methodologies.

Gonzalo Camiña
Gonzalo Camiña is a former member of the research group MuStE. He is responsible for designing and implementing the XML coding of the Coruña Corpus, and has co-edited and co-authored three of the subcorpora published so far by MuStE: astronomy and philosophy (the core of his PhD dissertation), and the current volume in life sciences. After two years as a teaching scholar at University of A Coruña, he has taught at University College Cork (Ireland), University of Ulster (UK) and since 2016 he has been teaching scientific and professional English at the Faculty of Sciences and Techniques at the University of Nantes (France). His fields of interest are word-formation in scientific English, computer science, and the combination of linguistics and computing. He is now engaged in developing online tools and contents for Second Language Acquisition in the domain of science, some of which can be viewed on his website caminha.org.

Begoña Crespo
Dr Begoña Crespo is a full professor at the University of A Coruña (Spain). She has published on lexicon enlargement in the History of English, genre/text-type taxonomy, the diachrony of English scientific writing and the discourse of science of female authors. She is currently supervising some PhD dissertations on various aspects of the language of science (conditionals, nominalisations, passive voice). She has published in journals such as *Studia Neophilologica, Nordic Journal of English Studies, Linguistica Pragensia, Neuphilologische Mitteilungen, English Studies, Anglica, SAP and Literary and Linguistic Computing*, among others, as well as book chapters in D.S Brewer, Rodopi, Peter Lang, Lincom Europa or John Benjamins.

María José Esteve-Ramos
María José Esteve-Ramos is a senior lecturer at University Jaume I in Castelló (Spain), where she teaches diachronic linguistics, Old and Middle English. She collaborates as a full member of MuStE since 2006, and is currently working in the *Coruña Corpus Project*. She has supervised a PhD on lexicology, and is supervising a thesis on medieval English. She is a member of the Board of the *Catalan Society for the History of Science* and of the *Spanish Society for Medieval English Language*

and Literature. Also, she has been director of IULMA (Instituto Interuniversitario de Lenguas Modernas Aplicadas) for six years and has been a Vice Dean of the Humanities Faculty until July 2018. She has been visiting the ASNC department in Cambridge as Honorary Researcher in 2016 and has spent a sabbatical year at the University of Glasgow in 2019–20. Her main research interests are the evolution of scientific English and its specialised lexicon, medieval texts and manuscripts, corpus linguistics and etymology.

Inés Lareo

Dr. Inés Lareo is currently a lecturer in the Department of English, French and German Philology of the University of Vigo. She previously worked as a researcher in the English Department of the University of A Coruña. She has also carried out research as a visiting scholar at the universities of Köln (Germany), Berkeley (USA), Aachen (Germany), Castelló (Spain), Cork (Ireland), Berlin (Germany), Valencia (Spain), Las Palmas (Spain) and Heidelberg (Germany). Her main research interests are lexicography, historical linguistics, corpus linguistics, and language teaching methods. As a member of the MuStE team that is compiling the *Coruña Corpus of English Scientific Writing* (CC), (https://www.udc.es/grupos/muste/), she has also focused her attention on grammatical and sociolinguistic aspects of the use of collocations in English scientific writing. She has already published a variety of articles on these topics in well-known journals and publications.

Margarita Mele Marrero

Margarita Mele Marrero is a tenured senior lecturer of the English and German Philology Department at the Universidad de La Laguna (ULL) since 2002. Her research interests have been mainly centred on English Historical Linguistics, being her most relevant publications on the Old and Early Modern English periods (2002, 2011, 2014, 2015, 2017, 2019). These have been approached from an initial Philological perspective moving to a more recent Pragmatics (stance and evidentiality) theoretical frame. She has collaborated with different research projects, being presently involved in the compilation of the *Coruña Corpus of English Scientific Writing* Coruña Corpus. She is also an active member of the interdisciplinary Institute IEMyR (Instituto de Estudios Medievales y Renacentistas at ULL) since its beginnings in 1990.

Dr. Mele serves in the editorial board of *LFE, Revista de Lenguas y Lingüística Aplicadas* and she is currently the Secretary of RCEI.

Leida Maria Monaco
Leida Maria Monaco obtained her doctorate in English Linguistics from the University of A Coruña with the MuStE team. Part of her research was carried out in the Université de Bretagne Occidentale and Lancaster University, where she studied modality and statistics applied to corpus linguistics. Later on, she worked as a predoctoral teaching assistant and, later on, as a research assistant in the English Department of her home institution. At this moment, she is an assistant lecturer in the University of Oviedo.

Isabel Moskowich
Isabel Moskowich is full professor of English Philology at the University of A Coruña where she leads the MuStE group. Her research covers different aspects of the History of English, mainly the Scandinavian lexical elements and sociolinguistic implications during the Middle Ages. However, since 2003, she has been involved in the compilation of the *Coruña Corpus of English Scientific Writing* and the study of the origins of ESP from a socio-pragmatic approach. She has published extensively in international journals such as *Folia Linguistica Historica, SELIM, Neuphilologische Mitteilungen, Studia Anglica Posnaniensia, Studia Neophilologica, Literary and Linguistic Computing, Australian Journal of Linguistics* or *English Studies*, and has edited or co-edited different volumes. She has been responsible for several funded research projects.

Elisa Narváez García
Elisa Narváez García holds a MA in Estudios Ibéricos e Iberoamericanos awarded by the Katholieke Universiteit Leuven (Belgium) and is a PhD candidate at the Institut für Übersetzen und Dolmetschen (Ruprecht-Karl Universität Heidelberg, Germany). She works as assistant editor at the Boletín de Lingüística (Universidad Central de Venezuela).

Luis Puente-Castelo
Luis Puente-Castelo (A Coruña, 1987) obtained a BA in English Philology (2009) and an MA in Secondary School Teaching (2010) from the University of A Coruña. He obtained his PhD in June 2017 with a dissertation about the use of conditional structures in English Scientific Writing from 1700 to 1900. He has been a visiting postdoctoral scholar at the University of Birmingham since from 2018 to mid 2021. His main research interests are corpus linguistics, discourse analysis, and scientific and pseudoscientific writing. Other research interests are new technologies, (social) media, and the popularisation of scientific knowledge.

CHAPTER 1

The making of the Corpus of English Life Sciences Texts (CELiST), a bunch of disciplines

Isabel Moskowich
University of A Coruña

1. Introduction: Biology or something else?

The Corpus of English Life Sciences Texts (CELiST) (Lareo *et al.* 2020) was originally conceived of as a collection of texts on Biology dating from the eighteenth and nineteenth centuries. However, it soon became clear that this was not a simple undertaking, given that the discipline of Biology did not exist as such when the works we intended to compile were written and published. The present chapter, then, aims to describe how CELiST developed and to give an account of its nature and place within the Coruña Corpus of English Scientific Writing (CC),[1] as well as to provide a description of its constituent texts.

The first recorded use of the term *biology* seems to date from 1686, this in a non-scientific sense, probably as the result of one of those strange new word-formation processes through which a new lexeme is obtained from the combination of two forms or "pieces of words" rather than from two complete ones. Such new formations are often found in the literature as (neo)classical compounds (Adams 1988; Kastovsky 2009) and are directly related to the *inkhorn* controversy (Millward and Hayes 2012: 226). According to the *Oxford English Dictionary* (OED, henceforth), this 1686 use is an early, isolated one, referring to "A biographical history of a person, place, etc.; a biography. Later also: the study of human life, character, or society. Now rare." It is not until the very end of the eighteenth century that the term is used to refer to what we understand nowadays as biology, defined by the OED as:

> The branch of science that deals with living organisms as objects of study, apart from any utilitarian value they may have, and now comprising more specialized disciplines such as zoology, botany, and bacteriology, chemical, human, marine, molecular, population, radiation biology, etc.: see the first element.

---

1. A complete account of the CC can be found at https://www.udc.es/grupos/muste/.

https://doi.org/10.1075/z.237.01mos

We can infer such a meaning in the following example, from the end of the eighteenth century:[2]

> Physiology therefore – or more strictly biology – by which I mean the doctrine of the living system in all its states, appears to be the foundation of ethics and pneumatology. (Beddoes 1799: 4)

This different taxonomy of the fields of science was one of the reasons why, when we began selecting samples in 2006, our searches in libraries and catalogues using the label *biology* for the period 1700–1800 resulted in all manner of texts, writings that we would never have classified as Biology from our twenty-first-century perspective. Thus, we found texts dealing with anatomy, descriptions of animal life, and the characteristics of plants. This was not completely unexpected, since something similar had occurred when collecting samples for the Corpus of English Texts on Astronomy, CETA (Moskowich *et al.* 2012). In that case, we soon came to realise that Astronomy was not seen as an independent field of knowledge in Late Modern English but was inextricably linked to Astrology, especially in the first period of our survey. In the same way, the keyword *biology* proved to be inadequate as a means of searching for the texts in the domain we intended to compile. Therefore, as already mentioned in several previous descriptions of the CC (Moskowich and Crespo 2007: 244; Crespo and Moskowich 2010; Puente-Castelo 2017; Moskowich 2019), we opted for an inclusive approach. A thorough revision of the UNESCO list of the fields of science and technology (1988) revealed that we needed to use a number of different labels in searching for texts. The UNESCO divides science and technology into six different fields, the first of which relates to the Natural Sciences.

Table 1 below sets out fields I and IV of the UNESCO taxonomy, the former called Natural Sciences and the latter with no specific name. Both were used for our compilation of CELiST.

Table 1. Fields of science and technology relating to Biology in the UNESCO taxonomy (UNESCO 1988)

I. Natural sciences
Astronomy, *bacteriology, biochemistry, biology, botany*, chemistry, *entomology*, geology, geophysics, mathematics, meteorology, mineralogy, computing, physical geography, physics, *zoology* and other allied subjects.
IV.
Agronomy, zootechnics, *fisheries*, forestry, *horticulture, veterinary medicine* and other allied subjects.

2. The term *biology* as such does not occur in the samples in CELiST. Related terms do not occur until the nineteenth century, thus coinciding with the dating of the term in the OED. The corpus contains two uses of *biological* (Spencer 1867 and Galton 1889) and one of *biologist* (Packard 1898).

The disciplines in italics are those for which we eventually found samples. They are certainly varied, although related to one another in some way, at least in that they all deal with the study of life and organisms. For this reason, we decided to group them all under the broader term Life Sciences.

Although we used this UNESCO classification as a starting point for our selection of texts in the different domains, we have not in fact followed it faithfully, since some of the divisions it proposes do not accurately reflect Late Modern notions of knowledge. Proof of this is that we have been able to select texts from disciplines in fields I and IV, and to group them together in one single, cohesive category. Also, as already noted, some other epistemological approaches consider there to be just two branches of the Natural Sciences, these being Life and Physical Sciences, and that they complement each other in the sense that the latter is concerned with non-living matter. We think our use of the term Life Sciences constitutes a better fit with the western tradition initiated by Alcmaeon of Croton in the fifth century BC (Malpas 2012), who, it has been argued, should be considered a philosopher rather than anything else (Barnes 1982). However, Alcmaeon was at once a biologist, anatomist, physician and philosopher, given that fields of knowledge in Classical times were not divided and specialised in the way that they are today. Indeed, such specialisation is very recent. Our twenty-first century prejudices lead us to think of people as either one thing or another, particularly when two areas of activity fall into what we consider opposite or unconnected fields of knowledge. Yet as late as the Renaissance, someone could be a painter, sculptor, anatomist and inventor simultaneously, a case in point being Leonardo da Vinci. Thus, there has been a long tradition, of some 2,000 years (if not longer), in which present-day epistemic compartments simply did not exist. Besides this, Naturalism is a Renaissance trend that values, enjoys and observes Nature, and this is something that would also last. With the arrival of Empiricism, the conjunction of nature and philosophy in what was termed Natural Philosophy came to its highest expression when authors such as David Hume argued for the need to base knowledge on the observation of natural phenomena and to simplify them as much as possible (Hume 1748: 4.1).

As compilers, we could have chosen to talk about *Natural Philosophy* rather than *Life Sciences*, but we felt that the former was somehow wider in its scope, in the sense that it could also refer to those aspects of nature that are non-living entities. That is, we considered that the term *Natural Philosophy* comprised both natural and physical sciences. It is true that some of the authors whose work we have included in the corpus are described as natural philosophers, but the works or parts of works chosen for compilation deal with living beings, although these same authors could occasionally write about other aspects of Natural Philosophy. The Late Modern period witnessed the birth of this new taxonomy of science, in which we can observe that scientists themselves are conscious of more specialised branches, although to nowhere near the extent of our present-day conceptions of this. The Scientific Revolution and the emergence of objectivity that accompanied it (Harari 2014) would not have taken place without the unification of humankind, that is, without the gradual consolidation of political organisations towards one global empire.[3] All these factors meant that new epistemological patterns appeared within a new tradition. It is this tradition that can be seen in the Corpus of English Life Sciences Texts, which will be described in what follows.

2. The structure of CELiST

The Corpus of English Life Sciences Texts (CELiST), as part of the Coruña Corpus is a specialised corpus since it is "delimited by a specific register, discourse domain or subject-matter" (De Beaugrande 2001: 11), that of Life Sciences. Also, as with other subcorpora within the CC, it has been compiled following strict principles regarding its shape, size and structure. In terms of size, we cannot say that it is small, in that corpora containing between 20,000 and 200,000 words are generally considered to be small (Aston 1997). As with its twin subcorpora in the CC, CELiST comprises two 10,000-word samples per decade, totalling 400,305 words. So, in this sense CELiST is indeed not small.[4] The texts are more or less equally distributed in the eighteenth (200,220 words) and nineteenth (200,085 words) centuries, thus guaranteeing balance.

Another characteristic of CELiST is that each text sample is accompanied by a file containing information about the author and the sample itself. Such information can be used to carry out searches, and is also shown in the Coruña Corpus Tool interface. More importantly, metadata files provide data for socio-historical contextualisation and can be used for linguistic research. Information from the whole CELiST that can be extracted from the two sections of metadata files is described in the following paragraphs.

3. Harari (2014) also claims that imperialism and science go hand in hand. He mentions Joseph Banks and Darwin (both in CELiST) and Green (in CETA) and establishes a close relationship between capitalism and Scientific Revolution.

4. It might be useful here to take into account Flowerdew's claim that "It may also be misleading to always equate *small* with *specialized* for the reason that while the notion of *small corpora* entails the notion of *specialized* (…) the term *specialized corpora* does not necessarily imply that the corpora in question are small" (Flowerdew 2004: 19–20).

3. The authors in CELiST

Following a further compilation principle, no author is ever repeated in any of the subcorpora of the CC. This means that the writers included in CELiST are not to be found in any other subcorpora, even though they may have written and published in other disciplines. The fact that we collect approximately the same amount of words from each author means that no individual linguistic idiosyncrasy from any one writer will ever skew the corpus.

Table 2. Authors and their works in CELiST

Date	Author	Title
1707	Douglas, James	*Myographiæ comparatæ specimen: or, a comparative description of all the muscles in a man and in a quadruped. … To which is added an account of the muscles peculiar to a woman. With an etymological table* … London printed by W. B. for G. Strachan.
1707	Sloane, Hans	*A Voyage to the islands Madera, Barbados, Nieves St Christophers and Jamaica; with the Natural History of the Herbs and trees, four-footed Beasts, Fishes, Birds, Insects, Reptiles, &c. of the last of those Islands; To which is prefix'd an introduction, wherein is an account of the inhabitants, air, waters, diseases, trade, &c. of that place, with some relations concerning the neighbouring continent and islands of America.* In Two Volumes. Vol. I. London: printed by M. B. for the author.
1717	Keill, James	*Essays on several parts of animal oeconomy.* Essay IV: Of Animal Secretion. London: printed for George Strahan.
1720	Gibson, William	*The Farriers new Guide: containing first, the anatomy of a horse, being an exact and compendious discription of all his parts; with their actions and uses: illustrated with figures curiously engrav'd on copper plates. Secondly, an account of all the diseases incident to horses, with their signs, causes, and methods of cure; wherein many defects in the farriers practice, are now carefully supply'd, their errors expos'd and amended, and the art greatly improv'd and advanc'd, according to the latest discoveries. The whole interspers'd with many curious and useful observations concerning feeding and exercise, &c.* London: printed for William Taylor.
1723	Blair, Patrick	*Pharmaco-Botanologia: or, an alphabetical and classical dissertation on all the British indigenous and garden plants of the new London Dispensatory. In which their genera, species, characteristik and distinctive notes are methodologically described; the botanical terms of art explained; their virtues, uses, and shop-preparations declared. With many curious and useful remarks from proper observation.* Decad I. London: printed for G. Strahan; W. and J. Innys; and W. Mears.

(*continued*)

Table 2. (*continued*)

Date	Author	Title
1730	Boreman, Thomas	*A description of three hundred animals; viz. beasts, birds, fishes, serpents, and insects. With a particular account of the whale-fishery. Extracted out of the best authors, and adapted to the use of all capacities; especially to allure children to read. …* London: printed by J. T. for Rich. Ware; Tho. Boreman, and Tho. Game.
1737	Blackwell, Elizabeth	*A Curious Herbal, containing five hundred cuts, of the most useful plants, which are now used in the practice of Physick. Engraved on folio copper plates after drawings, taken from the LIFE. To which is added a short description of ye plants and their common uses in PHYSICK.* In Two Volumes. Vol. I. London: printed for Samuel Harding.
1737	Brickell, John	*The Natural History of North-Carolina. With an account of the trade, manners, and customs of Christian and Indian inhabitants. Illustrated with copper-plates, whereon are curiously engraved the map of the country, several strange beasts, birds, fishes, snakes, insects, trees, and plants, &c.* Dublin: James Carson. For the author.
1743	Edwards, George	*A Natural history of uncommon birds. And of some other rare and undescribed animals, quadrupedes, fishes, reptiles, insects, &c. Exhibited in two hundred and ten copper-plates, from designs copied immediately from Nature, and curiously coloured after life. With a full and accurate description of each figure.* In Four Parts. Part I. London: printed for the author, at the College of Physicians in Warwick-Lane.
1750	Hughes, Griffith	*The Natural History of Barbados.* In Ten Books. London: printed for the author and sold by most booksellers in Great Britain and Ireland.
1752	Dodd, James Solas	*An Essay towards a Natural History of the Herring.* London: printed for T. Vincent. Near.
1758	Borlase, William	*The Natural History of Cornwall. The Air, Climate, Waters, Rivers, Lakes, Sea and Tides; Of the Stones, Semimetals, Metals, TIN, and the Manner of Mining; The Constitution of the Stannaries; Iron, Copper, Silver, lead, and Gold, found in Cornwall. Vegetables, Rare Birds, Fishes, Shells, Reptiles, and Quadrupeds: Of the Inhabitants, Their Manners, Customs, Plays or Interludes, Exercises, and Festivals; the Cornish Language, Trade, Tenures, and Arts.* Oxford: printed for the author by W. Jackson.
1766	Pennant, Thomas	*The British Zoology. Class I. Quadrupeds. II. Birds …* Vol. I. London: printed by J. and J. March for the society.
1769	Bancroft, Edward	*An essay on the Natural History of Guiana, in South America. Containing a description of many curious productions in the animal and vegetable systems of that country. Together with an account of the religion, manners and customs of several tribes of its Indian inhabitants. Interspersed with a variety of literary and medical observations. In several letters from a Gentleman of the Medical Faculty during his residence in that country.* London: printed for T. Becket and P. A. Hondt.

Table 2. (*continued*)

Date	Author	Title
1774	Goldsmith, Oliver	*An History of the Earth, and animated Nature:* In Eight Volumes. Vol VIII. London: printed for J. Nourse.
1776	Withering, William	*A botanical arrangement of all the vegetables, naturally growing in Great Britain. With the descriptions of the Genera and species, according to the system of the celebrated Linnaeus. Being an attempt to render them familiar to those who are unacquainted with the learned languages. Under each species are added, the most remarkable varieties, the natural places of growth, the duration, the time of flowering, the peculiarities of structure, the common English names; the names of Gerard, Parkinson, Ray and Baubine. The uses as medicines, or as poisons; as food for men, for brutes, and for insects. With their applications in oeconomy and din arts, with an easy introduction to the study of botany. Shewing the method of investigating plants, and directions how to dry and preserve specimens.* In Two Volumes. Vol. I. Birmingham: printed by M. Swinney for T. Cadel and P. Elmsley and G. Robinson.
1786	Speechly, William	*A Treatise on the Culture of the Pine Apple and the Management of the Hot-house. Together with a Description of every Species of Insect that infest Hot-houses, with effectual Methods of destroying them by William Speechly. To which is added A method to preserve peach and nectarine trees from mildew &c. by Robert Browne.* With plates. Book I. Dublin: printed for Luke White.
1789	Bolton, James	*An History of Fungusses, growing about Halifax. With forty-six copper-plates; or which are engraved sixty-four species of funguses, Including the Seven following GENERA, viz. CLATHRUS, HALVELLA, PEZIZA, CLAVARIA, LYCOPERDON, SPHAERIA, and MUCOR. Wherein their various appearances in the different stages of growth, are faithfully exhibited in about three hundred figures, copied with great care from the PLANTS, when newly gathered and in a state of perfection. With a particular DESCRIPTION of each SPECIES, in all its stages. From the first appearance to the utter decay of the plant; with the time when they were gathered; the soil and situation in which they grew; their duration; and the particular places mentioned, where all the new or rare species were found. The whole being a plain recital of FACTS, the result of more than twenty years observation.* In Three Volumes. Vol. III. Huddersfield: Printed by J. Brook for the author.
1794	Donovan, Edward	*Instructions for collecting and preserving various subjects of natural history; as animals, birds, reptiles, shells, corals, plants, &c. Together with a treatise on the management of insects in their several status: selected from the best authorities.* London: Printed for the Author.
1795	Smith, Sir James Edward	*English Botany; or coloured Figures of British Plants with their essential Characters, Synonyms, and Places of Growth.* In Thirty Six Volumes. Vol. IV. London: printed for the author by J. Davis.

(*continued*)

Table 2. (*continued*)

Date	Author	Title
1804	Jacson, Maria Elizabeth	*Botanical Lectures. By a Lady. Altered from "Botanical Dialogues for the use of schools", and adapted to the use of persons of all ages, by the same author*. London: printed for J. Johnson by T. Bensley.
1808	Wilson, Alexander	*American Ornithology; or, The Natural History of the Birds of the United States. Illustrated with plates and colored from the original drawings taken from Nature*. In Nine Volumes. Vol. I. Philadelphia: published by Bradford and Inskeep.
1816	Wakefield, Priscilla	*An introduction to the Natural History and Classification of Insects, in a series of familiar Letters*. With Illustrative Engravings. London: printed for Darton, Harvey and Darton.
1819	Lawrence, William	*Lectures on Physiology, Zoology, and the Natural History of Man, delivered at the Royal College of Surgeons*. London: printed for J. Callow, Medical bookseller.
1824	Jenner, Edward	*Some Observations on the Migration of Birds. By the late Edward Jenner, M.D. F.R.S.; with an Introductory Letter to Sir Humphry Davy, Bart. Pres. R.S. By the Rev. G. C. Jenner*. Philosophical Transactions of the Royal Society of London.
1828	Godman, John Davidson	*American Natural History*. In Three Volumes. Vol. III. Philadelphia: Carey Lea and Carey.
1832	Lincoln, Almira Hart Phelps	*Familiar Lectures on Botany, including practical and elementary Botany, with generic and specific Descriptions of the most common native and foreign Plants and a Vocabulary of botanical Terms. For the use of Higher schools and academies*. Hartford: J. F. Huntington.
1835	Jardine, Sir William	*The Naturalist's Library. In Forty Volumes. Mammalia Vol. III. Ruminantia. Part I. The Natural History of the Ruminating Animals containing Deer, Antelopes, Camels, &c. Illustrated by thirty-five Plates; with memoir and portrait of Camper*. Edinburgh: W. H. Lizards. London: Samuel Highley. Dublin: W. Curry.
1840	Pratt, Anne	*Flowers and their Associations*. London: Charles Knight and Co.
1848	Dalyell, Sir John Graham	*Rare and remarkable Animals of Scotland, represented from living Subjects: with practical Observations on their Nature*. In Two Volumes. Vol. II. London: John van Voorst.
1859	Agassiz, Elizabeth	*A First Lesson in Natural History*. Boston: Little, Brown and Company. London: Sampson Low, Son and Company.
1859	Darwin, Charles Robert	*On the Origin of Species by means of Natural Selection, or the Preservation of favoured Races in the Struggle for Life*. London: John Murray.
1863	Huxley, Thomas Henry	*On the Origin of Species, or, The Causes of the Phenomena of organic Nature. A Course of six Lectures to working Men*. New York: Appleton and Company.

Table 2. *(continued)*

Date	Author	Title
1867	Spencer, Herbert	*The Principles of Biology*. In Two Volumes. V. II. London/Edinburgh: Williams and Norgate.
1876	Macalister, Alexander	*An introduction to animal morphology and systematic Zoology*. London: Longmans, Green, and Co.
1879	Lankester, Phebe	*Wild Flowers worth Notice: A Selection of some of our Native Plants which are most attractive from their Beauty, Uses, or Associations*. London: David Bogue.
1880	Balfour, Francis M	*A Treatise on comparative Embryology*. In Two Volumes. Vol. I. London: Macmillan & Co.
1889	Galton, Francis	*Natural Inheritance*. London/New York: Macmillan & Co.
1895	Gregory, Emily Lovira	*Elements of Plant Anatomy*. Boston/London: Published by Ginn and company.
1898	Packard, Alpheus Spring	*A Text-book of Entomology including the Anatomy, Physiology, Embryology and Metamorphoses of Insects. For Use in agricultural and technical Schools and Colleges as well as by the working Entomologist*. New York/London: Macmillan & Co.

As can be seen in Table 2, there are twenty samples in the corpus to represent each century. In the following pages we will see how these samples are distributed in terms of sex of the authors and the places where they acquired their linguistic habits and style.

3.1 The sex of the authors

As widely noted (Whaley 2003; Watts 2007; Freire-Picos 2014; Crespo 2015), women were not readily welcome in learned circles outside the private spheres in this period. Taking the writer Margaret Cavendish, for instance, it is worth recalling that she was not a member of any of the existing societies of the time, nor was she a student or lecturer in a university (Rees 2003; Fyfe and Røstvik 2018) or that Mary Anning (1799–1847) was not allowed to the meetings where other (male) scientists discussed her discoveries on Palaeontology. This androcentric world notwithstanding, we have also seen that the Life Sciences constitute a field in which women were not regarded as complete strangers. The field was certainly wide, and our compiled texts include the description (and cooking) of herrings (in James Dodd's 1752 work *An Essay towards a Natural History of the Herring*) as well as texts on anatomy and other more technical aspects of biology (see Chapters 3 and 4), and women were welcome to write about certain topics that fell within the scope of horticulture or gardening. However, gardening and walks in the nature often provide women with the opportunity to produce observational texts which were a great deal more learned than mere diaries. Such is the case with Elizabeth Blackwell's herbal (1737), Maria Jacson's botanical lectures (1804), and Priscilla Wakefield's classification of insects (1816).

The 400,305 words in CELiST represent both male and female writers, but as Figure 1 below shows, only 20% of the total words were produced by women.

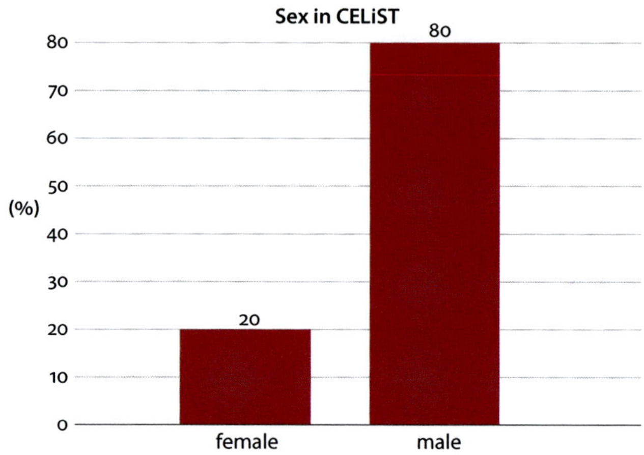

Figure 1. Overall distribution of words in CELiST by sex

We have, then, sought to reflect real life when compiling the corpus. After the disappearance of monasteries and nunneries in England under Henry VIII, women had nowhere to receive a formal education, and thus the Modern period was an especially poor one in this sense. It also exerted some influence on attitudes to women at the time. According to Whaley (2003) they were either considered as mentally and socially inferior to men, as equal but different, or as potentially equal in both their mental ability and their potential contribution to society. However, they were excluded from science and from formal education, on the prevailing grounds that women were emotional and incapable of objective reasoning[5] (Watts 2007: 63). One of the consequences of such views and ideas finds its echo in the CELiST. Women were discouraged from learning about the botanical classification proposed by Linnaeus because it was based on sexual features and reproduction, which might thus expose women to immoral lessons from nature. Taking all these factors into account, we opted to include only one sample of female writing in our eighteenth-century

5. On the analysis of linguistic behaviour supposedly caused by these female features, see Crespo and Moskowich (2015) on involvement and Monaco (2016) on the expression of abstract thinking.

subcorpus, a herbal by Elizabeth Blackwell, although in it she focuses more on the medical applications of herbs than on their morphological description.

For the following period, 1800 to 1900, we collected seven samples by female authors (of the twenty samples for this century) as a reflection of what was happening in society at the time. Exact word counts are set out in Figure 2 below:

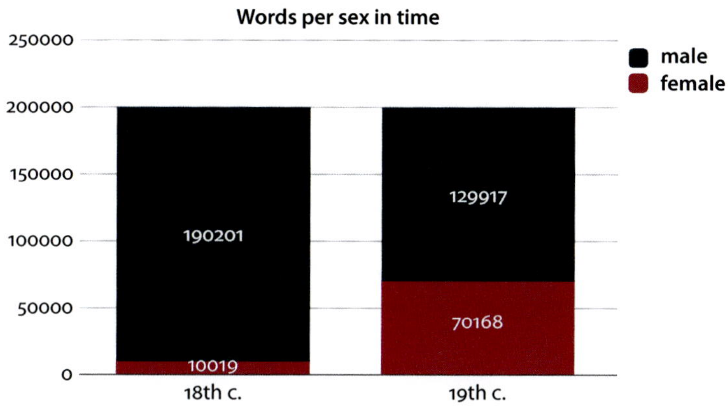

Figure 2. Distribution of words per sex in time

These seven women could perhaps have written about other topics, although they seem to prefer Botany in one form or another: Elizabeth Jacson (1804), Almira Lincoln (1832), Anne Pratt (1840), Phoebe Lankester (1839) and Emily Gregory (1895). Only Priscilla Wakefield (1816) writes about insects, and Elizabeth Agassiz, in her 1859 introductory lectures, writes about several general aspects of Life Sciences. Women were still excluded from formal scientific circles and institutions, and only began to gain access to education late in the century, and this can be seen in their production.

3.2 Geography

The Coruña Corpus metadata files (those files accompanying each text sample) also contain information about the geographical provenance of the authors, since this may be relevant for text selection when using the CC and for different studies. However, it is important to bear in mind that such information does not refer to where writers were born, but rather to where they learned how to write about science, that is, where they were educated and acquired their linguistic habits.

The samples in CELiST are extracts of texts written by authors who were educated in different parts of the anglophone world. Most were so in England (21), although we also have samples of authors educated in Scotland (8), Ireland (5) and North America, the territory today corresponding to the United States (6). The number of samples collected aims to be closely related to the quantity of work published in each place, this in an attempt to be representative of the language of the discipline at that time. Hence there are more samples from England than from anywhere else, with Scotland, which was experiencing the so-called Scottish Enlightenment, following; other areas, by contrast, were undergoing very different situations, such as war (USA) or famine (Ireland), and as a consequence fewer scientific texts were produced. In terms of word-counts, Figure 3 below shows that more than half the words in the corpus (210,041) were written by authors educated in England, followed by texts produced by Scottish writers (20%, 80,095 words), those acquiring their writing styles in North America (15%, 60,089 words), and those who did so in Ireland (13%, 50,090 words).

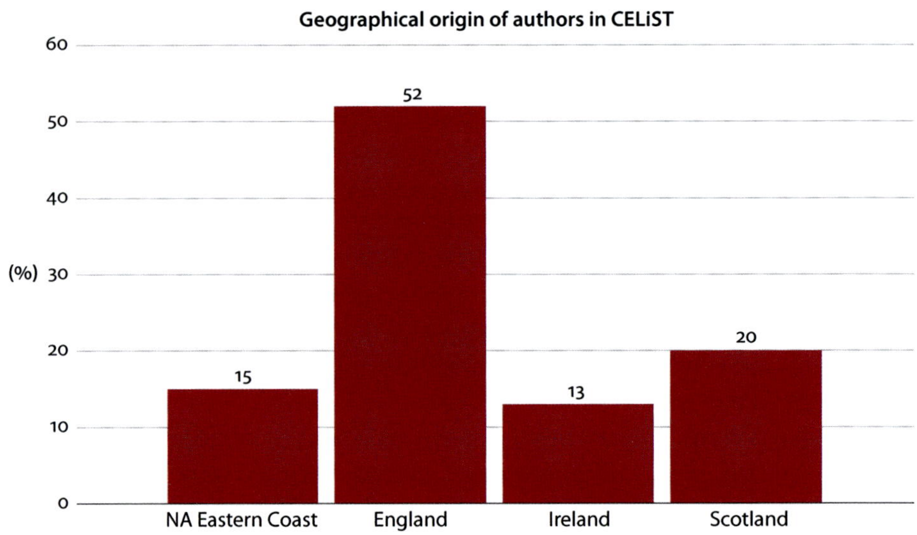

Figure 3. Overall geographical provenance of authors in CELiST

This overall distribution, however, is slightly different when we consider each subperiod separately. Thus, we can observe in Figures 4 and 5 that the nineteenth century witnesses a slight decrease in the number of words by authors educated in England, and a corresponding increase in those educated in Ireland, reflecting how conditions in the country had improved.

As regards the geographical distribution of the texts in the corpus, we can argue that it is certainly representative of the reality of Late Modern English scientific writing. In this respect, as with other corpora in the family, the principle of representativeness has been met.

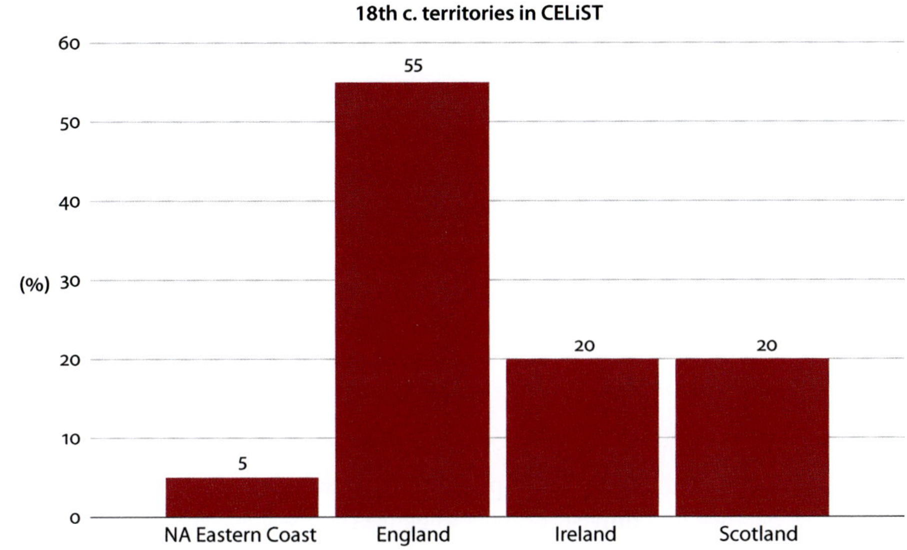

Figure 4. Distribution of geographical provenance of authors in the eighteenth century

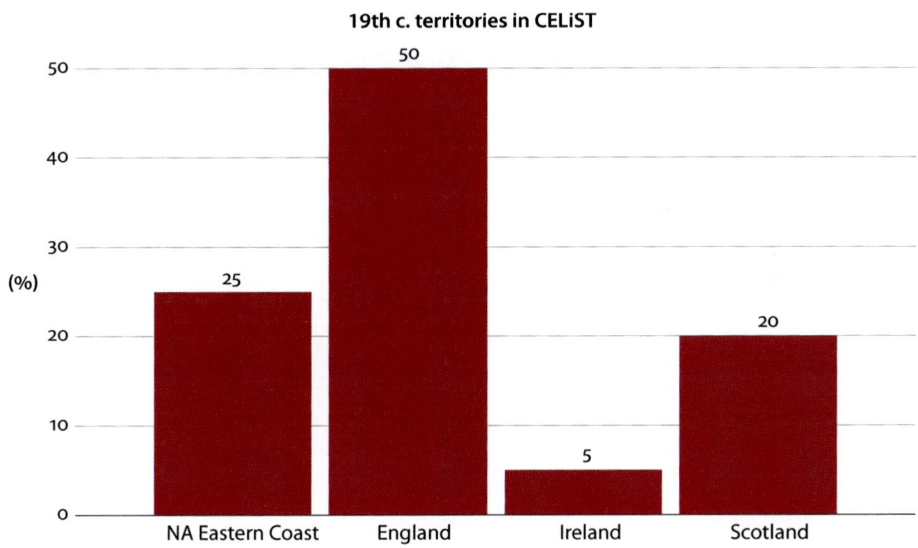

Figure 5. Distribution of geographical provenance of authors in the nineteenth century

4. The texts in CELiST

In the same way that the metadata files include certain information relating to authors, there are also several parameters that are directly connected to the texts sampled for each corpus, and again these are found in the metadata files. These include genre, date of publication, number of words contained in the sample, editorial details, etc. Among them, the genre to which a text belongs is perhaps the most relevant one for the purpose of this chapter, in that it serves to shed light on the evolution of the discipline.

4.1 The genres in CELiST

When selecting the samples for CELiST we wanted to offer as faithful a representation of the reality of the moment as possible. This implied that in our search for texts, genre balance was only secondary to genre representativeness. Thus, rather than collecting the same number of samples for each particular genre, we chose to gather more samples of those that were more popular within the field. Although some of the genres in CELiST are also to be found in other subcorpora of the CC, we found that some had not been seen in other sections.

Disciplines tend to develop and use their own genres, exclusive of each domain. Such is the case with guides and catalogues in CELiST. One of the definitions of *guide* in the OED is "A book of instruction or information for beginners or novices (in an art, etc.)". This use of the term first appeared in 1759 and seems to have been in use until 1892, thus roughly covering the years during which the various authors in CELiST were publishing their works. In order to assign a particular sample to a genre, we also had to take into account the ways in which authors themselves described and labelled their work (see Moskowich and Crespo 2016) and to consider what authors themselves meant when they used terms like handbook, manual and guidebook, since all of these genres can function as guides for "readers inspection of natural elements".[6]

We also found that catalogues are used to classify living entities, both animals and plants. In particular, we classified as catalogues all those samples that adhered to this definition, in use from 1667 until 1889:

> Now usually distinguished from a mere list or enumeration, by systematic or methodical arrangement, alphabetical or other order, and often by the addition of brief particulars, descriptive, or aiding identification, indicative of locality, position, date, price, or the like. (OED)

6. The work by James Douglas, *Myographiæ comparatæ specimen: or, a comparative description of all the muscles in a man and in a quadruped. ... To which is added an account of the muscles peculiar to a woman, etc. M.D.*, published in 1707, is the earliest guide contained in CELiST.

As a genre, Catalogue includes categories such as herbals, plus descriptions of flora and of animals. A good example of what we consider a catalogue is the text by the only woman included in the eighteenth-century section of the corpus, Elizabeth Blackwell, *A Curious Herbal, containing five hundred cuts, of the moſt uſeful plants, which are now uſed in the practice of Physick. Engraved on folio copper plates after drawings, taken from the LIFE. To which is added a short description of ye plants and their common uses in PHYSICK. In Two Volumes* (1737), of which we have sampled a fragment from volume I.

Apart from these two domain-specific genres, another six appear in CELiST, as can be seen in Figure 6 below:

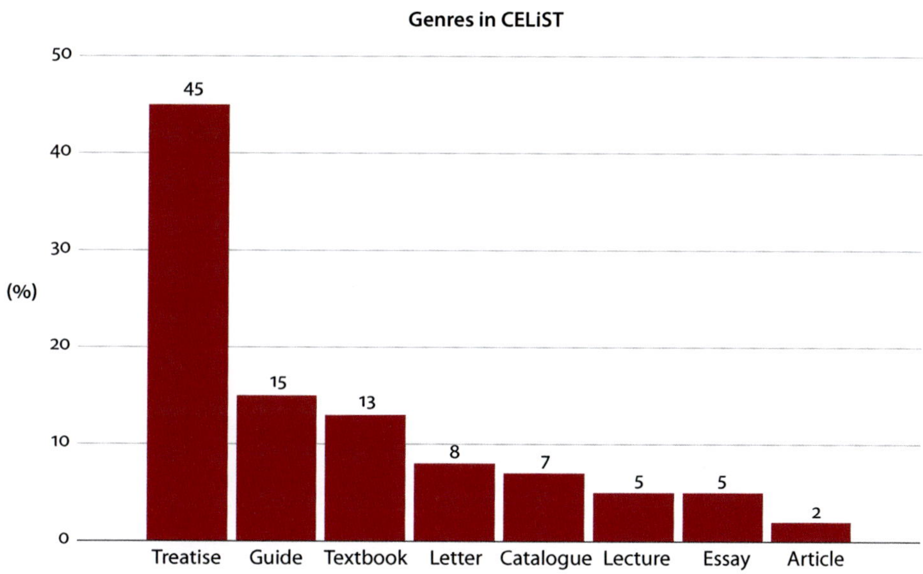

Figure 6. Overall distribution of genres in CELiST

Treatise is the genre to which most samples (45%) belong, with a total of 180,176 words. Very far from this popular format is the Guide, one of the domain-specific genres mentioned above. Guides constitute only 15% (60,068 words) of the total, almost the same as Textbooks (13%, 50,074 words). Letters, with 8% (30,052 words) occur at levels very close to Catalogue (7%, 30,045 words), the other domain-specific genre in CELiST. Lecture (20,099 words) and Essay (20,020 words) each constitute 5% of the total, with two samples per genre each, while Article (9,771 words), nowadays the king of genres in scientific communication, is represented by just one text (2% of the total words in CELiST). As compilers, we believe that this distribution is representative of the situation in Late Modern English science. However, looking separately at the distribution in the two centuries may give us an even deeper insight into the situation.

The way in which genres are distributed across time can be seen by comparing the figures for the two centuries. There we can see that each of them contains samples for six different genres, although those are not exactly the same. That is, Treatise, Letter, Textbook, and Guide appear throughout the period covered by CELiST; by contrast, Catalogue and Essay are recorded only for the eighteenth-century section, and Lecture and Article have been sampled only for the 1800–1900 period. This reflects tendencies in the discipline. So, for example, we know that although the research article appeared much earlier (Bazerman 1988; Atkinson 1999), it gained in popularity and prestige late in the nineteenth century as a means of conveying science.

Figure 7 shows that Treatise is the most popular genre in CELiST in the eighteenth century, as would also be the case in the following years (see also Figure 8) with almost half the total words. In fact, Treatise is the most popular genre in all the subcorpora of the CC. In CELiST it is followed by the discipline-specific Guide, containing four of the twenty samples, and Catalogue, with three samples. There are also two samples of Essay, with one extract from Edward Bancroft (1769), an example of Letter, and one, from Thomas Boreman (1730), corresponding to Textbook.

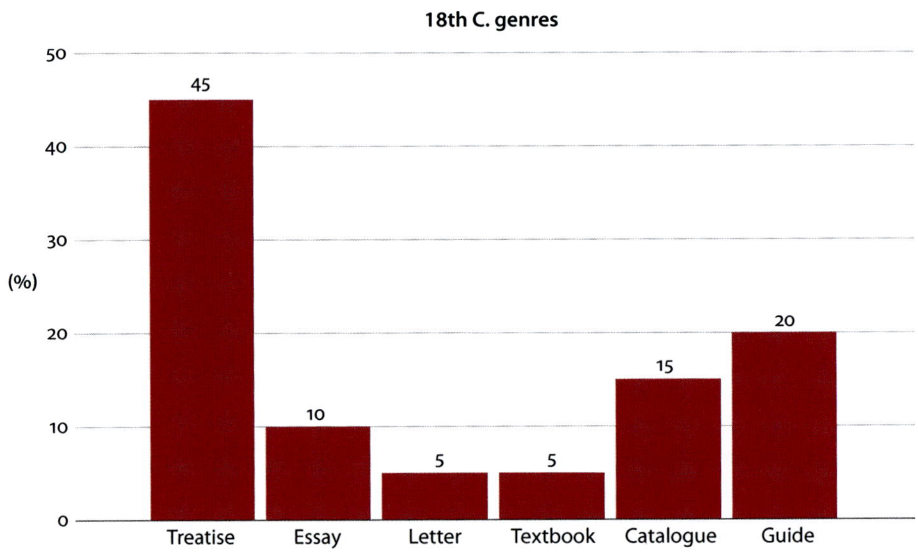

Figure 7. Genres in the eighteenth-century section of CELiST

The distribution of samples by their genre is slightly different in the nineteenth century, not only because of the formats that now began to appear but also because of the number of samples within each category. The genre Catalogue is not represented in this period, with only Guide surviving as a field-specific format. As mentioned above, almost half the words in this section belong to treatises, which are again the most popular way to transmit knowledge. Perhaps due to changes in the socio-historical situation, Textbook now attains 20% representation, and Lecture and Letter both reach 10%, the same as the field-specific format Guide. This new tendency may indicate that authors were interested in making knowledge available to a non-expert readership, in that more non-technical texts were now being published. As can be seen, technical texts written for members of an epistemic community (including a methodological framework, i.e.: treatises) are still the favoured format, and there is only one sample of a research Article (the one by Edward Jenner 1824). Apart from these, all the other texts seem to be addressed to non-expert readers.

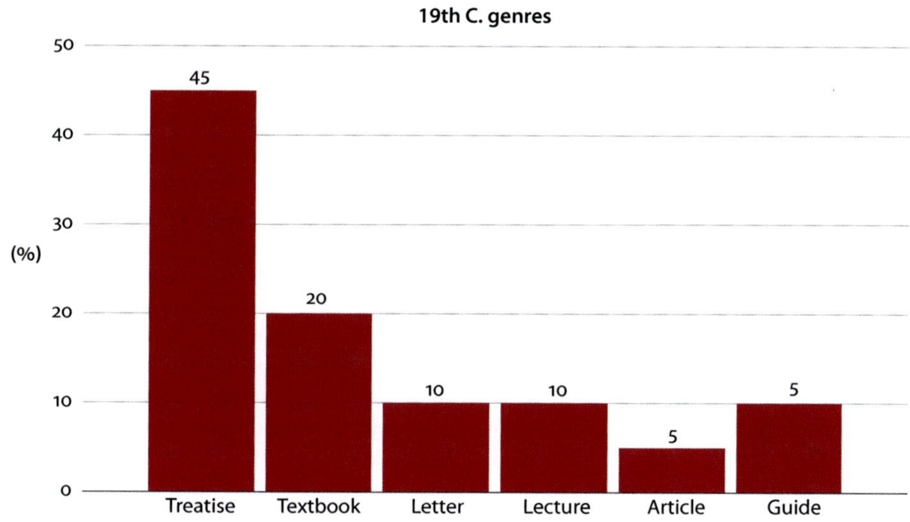

Figure 8. Genres in the nineteenth-century section of CELiST

Although both representativeness and balance are key elements in corpus compilation, the spread of genres as set out in the description above reinforces the idea that we, as compilers, are more concerned with the former, since it seems to better mirror the real situation than the latter. It is only in another sample-related variable, the number of samples, that we have tried to create balance, this by collecting the same number of words and the same number of samples per decade. The variables that relate to the authors of the texts compiled conform first and foremost to the principle of representativeness, this in an attempt to make of CELiST in particular, and the CC in general, a small-scale reflection of what was happening in the English-writing communities in the Late Modern English world.

Works cited

Adams, Valerie. 1988. *An Introduction to Modern English Word-Formation*. London: Routledge.

Aston, Guy. 1997. "Small and large corpora in language learning". In Lewandowska-Tomaszczyk, and Melia, Barbara and James (eds.), *Practical applications in language corpora*. Lodz: Lodz University Press. 51–62.

Atkinson, Dwight. 1999. *Scientific Discourse in Sociohistorical Context: the Philosophical Transactions of the Royal Society of London, 1675–1975*. Mahwah, New Jersey: Lawrence Erlbaum Associates.

Barnes, Jonathan. 1982. *The Presocratic Philosophers*. London: Routledge.

Bazerman, Charles. 1988. *Shaping Written Knowledge: The Genre and Activity of the Experimental Article in Science*. Wisconsin: The University of Wisconsin Press.

Beddoes, Thomas. 1799. *Contributions to physical and medical knowledge, principally from the West of England*, collected by Thomas Beddoes, M.D. Printed by Biggs & Cottle, for T. N. Longman and O. Rees, Paternoster-Row, London.

Crespo, Begoña. 2015. Women writing science in the eighteenth century: some hints about their language use. *Anglica. An International Journal of English Studies*, 24/2: 103–128.

Crespo, Begoña and Moskowich, Isabel. 2010. CETA in the Context of the Coruña Corpus. *Literary and Linguistic Computing*, 25/2: 153–164. https://doi.org/10.1093/llc/fqp038

Crespo, Begoña and Moskowich, Isabel. 2015. Involved in Writing Science: Nineteenth-Century Women in the Coruña Corpus. *International Journal of Language and Linguistics*, 2/5: 76–88.

De Beaugrande, Robert. 2001. "Large corpora, small corpora and the learning of language". In Ghadessy, Moshen; Henry, Alex and Roseberry, Robert L. (eds.), *Small corpus studies and ELT*. Amsterdam/ Philadelphia: John Benjamins. 3–28. https://doi.org/10.1075/scl.5.04bea

Flowerdew, Lynne. 2004. "The argument for using English specialized corpora to understand academic and professional language". In Connor, Ulla and Upton, Thomas A. (eds.), *Discourse in the Professions. Perspectives from corpus linguistics*. Amsterdam/Philadelphia: John Benjamins. 10–33. https://doi.org/10.1075/scl.16.02flo

Freire-Picos, María Angeles. 2014. "Hacia la igualdad de género en las evaluaciones de calidad científica". In Díaz López, Ana Jesús; González Penín, Anabel and Aguayo Lorenzo, Eva (eds.), *II Xornada Universitaria Galega en Xénero, XUGeX: roles de xénero nun mundo globalizado*. Universidade da Coruña, 2014, 57–60. Retrieved February 18, 2020, from http://hdl.handle.net/2183/15814

Fyfe, Aileen and Røstvik, Camilla Mørk. 2018. How female fellows fared at the Royal Society. *Nature*, 555: 159–161. https://doi.org/10.1038/d41586-018-02746-z

Harari, Yuval N. 2014. *Sapiens. A Brief History of Mankind*. London: Vintage Books.

Hume, David. 1748. *An Inquiry concerning Human Understanding*, published by Andrew Millar. London: Strand.

Kastovsky, Dieter. 2009. "Astronaut, astrology, astrophysics: About Combining Forms, Classical Compounds and Affixoids". In McConchie, Rod W.; Honkapohja, Alpo and Tyrkkö, Jukka (eds.), *Selected Proceedings of the 2008 Symposium on New Approaches in English Historical Lexis (HEL-LEX 2*. Somerville, MA: Cascadilla Proceedings Project. 1–13.

Lareo, Inés; Monaco, Leida Maria; Esteve-Ramos, María José and Moskowich, Isabel (comps.). 2020. *The Corpus of English Life Sciences Texts (CELiST)*. A Coruña: Universidade da Coruña. https://doi.org/10.17979/spudc.9788497497848

Malpas, Jeff. 2012. Donald Davidson. *The Standford Encyclopedia of Philosophy* (Winter 2012 Edition), Zalta, Edward N. (ed.). Retrieved February 17, 2020, from https://plato.standford.edu/archives/win2012/entries/davidson

Millward, Celia M. and Hayes, Mary. 2012. *A Biography of the English Language.* Boston: Wadsworth/Cengage Learning.

Monaco, Leida-Maria. 2016. "Abstractness as diachronic variation in CEPhiT: Biber's Dimension 5 applied". In Moskowich, Isabel and Crespo, Begoña (eds.), *'The Conditioned and the Unconditioned': Late Modern English Texts on Philosophy.* Amsterdam: John Benjamins. 99–121. https://doi.org/10.17979/spudc.9788497497077

Moskowich, Isabel. 2019. "An introduction to CHET, the Corpus of History English Texts". In Moskowich, Isabel; Crespo, Begoña; Puente-Castelo, Luis and Monaco, Leida-Maria (eds.), *Writing history in Late Modern English: Explorations of the Coruña Corpus.* Amsterdam/Philadelphia: John Benjamins. 41–58. https://doi.org/10.1075/z.225

Moskowich, Isabel and Crespo, Begoña. 2007. "Presenting the Coruña Corpus: A Collection of Samples for the Historical Study of English Scientific Writing". In Pérez Guerra, Javier et al. (eds.), *Of Varying Language and Opposing Creed: New Insights into Late Modern English.* Bern: Peter Lang. 341–357.

Moskowich, Isabel and Crespo, Begoña. 2016. Classifying communicative formats in CHET, CECHeT and others. *EPiC Series in Language and Linguistics*, 1: 308–320. https://doi.org/10.29007/kvjx

Moskowich, Isabel; Lareo, Inés; Camiña, Gonzalo and Crespo, Begoña (comps.). 2012. *Corpus of English Texts on Astronomy.* Amsterdam: John Benjamins. https://doi.org/10.17979/spudc.9788497497084

OED. *Oxford English Dictionary.* Retrieved February 11, 2020, from http://www.oed.com

Puente-Castelo, Luis. 2017. On conditionality: A corpus-based study of conditional structures in Late Modern English scientific texts. Unpublished PhD Dissertation. A Coruña: Universidade da Coruña.

Rees, Emma L. E. 2003. *Margaret Cavendish: gender, genre, exile.* Manchester: Manchester University Press.

UNESCO. 1988. *Proposed International Standard Nomenclature for Fields of Science and Technology.* UNESCO/ROU257 rev. 1. Paris.

Watts, Ruth. 2007. *Women in Science: A Social and Cultural History.* London and New York: Routledge.

Whaley, Leigh Ann. 2003. *Women's History as Scientists.* Santa Barbara, California: ABC-CLIO, INC.

CHAPTER 2

Editorial policy in the Corpus of English Life Sciences Texts

Criteria, conventions, encoding and editorial marks

Gonzalo Camiña
University of Nantes

1. General remarks

The Corpus of English Life Sciences Texts (CELiST) contains forty samples, twenty of which are from the eighteenth century and twenty from the nineteenth century. These samples come from images of the original texts subsequently converted into digital files, so that specialised software tools can manage them. One of such tools, the Coruña Corpus Tool (CCT), can be downloaded from https://ruc.udc.es/dspace/handle/2183/21850.

Converting original texts into a digital format poses some technical, linguistic and extralinguistic challenges that need to be addressed. This chapter briefly outlines the challenges we have faced when creating this corpus and explains the decisions we have made to tackle them in our effort to make the work of researchers easier.

For example, while the works published in the nineteenth century allowed for scanning and Optical Character Recognition (OCR) to produce a relatively faithful digital copy of the original text, those published in the eighteenth century (and most of the ones published in the nineteenth) required manual transcription, firstly because they contained smeared words, ink blots, worn-out or torn paper, and secondly because of the presence of non-standard characters such as <ſ> (long <s>), <*ſ*> (italicised long <s>) and the ligatured digraph <ct>, which made OCR impossible (view Figure 1).

Therefore, building our corpus required the use of both OCR and typing, followed by encoding and three manual revisions to minimise the chances of gross errors, though some inconsistencies may remain.

Some years ago in 2004, when we established the criteria for the Coruña Corpus, we opted for the eXtensible Mark-up Language (XML) to encode our files, because it is free, versatile and offers a great potential to further develop, change and export the encoded files, as explained in Section 2 below. Figure 2 illustrates the conversion of the previous original into its digital counterpart.

https://doi.org/10.1075/z.237.02cam

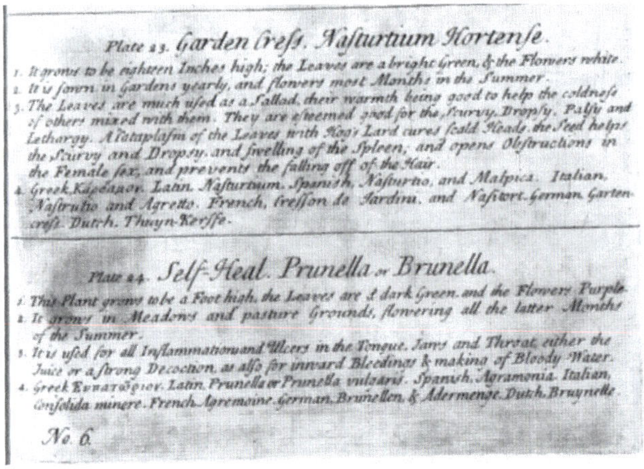

Plate 23. Garden Creſs. Naſturtium Hortenſe.

1. It grows to be eighteen Inches high; the Leaves are a bright Green, & the Flowers white.
2. It is ſown in Gardens yearly, and flowers most Months in the Summer.
3. The Leaves are much uſed as a Sallad, their warmth being good to help the coldneſs of others mixed with them. They are eſteemed good for the Scurvy, Dropſy, Palſy and Lethargy. A Cataplaſm of the Leaves with Hog's Lard cures ſcald Heads. the Seed helps the Scurvy and Dropsy, and ſwelling of the Spleen, and opens Obſtructions in the Female ſex, and prevents the falling off of the Hair.
4. Greek Κάρδαμον. Latin. Naſturtium. Spanish, Naſturtio, and Malpica. Italian, Naſtrutio and Agretto. French, Creſſon de Jardini, and Naſitort. German, Garten-creſs. Dutch. Thuyn-Kerſſe.

Plate 24. Self-Heal. Prunella or Brunella.

1. This Plant grows to be a Foot high, the Leaves are a dark Green, and the Flowers Purple.
2. It grows in Meadows and pasture Grounds, flowering all the latter Months of the Summer.
3. It is uſed for all Inflammations and Ulcers in the Tongue. Jaws and Throat, either the Juice or a ſtrong Decoction, as alſo for inward Bleedings & making of Bloody Water.
4. Greek Ευπατώριον. Latin. Prunella or Prunella vulgaris. Spanish. Agramonia. Italian, Conſolida minore. French Agremoine. German. Brunellen. & Adermenge. Dutch. Bruynelle.

No. 6.

Figure 1. Original text (1737)

Plate 23. Garden Crefs. Nafturtium Hortenfe.

1. It grows to be eighteen Inches high; the Leaves are a bright Green, & the Flowers white.

2. It is fown in Gardens yearly, and flowers most Months in the Summer.

3. The Leaves are much ufed as a Sallad, their warmth being good to help the coldnefs of others mixed with them. They are efteemed good for the Scurvy, Dropfy, Palfy and Lethargy. A Cataplafm of the Leaves with Hog's Lard cures fcald Heads. The Seed helps the Scurvy and Dropsy, and fwelling of the Spleen, and opens Obftructions in the Female fex, and prevents the falling off of the Hair.

4. Greek, [fragmentgreek]. Latin, Nafturtium. Spanish, Nafturtio, and Malpica. Italian, Naftrutio and Agretto. French, Creffon de Jardini and Nafitort. German, Gartencrefs, Dutch, Thuyn-Kerffe.

Plate 24. Self-Heal. Prunella or Brunella.

1. This plant grows to be a Foot high, the Leaves are a dark Green, and the Flowers Purple.

2. It grows in Meadows and pasture Grounds, flowering all the latter Months of the Summer.

3. It is ufed for all Inflammations and Ulcers in the Tongue. Jaws and Throat, either the Juice or a ftrong Decoction, as alfo for inward Bleedings & making of Bloody Water.

4. Greek, [fragmentgreek]. Latin, Prunella or Prunella vulgaris. Spanish, Agramonia. Italian, Confolida minore. French, Agremoine. German, Brunellen, & Adermenge. Dutch, Bruynelle.

Figure 2. Digital sample (1737)

Technical questions aside, several other issues needed to be addressed, such as spelling variation, truncated words, the distinction between the roman number "I" and the personal pronoun *I*, abbreviations in general, the use of hyphens across texts, quotations, quotes and fragments in languages other than English, and so forth.

Despite the logical differences between an original typeset file and its digital output, we have tried to keep the digital copy as true to the original as possible. Thus, the encoded sample has become a third step in text production, following the first and second steps executed by the author and the printer, respectively.

In converting the texts, our primary goal has been to keep the balance between two options: (a) showing the text the way it originally was, and (b) offering researchers unrestricted access to the data in the texts in an open, flexible and productive way. To achieve this, we have made some editorial decisions that will be explained in Section 3 below.

1.1 Headers

Each sample is preceded by a header containing relevant information about the file. The CCT displays this information in the Info Display window in a box right above the searchable text (please check the CCT Manual for further reference). Figure 3 below shows the elements included in headers.

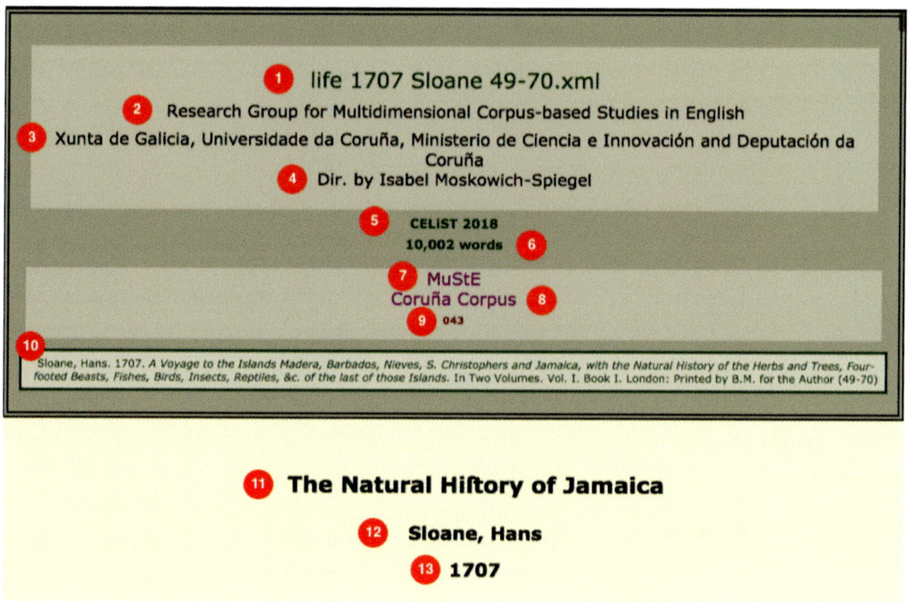

Figure 3. Header

Information provided inside the header box:

(1) Name of the file in the format [discipline, year, author, pages] with its corresponding .xml extension.

(2) Full name of the research group involved in the corpus.

(3) Institutions sponsoring the project.

(4) Director of the project, Isabel Moskowich.

(5) Name of the sub-corpus, CELiST in this case.
(6) Year we encoded the sample.
(7) Number of analysable items included in the file.
(8) Official name of the research project carried out at Universidade da Coruña.
(9) Reference number of the file within the Coruña Corpus.
(10) Bibliographic references of the sampled extract.

Information provided outside the header box:

(11) Reduced version of the title.
(12) Author of the text sampled.
(13) Year of publication.

The reason for providing a reduced version of the title here is that some titles are so long that they nearly cover a full page. Users can view the long version of titles in the Metadata section of the CCT.

1.2 Fonts

Original texts in CELiST were printed using diverse font families and sizes, but the digital samples have been encoded using only one font throughout the corpus: Arial Unicode MS. We have opted for this font because it is supported by most internet browsers, and the Unicode standard ensures interoperability and portability across platforms. Since the corpus complies with Unicode guidelines, users experiencing display issues should resort to a different browser, as some of them are not fully compatible with the Unicode standard, even if they claim so. Also, we have used a generous 13-point font size, to enhance readability on smaller mobile devices (see Figure 4):

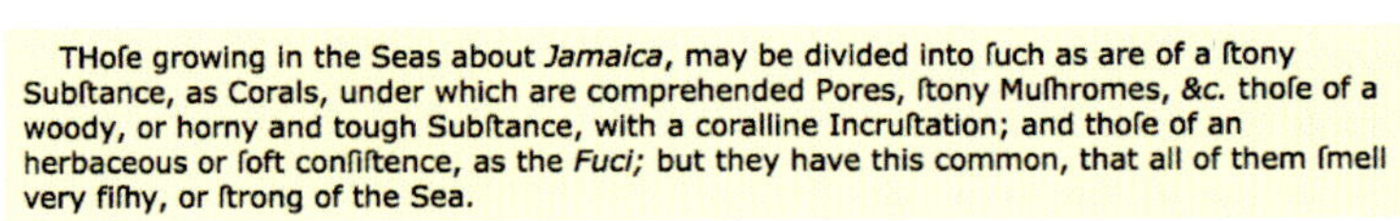

THoſe growing in the Seas about *Jamaica*, may be divided into ſuch as are of a ſtony Subſtance, as Corals, under which are comprehended Pores, ſtony Muſhromes, &c. thoſe of a woody, or horny and tough Subſtance, with a coralline Incruſtation; and thoſe of an herbaceous or ſoft conſiſtence, as the *Fuci;* but they have this common, that all of them ſmell very fiſhy, or ſtrong of the Sea.

Figure 4. Arial unicode MS displayed on Mozilla Brave

1.3 Page numbers

We have kept the original numbering in the texts, even when they do not start on page one. For the sake of consistency all page numbers appear centre-aligned, in bold type, and preceded and followed by blank lines. Figure 5 illustrates how page numbers are displayed in the final layout:

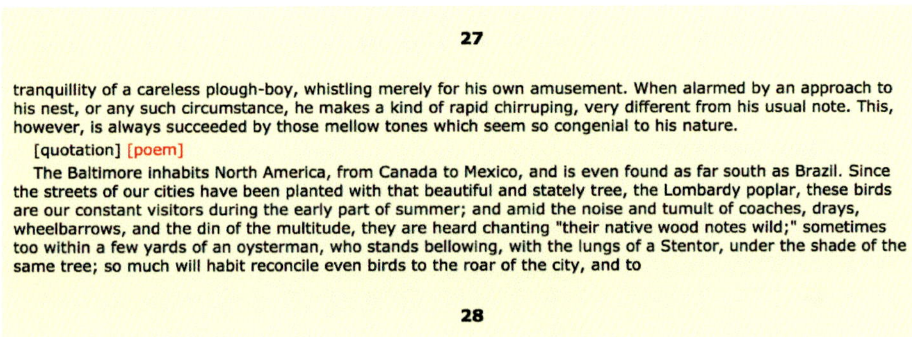

27

tranquillity of a careless plough-boy, whistling merely for his own amusement. When alarmed by an approach to his nest, or any such circumstance, he makes a kind of rapid chirruping, very different from his usual note. This, however, is always succeeded by those mellow tones which seem so congenial to his nature.

[quotation] [poem]

The Baltimore inhabits North America, from Canada to Mexico, and is even found as far south as Brazil. Since the streets of our cities have been planted with that beautiful and stately tree, the Lombardy poplar, these birds are our constant visitors during the early part of summer; and amid the noise and tumult of coaches, drays, wheelbarrows, and the din of the multitude, they are heard chanting "their native wood notes wild;" sometimes too within a few yards of an oysterman, who stands bellowing, with the lungs of a Stentor, under the shade of the same tree; so much will habit reconcile even birds to the roar of the city, and to

28

Figure 5. Page numbers

1.4 Chapter and section titles

The Info Display window of the CCT shows chapter and section titles centred on the screen in a larger, blue, bold font, to make the visual revision of texts easier, as seen in Figure 6 below.

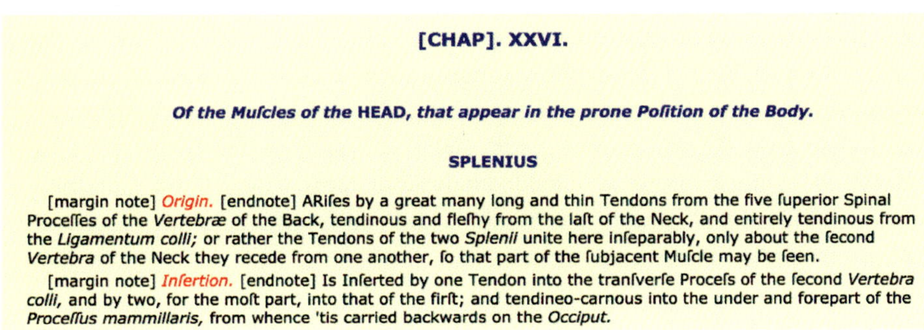

[CHAP]. XXVI.

Of the Muſcles of the* HEAD, *that appear in the prone Poſition of the Body.

SPLENIUS

[margin note] *Origin.* [endnote] ARiſes by a great many long and thin Tendons from the five ſuperior Spinal Proceſſes of the *Vertebræ* of the Back, tendinous and fleſhy from the laſt of the Neck, and entirely tendinous from the *Ligamentum colli;* or rather the Tendons of the two *Splenii* unite here inſeparably, only about the ſecond *Vertebra* of the Neck they recede from one another, ſo that part of the ſubjacent Muſcle may be ſeen.

[margin note] *Inſertion.* [endnote] Is Inſerted by one Tendon into the tranſverſe Proceſs of the ſecond *Vertebra colli,* and by two, for the moſt part, into that of the firſt; and tendineo-carnous into the under and forepart of the *Proceſſus mammillaris,* from whence 'tis carried backwards on the *Occiput.*

Figure 6. Titles

1.5 Paragraphs and lines

Digital paragraphs preserve the structure of the original versions, and the same applies to line length as shown in the Search Window. However, lines will automatically adjust to the width of the Info Display window. This is so for two reasons: first, the CCT can get rid of unnecessary spaces at the end of lines, and second, users can shrink the active window to have two or more windows open on their desktop at the same time. Figure 7 illustrates the previous example after shrinking:

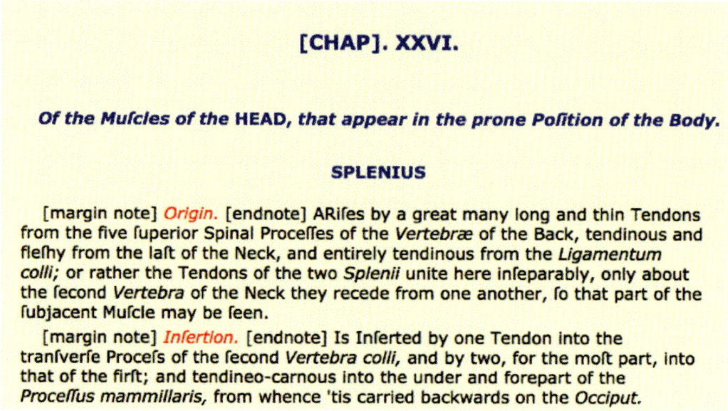
[CHAP]. XXVI.

Of the Muſcles of the **HEAD,** *that appear in the prone Poſition of the Body.*

SPLENIUS

[margin note] *Origin.* [endnote] ARiſes by a great many long and thin Tendons from the five ſuperior Spinal Proceſſes of the *Vertebræ* of the Back, tendinous and fleſhy from the laſt of the Neck, and entirely tendinous from the *Ligamentum colli;* or rather the Tendons of the two *Splenii* unite here inſeparably, only about the ſecond *Vertebra* of the Neck they recede from one another, ſo that part of the ſubjacent Muſcle may be ſeen.

[margin note] *Inſertion.* [endnote] Is Inſerted by one Tendon into the tranſverſe Proceſs of the ſecond *Vertebra colli,* and by two, for the moſt part, into that of the firſt; and tendineo-carnous into the under and forepart of the *Proceſſus mammillaris,* from whence 'tis carried backwards on the *Occiput.*

Figure 7. Lines in the info display window

1.6 Analysable items

The Info Display window of the CCT shows all items contained in the samples, even those that cannot be analysed from a syntactic point of view. For instance, isolated numbers, symbols, and items not belonging to an author's own writing, such as quotations and marginal notes, will appear in red, but they are excluded from word count. Also, the CCT will show our editorial marks, such as [fragment], [note] and [quotation], among others, alongside analysable items. This will be explained in Section 3. Figure 8 illustrates how symbols and text fragments quoted from other authors are shown in the final layout:

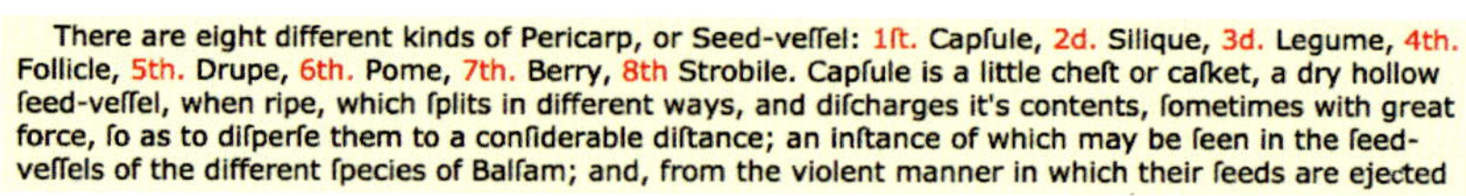
There are eight different kinds of Pericarp, or Seed-veſſel: 1ſt. Capſule, 2d. Silique, 3d. Legume, 4th. Follicle, 5th. Drupe, 6th. Pome, 7th. Berry, 8th Strobile. Capſule is a little cheſt or caſket, a dry hollow ſeed-veſſel, when ripe, which ſplits in different ways, and diſcharges it's contents, ſometimes with great force, ſo as to diſperſe them to a conſiderable diſtance; an inſtance of which may be ſeen in the ſeed-veſſels of the different ſpecies of Balſam; and, from the violent manner in which their ſeeds are ejected

Figure 8. Non-analysable elements

1.7 Omissions and amendments

Despite our commitment to respect the original versions of the samples as much as possible, we have omitted some editorial material such as page headers, footers and some marginal notes, as these are elements added by the printer and therefore do not represent the author's linguistic habits.

Older texts include an extra blank space before colons and semi-colons. For the sake of consistency, we have removed these spaces, because even though they do not affect the printed text, they carry meaning in the digital version. Please note that punctuation signs do not appear in the Search Window of the CCT.

Besides, some spelling errors have been corrected because they are likely to have occurred during the printing process. Finally, we have taken into account the different spellings over time and checked all items in The *Oxford English Dictionary*.

2. Mark-up language

The language used to encode CELiST is XML, a subset of the Standard Generalized Mark-up Language (SGML). We have opted for this language because it is easier to implement than SGML. For this reason, it is widely used nowadays for driving document specifications and building general-purpose applications, such as those for electronic commerce, RSS, XHTML and so on. It has been used to create the second edition of the *Oxford English Dictionary*, for example.

We pioneered the use of XML when we decided to use this language to encode the samples contained in the Coruña Corpus at a time when the prevailing trend was using raw text (TXT), even among the most reputed corpora. Our decision implied much more work, as coding in XML is obviously more time-consuming, but it was worth the effort, because it is also much more powerful when it comes to complex searches.

The most significant advantages of XML are ease of use, flexibility and cross-platform operability. This means that any text encoded with XML can be stored, modified, expanded, viewed and displayed on any computer running any operating system.

To tag our texts, we have followed the guidelines provided by the Text Encoding Initiative (TEI). In our corpus, we have incorporated some tags proposed by TEI to provide CELiST users with a structure that will allow them to carry out research accurately. The Unicode Consortium, developer of the Unicode Standard, supplies the characters necessary to represent most symbols, including mathematical and astronomical, which allows CELiST to offer a representation close to the original text.

All texts have been tagged using the free software text editor Emacs (see Figure 9), because it is open-source, powerful, extensible and highly customisable.

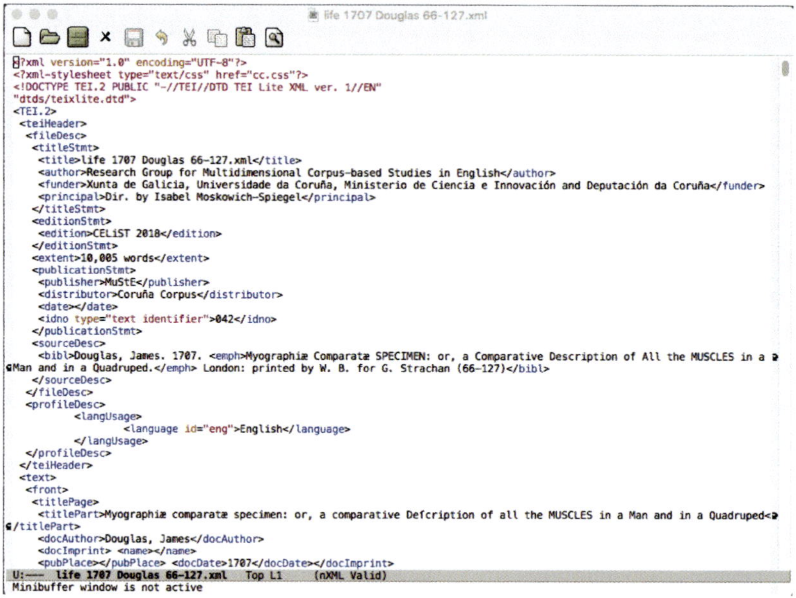

Figure 9. The Emacs editor (MacOS version)

2.1 Tags in CELiST

The header shown in Section 1 above (Figure 2) contains a significant number of TEI tags that will not be explained in this volume because of length constraints, but we suggest that all those interested in the topic visit the TEI webpage at http://www.tei-c.org. In this chapter we will focus on the TEI tags present in the main body of the xml file only, as seen in the following list:

TAG	MEANING
<div>	Chapter and section division
<head>	Chapter and section title
<p>	Paragraph
<l>	Line
<pb>	Page break
<abbr>	Abbreviation
<emph>	Emphasis. Used for italics only
<del>	Deleted element. The CCT does not analyse items delimited by the tags <del> and </del>, but they will still appear in red in the Info Display window to preserve the context
<note>	Footnote

Apart from these few TEI tags we have included a set of editorial marks to make analysing texts with the CCT more straightforward. They are discussed in the following sections.

3. Editorial marks and decisions

In line with the criteria laid out for all the sub-corpora that make up the Coruña Corpus, we have selected samples of around 10,000 words taken from various parts of the works included in CELiST. Although our samples try to stay true to the original texts, some changes have been necessary to make searching on the CCT more comprehensible. This section will deal with the specific markers that appear in each part of the text: page, paragraph and word.

Using editorial marks is one of the most controversial decisions taken by the creators of CELiST, as some members were in favour of only using tags that conform to the XML standard. Despite this, we all agree that the CCT can handle these markers seamlessly, and in all fairness the editorial information they contain should make the work of researchers easier. The characters we have selected to represent this information are square brackets [ ], because authors seldom use them in the original texts (see Figure 10 below). Evidently, this choice compelled us to replace the (very few) square brackets present in our corpus with parentheses to avoid confusion. Square brackets serve two purposes: first, they enclose data such as quotations, figures and formulae; and second, they disambiguate homographic forms that the CCT might otherwise consider the same word.

For instance, the Roman number one appears between brackets [I], to avoid miscounting the number of occurrences of the personal pronoun *I* (see Figure 11 below). Hence, if we generate a full word list for CELiST in the CCT, the first personal pronoun will display two different types: i-1216, showing that authors have used the first-person pronoun 1,216 times, and [i]-102, because the Roman number one appears 102 times. Apart from Roman numbers, brackets also enclose abbreviations and editorial marks. For instance, the abbreviation for number, No, will appear as [no] to distinguish it from the negative particle *no*, etc. Editorial marks will be described in more depth in Section 3.

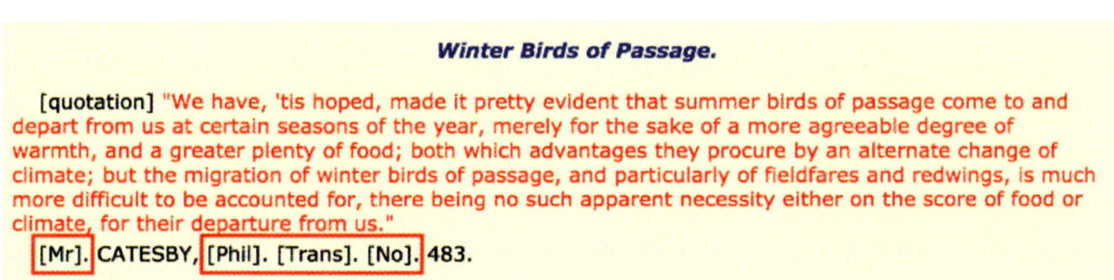
Winter Birds of Passage.

[quotation] "We have, 'tis hoped, made it pretty evident that summer birds of passage come to and depart from us at certain seasons of the year, merely for the sake of a more agreeable degree of warmth, and a greater plenty of food; both which advantages they procure by an alternate change of climate; but the migration of winter birds of passage, and particularly of fieldfares and redwings, is much more difficult to be accounted for, there being no such apparent necessity either on the score of food or climate, for their departure from us."
[Mr]. CATESBY, [Phil]. [Trans]. [No]. 483.

Figure 10. Use of square brackets with abbreviations

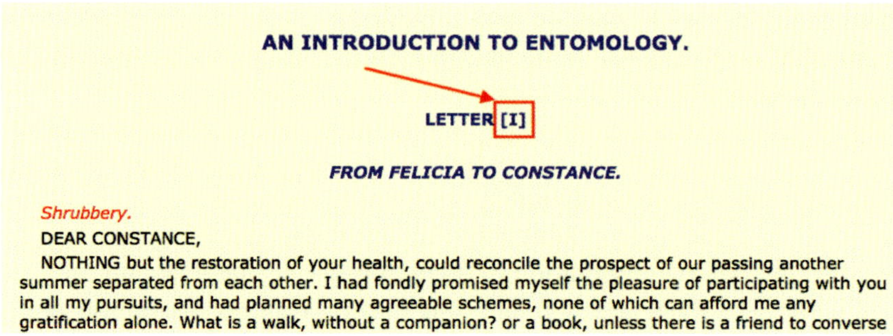

Figure 11. Roman number "I"

Users need to remember that the CCT will display editorial marks both in the Info Display window and the Search window, but they are not taken into account for the final total number of items/words of each sample.

3.1 Pages

Samples compiled for CELiST respect original page numbers and contents. Tables, figures and formulae have been tagged using xml code in the digital copy to mark their location in the original texts. However, the contents of these non-text elements have been omitted.

Another interesting visual aid in eighteenth-century books is that they show the first word of the next page at the bottom of the current page, isolated from the body of the text. Despite its alleged usefulness, this visual aid represents an unnecessary repetition of words that should not appear in the digital sample, so all of them have been omitted from CELiST samples.

3.2 Lines and paragraphs

Lines are automatically resized to the width of the active window (Info Display Window only), so we have joined truncated words at the end of lines. Conversely, the structure of original paragraphs has been preserved. The only exception to this is footnotes, as TEI restricts the number of possible paragraphs to one. Accordingly, the information included in footnotes is displayed as one paragraph.

These restrictions also apply to the place where footnotes may appear in the Info Display window of the CCT. TEI guidelines state that footnotes must follow the word they refer to, but this may lead to a cluttered text difficult to read on screen. To prevent this, we have encoded footnotes so they will be displayed as a separate chunk of text in a smaller font size in the Info Display window. This makes the electronic text easier to interpret. (see Figure 12).

Another decision we have made regarding notes is eliminating those considered editorial notes, because they have been added by the printer and therefore do not represent the author's language. Similarly, we have added the following editorial marks to our samples to provide information to the final file and make researchers' work easier:

- **[note] / [endnote]**
  As mentioned above, footnotes appear in the Info Display window in a separate text block after the reference mark used by the author (a number, a letter, a symbol, etc.).

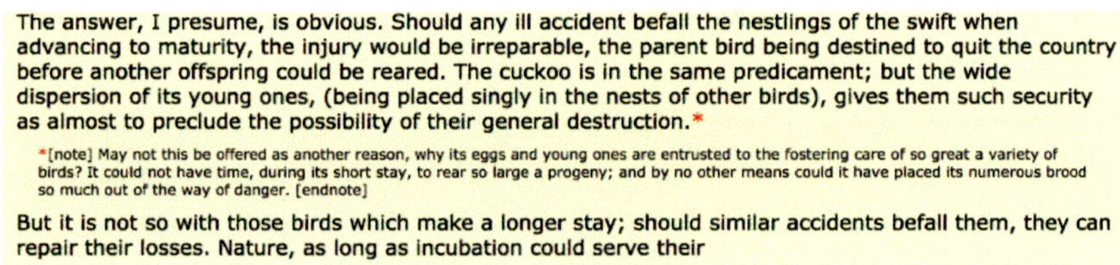

Figure 12. Info display window. Editorial mark [note]

Yet, when working in the Search window, notes are more difficult to identify at first sight since they are inserted in the main body of the text due to TEI restrictions. To avoid this, we have included these two editorial marks to identify the beginning [note] and the end of a note [endnote] (see Figure 13):

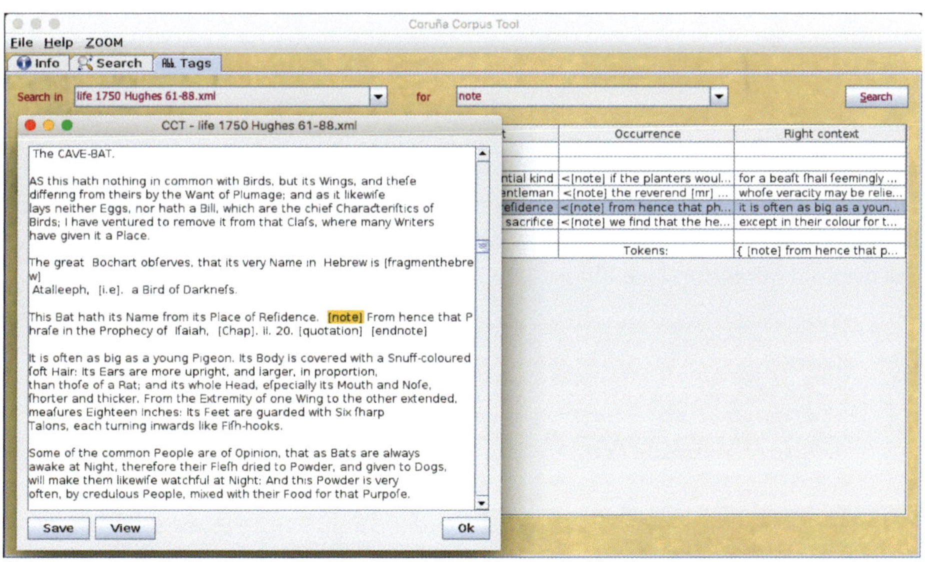

Figure 13. Search window. Editorial mark [note]

- **[margin note] / [endnote]**
  Although margin notes may have been a convenient visual aid for eighteenth-century text readers, these notes do not reflect the author's language. Therefore, they have been excluded from the word count and, since they do not refer to a marker in the text, they have been marked in red and inserted at the beginning of the paragraph they point to.
- **[quotation]**
  As mentioned above, the corpus only includes the words reflecting the language of the authors. Therefore, we have deleted quotations from the original texts. The mark [quotation] is used instead of the original words for easier understanding of the CCT Search window (see Figure 14):

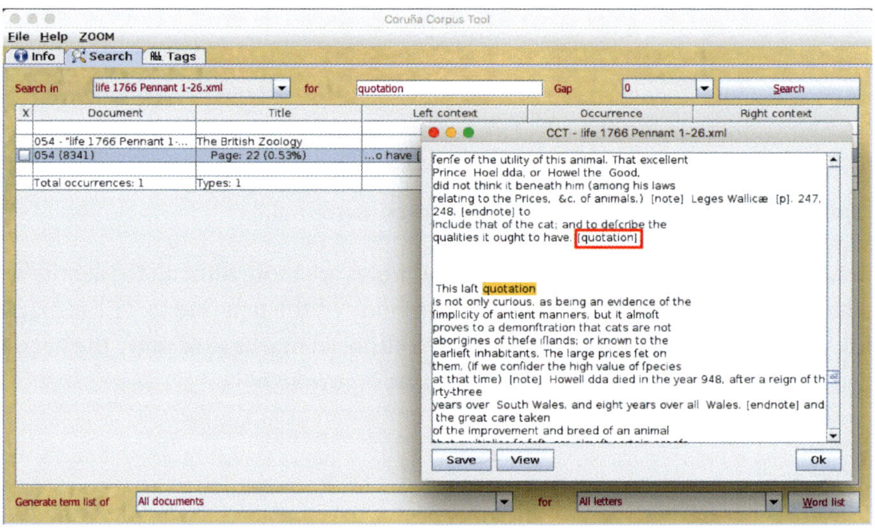

Figure 14. CCT Search window. Editorial mark [quotation]

However, when considered relevant for the context, many deleted quotations can be seen in red in the Info Display window to the right where this mark precedes the original quotation (see Figure 15):

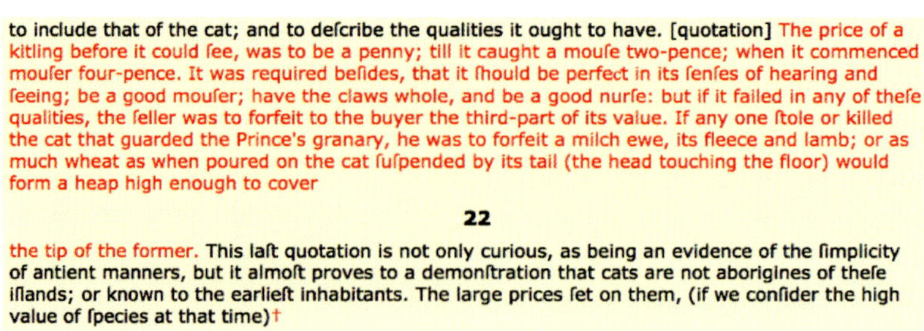
to include that of the cat; and to deſcribe the qualities it ought to have. [quotation] The price of a kitling before it could ſee, was to be a penny; till it caught a mouſe two-pence; when it commenced mouſer four-pence. It was required beſides, that it ſhould be perfect in its ſenſes of hearing and ſeeing; be a good mouſer; have the claws whole, and be a good nurſe: but if it failed in any of theſe qualities, the ſeller was to forfeit to the buyer the third-part of its value. If any one ſtole or killed the cat that guarded the Prince's granary, he was to forfeit a milch ewe, its fleece and lamb; or as much wheat as when poured on the cat ſuſpended by its tail (the head touching the floor) would form a heap high enough to cover

22

the tip of the former. This laſt quotation is not only curious, as being an evidence of the ſimplicity of antient manners, but it almoſt proves to a demonſtration that cats are not aborigines of theſe iſlands; or known to the earlieſt inhabitants. The large prices ſet on them, (if we conſider the high value of ſpecies at that time)†

Figure 15. CCT Info display window. Editorial mark [quotation]

When the quotation is presented in direct speech, the mark [quotation] is used twice; one before the words preceding forms such as *said he, includes she, mentioned they*, and so on, followed by the corresponding part of the original quotation in red (meaning deleted) and another one after the same forms and before the following part of the quotation, as seen in this sample of Collins (1774: 126):

> It often happens that thefe infects quit their retreat in a body, and go in queft of adventures. [quotation] "During my ftay," fays Smith, [quotation] "at Cape Corfe Caftle, a body of thefe ants came to pay us a vifit in our fortification. It was about day-break when the advanced guard of this famifhed crew entered the chapel, where fome negroe fervants were afleep upon the floor [...]".

But sometimes, when the quotation is unnecessary to understand the text its content is omitted, and will not even appear on the Info Display window, which only shows the editorial mark [quotation].

- **[quot(lang)]**
  The previous mark [quotation] implies that the quoted text is written in English, but what happens when authors quote using languages other than English? In these cases, we have added the editorial mark [quot(lang)], in which (lang) stands for the corresponding language. Therefore, users will encounter markers such as:
  [quotlat], when the original language is Latin.
  [quotgreek], when the original language is Greek.
  [quotfrench], when the original language is French, etc.

Including these text marks alongside the original quotation, which can be viewed in the Info Display window, will allow researchers to study intertextuality and language variation in scientific writing through specific searches in the Search window (for more information about searches please check CCT Manual here https://ruc.udc.es/dspace/handle/2183/21850).

- **[fragment]**
  This mark is used to identify other deleted parts of the original text. For instance, the beginning or the end of the sample may have been omitted because sentences began on the previous page or continued on the following page (see Figure 16). This does not apply to quotations, as they have their own identifying mark.

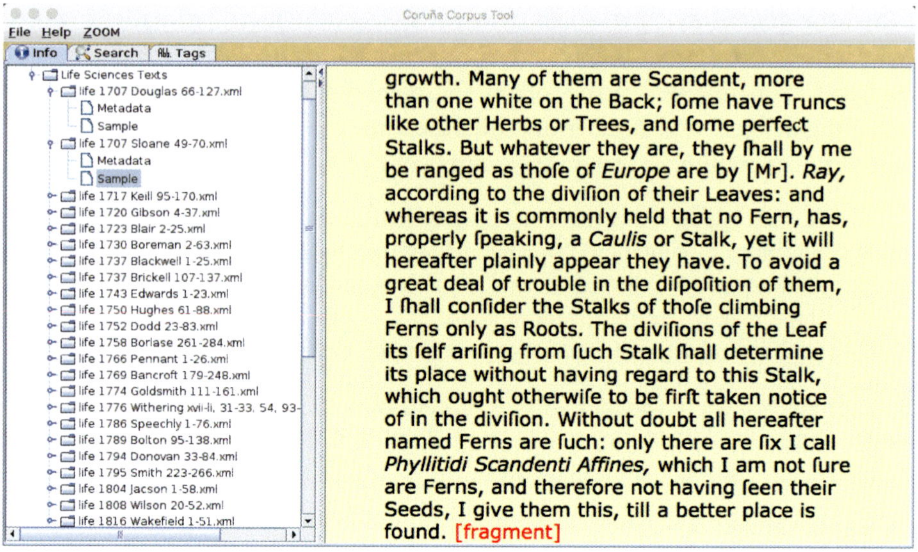

Figure 16. Editorial mark [fragment]

This mark is also used when a word or fragment is written in a non-Latin alphabet in the original text. Thus, words that cannot be represented by the Latin alphabet are marked as [fragmentgreek] or [fragment, followed by the corresponding language] (see Figure 17):

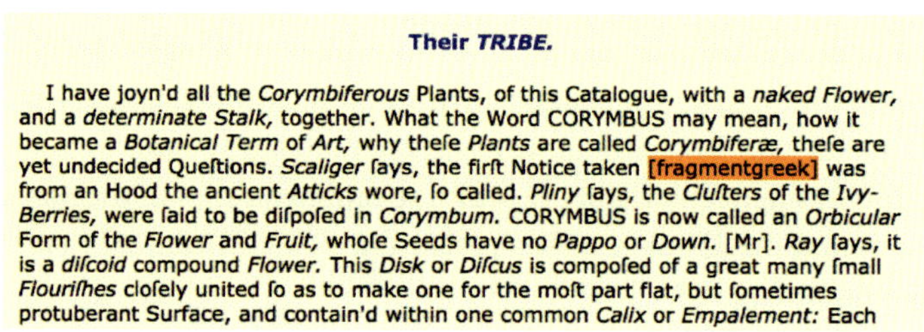
Their *TRIBE.*

I have joyn'd all the *Corymbiferous* Plants, of this Catalogue, with a *naked Flower,* and a *determinate Stalk,* together. What the Word CORYMBUS may mean, how it became a *Botanical Term* of *Art,* why theſe *Plants* are called *Corymbiferæ,* theſe are yet undecided Queſtions. *Scaliger* ſays, the firſt Notice taken [fragmentgreek] was from an Hood the ancient *Atticks* wore, ſo called. *Pliny* ſays, the *Cluſters* of the *Ivy-Berries,* were ſaid to be diſpoſed in *Corymbum.* CORYMBUS is now called an *Orbicular* Form of the *Flower* and *Fruit,* whoſe Seeds have no *Pappo* or *Down.* [Mr]. *Ray* ſays, it is a *diſcoid* compound *Flower.* This *Disk* or *Diſcus* is compoſed of a great many ſmall *Flouriſhes* cloſely united ſo as to make one for the moſt part flat, but ſometimes protuberant Surface, and contain'd within one common *Calix* or *Empalement:* Each

Figure 17. Editorial mark [fragmentgreek]

3.3 Words

One of the most interesting characteristics of CELiST is that it contains texts from two different centuries, which enables researchers to conduct diachronic studies. It has also urged compilers to keep original word spellings, e.g., the ligatured digraph <ct> and the long <ſ> mentioned earlier in this chapter (see Figure 18):

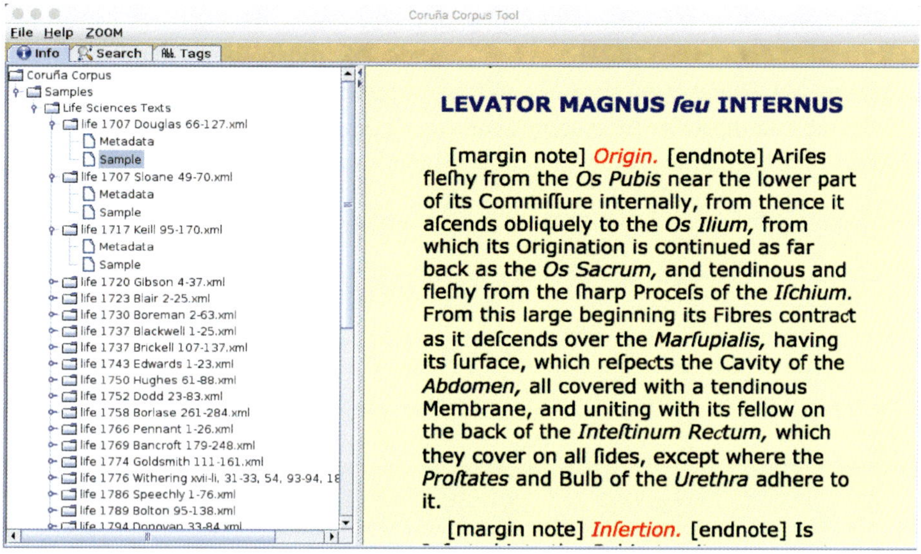

Figure 18. Spelling variants

We have developed the CCT taking these spelling variants into consideration. Therefore, when a searched word has two or more different spellings, they will be considered and displayed as distinct types (see Figure 19, bottom row):

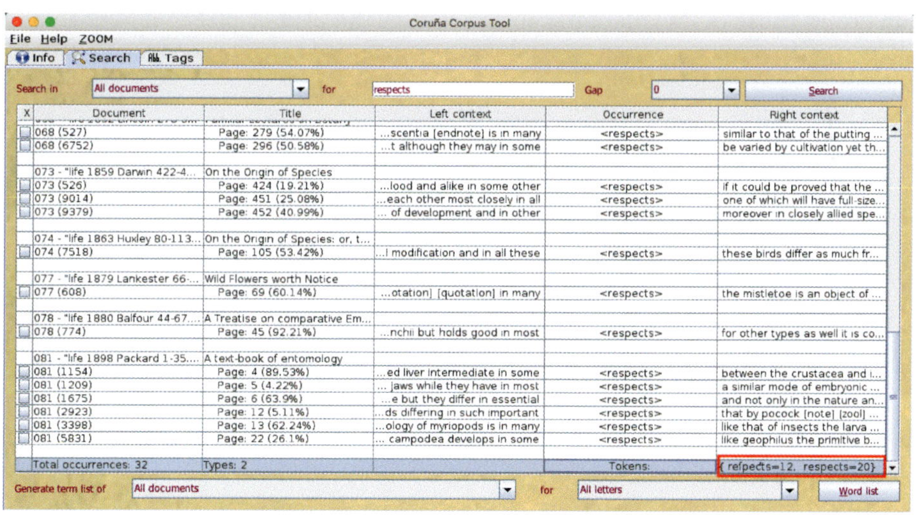

Figure 19. Search window. Spelling variants

CELiST samples keep original line breaks, but we have joined the truncated words at the end of lines into a single segment, and deleted the hyphen separating them. This serves a particular purpose, as hyphens are reserved for hyphenated compound words that can therefore be searched for as well (see Figure 20):

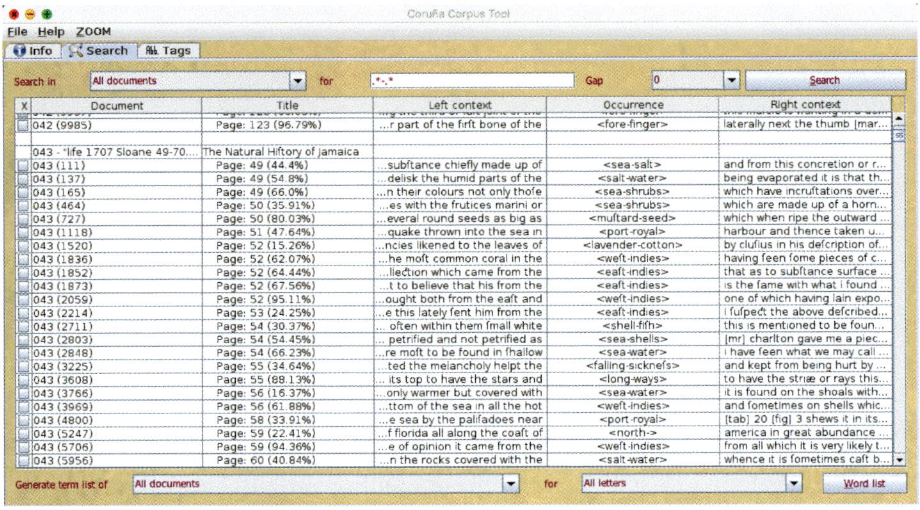

X	Document	Title	Left context	Occurrence	Right context
	042 (9985)	Page: 123 (96.79%)	...r part of the firft bone of the	<fore-finger>	laterally next the thumb [mar...
	043 - "life 1707 Sloane 49-70....	The Natural Hiftory of Jamaica			
	043 (111)	Page: 49 (44.4%)	...fubftance chiefly made up of	<sea-salt>	and from this concretion or r...
	043 (137)	Page: 49 (54.8%)	...delisk the humid parts of the	<salt-water>	being evaporated it is that th...
	043 (165)	Page: 49 (66.0%)	...n their colours not only thofe	<sea-shrubs>	which have incruftations over...
	043 (464)	Page: 50 (35.91%)	...es with the frutices marini or	<sea-shrubs>	which are made up of a horn...
	043 (727)	Page: 50 (80.03%)	...everal round seeds as big as	<muftard-seed>	which when ripe the outward ...
	043 (1118)	Page: 51 (47.64%)	...quake thrown into the sea in	<port-royal>	harbour and thence taken u...
	043 (1520)	Page: 52 (15.26%)	...ncies likened to the leaves of	<lavender-cotton>	by clufius in his defcription of...
	043 (1836)	Page: 52 (62.07%)	...he moft common coral in the	<weft-indies>	having feen fome pieces of c...
	043 (1852)	Page: 52 (64.44%)	...llection which came from the	<eaft-indies>	that as to fubftance surface ...
	043 (1873)	Page: 52 (67.56%)	...t to believe that his from the	<eaft-indies>	is the fame with what i found ...
	043 (2059)	Page: 52 (95.11%)	...ought both from the eaft and	<weft-indies>	one of which having lain expo...
	043 (2214)	Page: 53 (24.25%)	...e this lately fent him from the	<eaft-indies>	i fufpect the above defcribed...
	043 (2711)	Page: 54 (30.37%)	... often within them fmall white	<shell-fifh>	this is mentioned to be foun...
	043 (2803)	Page: 54 (54.45%)	... petrified and not petrified as	<sea-shells>	[mr] charlton gave me a piec...
	043 (2848)	Page: 54 (66.23%)	...re moft to be found in fhallow	<sea-water>	i have feen what we may call ...
	043 (3225)	Page: 55 (34.64%)	...ted the melancholy helpt the	<falling-sicknefs>	and kept from being hurt by ...
	043 (3608)	Page: 55 (88.13%)	... its top to have the stars and	<long-ways>	to have the striæ or rays this...
	043 (3766)	Page: 56 (16.37%)	...only warmer but covered with	<sea-water>	it is found on the shoals with...
	043 (3969)	Page: 56 (61.88%)	...ttom of the sea in all the hot	<weft-indies>	and fometimes on shells whic...
	043 (4800)	Page: 58 (33.91%)	...e sea by the palifadoes near	<port-royal>	[tab] 20 [fig] 3 shews it in its...
	043 (5247)	Page: 59 (22.41%)	...f florida all along the coaft of	<north->	america in great abundance ...
	043 (5706)	Page: 59 (94.36%)	...e of opinion it came from the	<weft-indies>	from all which it is very likely t...
	043 (5956)	Page: 60 (40.84%)	...n the rocks covered with the	<salt-water>	whence it is fometimes caft b...

Figure 20. Hyphenated compound words

Finally, in those cases where the author and/or the printer placed a hyphen as a layout mark in the text, we have replaced it with an EM-dash (see Figure 21):

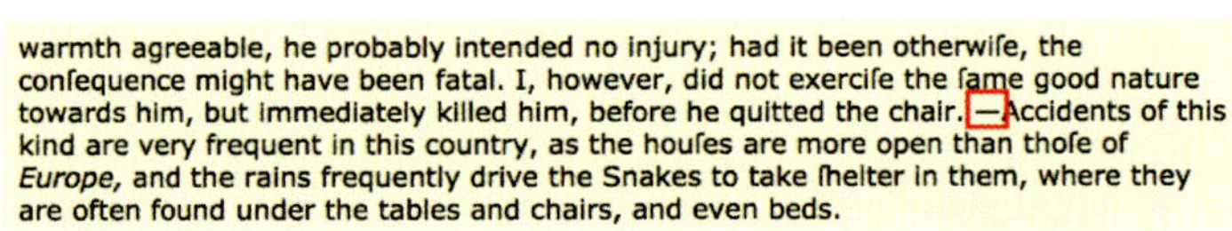
warmth agreeable, he probably intended no injury; had it been otherwife, the confequence might have been fatal. I, however, did not exercife the fame good nature towards him, but immediately killed him, before he quitted the chair.—Accidents of this kind are very frequent in this country, as the houfes are more open than thofe of *Europe*, and the rains frequently drive the Snakes to take fhelter in them, where they are often found under the tables and chairs, and even beds.

Figure 21. Use of EM-dashes

4. List of editorial marks used in CELiST

Below is a list of the editorial marks used in CELiST by alphabetical order, and the samples containing them:

[figure]

In Keill (1717), Boreman (1730), Brickell (1737), Jackson (1804), Jardine (1835), Agassiz (1859), Balfour (1880), Galton (1889), Gregory (1895) and Packard (1898).

[formula]

In Keill (1717) and Macalister (1876).

[fragment]
In all samples except Jenner (1824) and Agassiz (1859).
[fragmentgreek]
In Blair (1723), Blackwell (1737), Dodd (1752), Godman (1828) and Jardine (1835).
[fragmenthebrew]
In Hughes (1750).
[fragmentlatin]
In Dodd (1752).
[margin note] / [endnote]
In Douglas (1707), Sloane (1707), Keill (1717), Gibson (1720), Borlase (1758), Withering (1776) and Bolton (1789).
[note] / [endnote]
In Sloane (1707), Hughes (1750), Dodd (1752), Borlase (1758), Pennant (1766), Bancroft (1769), Goldsmith (1774), Withering (1776), Speechly (1786), Donovan (1794), Jacson (1804), Wilson (1808), Wakefield (1816), Lawrence (1819), Jenner (1824), Godman (1828), Lincoln (1832), Jardine (1835), Agassiz (1859), Huxley (1863), Spencer (1867), Macalister (1876), Lankester (1879), Balfour (1880), Galton (1889), Gregory (1895) and Packard (1898).
[poem]
In Hughes (1750), Bancroft (1769), Wilson (1808), Wakefield (1816), Godman (1828), Lincoln (1832), and Pratt (1840)
[quotation]
In Keill (1717), Hughes (1750), Dodd (1752), Borlase (1758), Pennant (1766), Bancroft (1769), Goldsmith (1774), Bolton (1789), Donovan (1794), Smith (1795), Jacson (1804), Wilson (1808), Wakefield (1816), Lawrence (1819), Jenner (1824), Godman (1828), Lincoln (1832), Jardine (1835), Pratt (1840), Agassiz (1859), Darwin (1859), Huxley (1863), Spencer (1867), Lankester (1879), Balfour (1880), Galton (1889) and Packard (1898).
[quotfrench]
In Bancroft (1769) and Lawrence (1819).
[quotgreek]
In Hughes (1750) and Pennant (1766).
[quotlat]
In Sloane (1707), Blair (1723), Hughes (1750), Dodd (1752), Borlase (1758), Pennant (1766), Bancroft (1769), Wilson (1808), Lawrence (1819), Godman (1828) and Lincoln (1832).
[quotwelsh]
In Pennant (1766).
[table]
In Keill (1717), Gibson (1720), Withering (1776), Speechly (1786), Lawrence (1819), Godman (1828), Lincoln (1832), Macalister (1876) and Packard (1898).

5. Concluding remarks

The criteria and conventions explained in this chapter have been discussed at length. At MuStE we have made our best efforts to come up with solutions to the challenges posed by TEI constraints and omissions. From the early 2000s, our primary purpose has been to keep these criteria and conventions simple and as closely connected to the standards as possible, to offer CELiST users the best tools to conduct their research with the least difficulty possible. To achieve this goal, some editorial decisions had to be made in particular situations so that we can guarantee the exploitation of the corpus at its full potential, given the particularities of the texts included in the corpus, especially those dating from an earlier period.

We assume the need to reconsider the use of editorial marks in CELiST, as further implementation of XML may make these marks unnecessary. Apart from this, we are open to suggestions from users and would welcome their feedback on our corpus and tool. In a nutshell, we believe that the Coruña Corpus in general, and the CELiST sub-corpus in particular, along with the CCT and the criteria adopted here may provide solid and manageable grounds to investigate the scientific register of the English language in the eighteenth and nineteenth centuries.

Acknowledgements

We would like to acknowledge the work carried out by the Emacs developers, the TEI Consortium and the Unicode Co*ns*ortium, which has been of significant help to our project.

CHAPTER 3

A look beyond the texts

The samples in the eighteenth-century Corpus of English Life Sciences Texts

Inés Lareo and Isabel Moskowich
University of Vigo / University of A Coruña

1. Introduction

This chapter provides bibliographical information on the eighteenth-century section of the Corpus of English Life Sciences Texts (CELiST) (Lareo *et al.* 2020). We also describe the editorial changes made by compilers in the text samples that are shown in the Coruña Corpus Tool (CCT). Although the metadata files for each sample displayed in the CCT's info window contain information on both the author and the text, the present chapter offers a more detailed account, with data taken from the works' prefatory material and other sources used to edit and tag the samples.

Thus, we will comment on the way each work was published and the role of publishers and sellers as the field of science developed. Raven distinguishes three types of booksellers:

> those who invested and dealt the ownership of the copyright to publication [...], and those who either printed, sold or distributed books for existing copyright-owners, or who traded entirely outside the bounds of copyright materials.
>
> (Raven 2005: 22)

We will see that some of the works in CELiST were published by subscribers, but most were printed for the author, that is, "to publish at his or her own risk" (Downie 2013: 61). This was a common option in the eighteenth century when authors wanted to see their works in print and to retain copyright, so common indeed that Sher (2006), discussing the situation in Scotland, notes that "in the second half of the century, as the enlightenment and the book reveals, a number of Scottish authors decided to retain the copyright of their works and take the risk of publishing for themselves" (Sher 2006: 72). The Appendix summarises this information,

https://doi.org/10.1075/z.237.03lar

as well as data about publishers, printers and copyright-holders involved in the development of science from 1700 to 1800.

The fullest imprint of each work has been included in the references to this chapter, because this has been used to identify the people involved in the publication process. Following the place of publication, the imprints typically name three groups of people (Foxon and McLaverty 1991: 2): the printer, the copyright-holding bookseller for whom the book was printed, and the publisher who actually distributed or sold the work. Thus: "London: printed by X (the printer), for Y (the bookseller who owned the copyright), and sold by Z". Foxon and McLaverty (1991: 2) explain that a publisher was, "one who distributes books without having any other responsibility – he does not own the copyright or employ a printer, or even know the author". Feather (1985: 60) also states that X is the printer, but that Y is the owner of the copyright, and Z the distributor or wholesaler. If the copyright holder and the distributor were the same person, we usually find, "Printed for Y, or: Printed for and sold by Y, or: Printed and sold by Y".

The column DESCRIPTION in the Appendix includes the different formulae used in these corpus samples. The COPYRIGHT column illustrates that copyrights were quite often sold by authors. Wigelsworth (2010: 156) notes that "the sale of the copyright to a printer or bookseller represented the extent of authorial income in the majority of cases in eighteenth-century London". It was very common to sell copyrights, normally to the publisher or bookseller, who kept them forever under the Common Law; although the Copyright Act was introduced in 1710, it was not immediately applied (Raven 2007: 128).

So, authors who wished to publish their work and make money from it had three paths open to them (Prescott 2003: 103): to be engaged directly with a bookseller, to seek an influential and wealthy patron, or to publish a work by subscription or commission. The advantage of this final option was authorial independence, but the author might also have needed to satisfy the wishes and expectations of the patron-subscribers, which would not always have been shared by the author. Nevertheless, this avenue did ensure payment in advance, thus making money from the work without the author risking their own funds. Some authors in CELiST chose this option. Despite this, selling the copyright to the bookseller was the most obvious way of making money from writing at the time (Prescott 2003).

The Licensing Act of 1695 effectively meant that printing was no longer restricted to London, the university towns of Oxford and Cambridge, plus York, and the book trade now gradually spread to the provinces (Belanger 1982: 6). Yet this was also the moment when the first associations of London booksellers appeared, which would go on to dominate the English book trade through the eighteenth century. This London-centric situation is reflected in the eighteenth-century CELiST (see column PLACE PUB in Appendix).

Dedications were so common from the beginning of the printing era that they constitute an object of study in themselves. Brown (1913) distinguishes some thirty categories, including dedications to the Virgin Mary, royalty, dignitaries, institutions, friends, etc.; meanwhile, Rogers' (1993) study examines 933 dedications printed in a variety of books during the eighteenth century. Griffin (1996: 24) mentions the extra value of dedications, noting that "if a work comes complete with a dedication to a noble lord who has in some sense authorized it or given his stamp of approval, it will tend to sell better". The naming of a patron or dedicatee effectively attached their good reputation to a book, and hence booksellers assumed less risk in buying the property or copyright; the success of a book also showed the named patron "as an arbiter of taste" (Griffin 1996: 25), thus increasing their social prestige or recognition. However, dedications began to lose their initial effect and were "by 1750 a mere formality, signifying no more than a fine leather binding" (Griffin 1996: 268). The column DEDI in the Appendix shows that eleven of the twenty works in the eighteenth-century CELiST include a dedication of some kind.

In what follows, we will describe each sample in the eighteenth-century CELiST, providing its shortened title, the pages sampled, and the number of words. We will also provide bibliographical information such as the number of chapters and plates, list of subscribers, dedication, and any information on editions available. This will be followed by data extracted from the prefaces, with attention to comments made by authors regarding the expected audience, language used, genre mentioned, plus any other relevant comments. We will also note the level of technicality of the sample and any changes made during the work's edition process.

2. The eighteenth-century samples in CELiST

1707. James Douglas, M. D.

Myographiæ Comparatæ Specimen (Pages 66–127, 10,005 words)

This work, extending to 267 pages, is divided into forty-six chapters and an Appendix, and describes the muscles in humans and quadrupeds. The descriptions are generally brief. The sixty-one pages sampled are taken from chapters XXII to XXXVII. Contrary to the customary practice at the time, the book contains no plates or figures, dedication, or list of subscribers. It was printed by William

Bowyer[1] for George Strachan,[2] an eminent Scottish bookseller; as noted above, this was one of the typical methods of seeing one's work in print without any personal outlay.

In his preface the author describes the book as a Treatise in which the muscles of a dog and a man are described and compared, this for two reasons. First, to show the use of the muscles during the motions of a dog and to highlight the differences in respect of origin and insertion. Second, "for the benefit and convenience of the young students of anatomy, being concerted in the most easy and distinct way" (Douglas 1707: viii). The author also mentions that "he has purposely omitted the Anatomical Administration, or the manner of raising the muscles, upon the consideration, that it would have inlarged the bulk of this treatise, which is designed for a manual, fit to be carried about to publick dissections, and would have inflamed its price, without doing the young students any additional service" (Douglas 1707: x). In a separate chapter, Douglas provides an introduction with a short account of the structure and composition of muscles in general, explaining the core vocabulary used, for younger readers and beginning students. That is, scientific advances and the dissemination of knowledge are at the forefront of the author's aims throughout.

Although Douglas refers to the book as a Treatise, the sample was classified as a Guide in CELiST because it is concise and designed as a conveniently sized manual rather than as a learned volume including extensive methodological details. Given its intended audience (young students attending dissections), the degree of technicality of the work is low.

The work's layout speaks of its nature: margin notes are used to show the exact paragraph where topics such as "origin, insertion and use" are described for the ease of students. For compilation, these margin notes have been relocated and marked with the editorial mark [margin note], either at the beginning or at the end

1. William Bowyer (1663–1737) was a London printer admitted to the Company of Stationers in 1686. In 1713 a fire destroyed his printing office and dwelling and killed one member of the family. Many valuable works and manuscripts were lost. Contributions from friends and subscriptions from his own professional fraternity allowed him to continue his business and return to White Friars, where he became the foremost printer of his day, until he was overshadowed by the fame of his learned son. The latter was taken into partnership in 1722 to correct the press, while his father retained an executive role until his death, the imprint continuing to be "Printed by William Bowyer" (ODNB).

2. George Strachan (or Strahan) was a leading Scottish bookseller and publisher in London. He arrived there from Edinburgh in 1737 and was one of the six booksellers that had contracts with the Society for the Encouragement of Learning. As Raven (2007: 164) notes, "he eclipsed even the Bowyers and Nichols". His wealth derived from investment in copyrights. He was the printer and publisher of Johnson's *Dictionary* and of many other well-known writers, among them Hume, Robertson and Adam Smith (Plomer 1900, 1922; Raven 2007).

of a sentence, this to avoid the inclusion of non-analysable material in the middle of sentences and hence making the CCT more useful for linguistic research. This work was very popular and new editions appeared in 1729 (in Latin), 1750, 1755, 1760, 1763, 1775 and 1777.

1707. Sir Hans Sloane

A Voyage to the Islands Madera, Barbados, Nieves, S. Christophers and Jamaica ... (Pages 49–70, 10,002 words)

This work extends to 418 pages divided into seventeen chapters, an introduction with independent numbering, and around 155 copperplates at the end. As the author explains on the front page, images of those things described in the book which had not been previously engraved are placed at the end of the book "as big as the life". The work itself provides an account of the travels and observations of Sloane while he was in Jamaica. This first volume is concerned mainly with plants, and the second (published in 1725) covered Jamaican zoology. Both were illustrated with engravings by Michael van der Gucht, one of the best in his field, based on original drawings by Reverend Moore in Jamaica and (usually) by Everhardus Kickius back in England. Although Sloane published little else of a scientific nature, these volumes were so well received that the Royal Society elected him President in 1727.

The book includes a dedication to the Queen and was printed by Benjamin Motte[3] "for the author", one of the ways of publishing works quickly while keeping copyrights.

In the preface (without page numbers) the author explains his reasons for travelling to Jamaica and his desire to describe its nature and to compare it to the plants cultivated in England. He describes very humbly how this work came to light after being corrected by Mr. Ray, whose method of classification, presented in his *History of Plants*, is followed by Sloane here. In his *captatio benevolentiae*, he also apologises "for the great many faults, nor only in hypotheses or opinions, but also for the imperfections in the language" (Sloane 1707). He claims that knowledge of Natural History is based on the observation of facts and is therefore more certain than other sciences. He mentions three uses of his *History*: to contribute to the spread of knowledge about nature, to teach the inhabitants of these territories where specific

3. Benjamin Motte (d. 1710) was a prosperous London printer of presumed French ancestry. His name is associated with the publication of Samuel Wesley's works. He was the publisher of *Gulliver's Travels* and several textbooks, classical works, translations and a number of illustrated books. After his death, his widow Anne Clarke continued the printing business, and even though their son Benjamin took over the family firm in 1715, she remained active in the trade until at least 1731 (ODNB; Plomer 1922).

plants grow and how to recognise, cultivate and use them, and finally to know more about certain plants and words through providing their history in the catalogue, which includes citations from earlier sources of information. Sloane aimed to avoid any confusion in nomenclature, in that his intended readership was the population of Jamaica as well as readers with a general interest in nature. For this reason, and although the author uses specialised terminology, descriptions are simple and easy to understand by a novice reader, the degree of technicality being low or medium.

The work was classified as a Treatise in CELiST. However, together with those by Brickell (1737); Hughes (1750) and Bancroft (1769) it could also be considered as belonging to the "discovery voyages" category, so important and frequent at a time when scientists were searching for new orders in the natural world.

As regards the preparation of the sample, we have used delete tags to mark description and figures included at the beginning of each explanation. In this way, such words, although excluded from the search index, will nevertheless appear in the Info Window so that no context is lost. We have also made sure that certain forms were not printing mistakes, such as *sometimes* and *somtimes* on page 51, variants that were used interchangeably at the time.

1717. James Keill

Essays on Several Parts of Animal Oeconomy (Pages 95–170, 10,014 words)

Running to some 233 pages, this work is divided into five essays, the sample in CELiST being the longest of these. In 1708 Keill published *An Account of Animal Secretion: the Quantity of Blood in the Humane Body, and Muscular Motion*, containing three of these chapters; meanwhile, Essays II and III, "Of the velocity of the blood" and "Of the force of the heart in driving the blood through the whole body", are new material for this 1717 edition. In the preface the author describes animal economy as being a considerable part of natural philosophy (Keill 1717: xviii) and notes the importance of physicians making anatomical experiments and observations on the structure of the body, the nature of blood, and secretions, since "if the animal economy were perfectly understood, and the histories of diseases exactly known, the right method for curing each disease might be evidently and certainly deduced" (Keill 1717: xvii).

In Keill we observe an inconsistency often found in other authors in the Coruña Corpus (CC): each of the chapters is called an "Essay", but in his preface the five different parts are referred to as "Treatises" (Keill 1717: xxiv). Later in the preface we read: "any one with a moderate skill in the mathematicks may understand these discourses" (Keill 1717: xxviii). Given that the text contains explanations of experiments, it has been classified as an Essay (experimental Essay) in CELiST.

The book was printed for George Strahan (see note 2). It does not include any dedication or advertisement although we can infer that its intended readership included physicians and surgeons with some skill in mathematics. The topic, vocabulary, formulae and explanations are all quite specialised, and the sample can be classified as technical. A fourth edition was printed in London in 1738, which testifies to its popularity.

1720. William Gibson

The Farriers new Guide (Pages 4–37, 10,000 words)

The book (303 pages plus plates) is divided into 94 chapters and was published for William Taylor.[4] It was considered the best and most accurate book on the treatment of equine diseases of horses written in English, and indeed a further eight editions appeared before the end of the 1730s. Although it is referred to as a Treatise throughout the preface, the book is addressed to farriers,[5] a specialised audience, and the word *guide* is used in the very title. Given that it is a concise and basic introduction, it was classified as a Guide in CELiST.

In his preface Gibson reviews some existing works dealing with horses and disease, explaining that he writes this book "because all the advantages made in this field and the new methods employed have not been communicated to the public, and because of the frequent complains of those who are the greatest sufferers by the deficiency of this art" (Gibson 1720: ix). A military surgeon, Gibson used his army experience to treat horses in London for almost thirty years, and set out all that he had learnt in this first book. The work contains a description of equine anatomy, diseases, and methods of treatment, in which the author has "wholly study'd the Benefit of such as are unacquainted with the Subject" (Gibson 1720: ix). When he mentions the work of two of the most famous farriers to Charles II, Signior de Ruini and Mr. Snape, he criticises the former for not writing in English and the latter for

4. William Taylor, known as a man of capital, respectability and of a certain social position, died in 1724 after having traded under The Sign of the Ship. Thomas Longman bought the business from Taylor's executors, John Osborne and William Innys (Briggs 2008).

5. Farriers specialised in shoeing and treating horses, and acted as veterinarians. Veterinary colleges were intended to mark the difference between the educated vets and the farrier, but many horse doctors continued to call themselves farriers until the late nineteenth century. Although veterinarians might have wished to be compared to doctors of humans, due to their formal training, their profession was classified as being at the same level as farriers. In turn, professional farriers were not illiterate: evidence from their bills, receipts and writings implies that they could have learned their skills through apprenticeship and the reading of popular veterinary texts like Gibson's (Derry 2006; Heymering 2009; Lane 1993; Porter 1993; Swabe 1999).

dealing only with anatomy. Yet Gibson, aware of the goal of his work (to be used as a guide), also refers to the functions of all the parts of a horse, this written "in as short and concise a manner as possible" and with the necessary explanations to render everything intelligible and useful. In fact, the book begins with a basic introduction and description of the core vocabulary used in the work.

1723. Patrick Blair

Pharmaco-Botanologia: Or, An Alphabetical and Classical Dissertation...
(Pages 2–25, 10,018 words)

This volume is divided into seven decades[6] (each containing the description of forty plants), the last six of which have a separate title page and preface. The whole work was published between 1723 and 1728[7] for reasons explained in the preface, with continuous page numbering (until page 343). It was printed for well-known booksellers such as George Strahan (see note 2), William and John Innys[8] and William Mears.[9]

Plants are presented in alphabetical order and classified in tribes, but the author died when he had reached the letter H. In his preface, Blair explains that he has arranged the plants in alphabetical order "in this catalogue in a botanical as well as physical respect; that so he who will study botany, may attain to the knowledge of plants, without regard to their virtues; and the young beginner in physick, or pharmacy, may be taught their virtues, without prying into the niceties of botany" (Blair 1723: i). Although he uses the word *catalogue* in the preface, the work was classified as a Treatise in CELiST because the subtitle of each the seven parts includes this term (*Pharmaco-botanologia or, a treatise of dispensatory plants, alphabetically*

6. The word *decad* is an old form of *decade* defined in the OED as the division of a literary work, containing ten books or parts.

7. Decad I (1723: 1–48), Decad II (1724: 49–96), Decad III (1725: 97–144), Decad IV (1727: 145–188), Decad V (1727: 189–236); Decad VI (1727: 237–296), Decad VII (1728: 297–343).

8. They were printers to the Royal Society specialising in natural and experimental philosophy. William, one of the leading booksellers in London in the first quarter of the eighteenth century, published at The Sign of the Princess Arms and was a leading specialist in science and medicine, associated with large-scale publishing projects. He married William Taylor's widow, whose bookselling and publishing business was sold to Thomas Longman on her husband's death. William also contributed to Bowyer's fund and joined John. Together they published many of Newton's works (Briggs 2008; Plomer 1922).

9. He was a London bookseller who offered Bradley's *A Philosophical Account of the Works of Nature* by subscription. The names of subscribers were recorded for posterity. He was also the publisher of Defoe's *New Voyage around the World* (Plomer 1922; Wigelsworth 2010).

and classically disposed) and because it contains a methodological part and more information than a catalogue.

Although its primary readership is of young medicine and pharmacy students, it is also addressed to the general public with an interest in learning more about Botany. Blair's intention is to offer a concise and well-arranged compendium of plants, including their description and uses.

It is notable that the author apologises for using English instead of Latin: "instead of obliging the unlearn'd, to acquire such a competency of the universal languages, as to make them capable to understand these, I am forced to convert them into the mother tongue, by which the most impudent pretender to physick, as well as the most regularly educated in that delightful and useful science, may become equally knowing" (Blair 1723: vii). After criticising apothecaries and herb-women, he states that he had to adapt his discourse to those of the "meanest capacity", including adaptations of Greek and Latin terms in that their difficulty for the average reader has been the cause of the neglect of Botany. His aim is to reach a wide audience. This is made clear when he writes: "so delightful a science so easy, that it may be universally known instead of being a hidden treasure in the hands of a very few" (Blair 1723: viii).

Blair also acknowledges that he opted for a monthly publication, since doing otherwise would have implied too great a workload. Besides, this also constituted a smaller financial risk for the project, a lower cost to buyers, and it also meant that readers could communicate to him their observations as he advanced in his work.

In the fragment sampled in CELiST we have omitted the first section for each plant description, where the author includes Latin and Greek terms and gives previous references to the plant itself. Since most of these words are abbreviations and numbers, they can only be seen in the XML file in the (CCT) info window. The level of technicality of this work, as the author says, is low to make it accessible to a wide audience.

1730. Thomas Boreman

A description of three hundred animals (Pages 2–63, 10,017 words)

Thomas Boreman was a bookseller and printer who owned several bookstalls in London. He was one of the first to write for children and to specialise in children's books and games. The work selected here (which is 212 pages long) is divided into four books or parts and is considered the first children's work on zoology (Ritvo 1985). Boreman includes an address to the reader (those instructing children) in which he notes the importance of children's instruction and his method for attracting their attention. He also explains that he has adapted the information for those of all abilities, and especially to encourage children to read as well as for their teaching with entertainment. This intended readership was formed by middle-class children who would be part of the future commercial society, one that would consider Natural History to be one of the most favoured and respected fields of knowledge (Ritvo 1985). Boreman was right in thinking that children deserved some attention as readers, and his choice of animal life was a successful one. As science became more sophisticated, the very term *natural history*, which had an aura of amateurism and speculation, gave way to soberer, more precise terms like *zoology* (Ritvo 1985: 5).

Although Boreman refers to his work as a brief Essay, perhaps using one of the senses first recorded in the OED ("a composition of moderate length on any particular subject"), we have classified the sample in CELiST as a Textbook since we consider that an Essay is "a composition more or less elaborate in style, though limited in range".

The book was published for three booksellers: the author himself, Richard Ware, another well-known bookseller and stationer in Ludgate-Hill who was not afraid of taking risks with new publications (Nichols 1814: 485), and Thomas Game, a bookseller in the Strand. The work was so popular that by the end of the nineteenth century thirty-eight editions had appeared, mostly in London, although with one in Dublin in 1762 and another in Edinburgh in 1799. This work was followed by similar ones published by Boreman alone.

1737. Elizabeth Blackwell

Blackwell, Elizabeth. 1737. A Curious Herbal… (Pages 1–25, 10,019 words)

Elizabeth Blackwell is the first female author whose work (in two volumes) has been sampled for CELiST. In fact, she seems to have been the first woman to publish works on Botany in English. Her own financial difficulties seemed to persuade her to overcome a fear of publishing in a men's world. Like many well-situated women, she had received tuition in drawing and painting, and her herbal contains plates that she had painted. Her work was initially published weekly, four plates depicting plants and their descriptions in each instalment, until she had produced some 500 images. The volume sampled includes the description of 152 plates in forty pages of descriptions and an index. The descriptions are so concise and organised in such a way that the sample has been classified as a Catalogue.[10]

10. Raven (2005: 16), however, considers that "the term *catalogues* issued by booksellers between 1668 and 1709 provides (sic) a very general sense of market interests", giving the impression that this was not a genre.

Blackwell's intention was to offer a reference book for apothecaries, one that included the species discovered in America. The first copies were published at the author's expenses, but given the high cost and Blackwell's financial difficulties, Samuel Harding soon became the first bookseller (Feather 1985). The book enjoyed considerable success at the time, even among physicians and apothecaries, and a second volume followed in 1738 or early 1739; a new edition also appeared in London in 1782. The book was published by subscription, and Blackwell used the income to pay off her husband's debts and release him from debtors' prison. However, the work had a complex history. Blackwell eventually had to sell part of her publication rights to Samuel Harding (see the imprint in the References section) as a means of raising money for her husband's new debts. In turn, Harding approached John Nourse, who was in a better situation to advertise and distribute the book, and later copies were printed with his name. In 1739 Nourse signed an agreement with Alexander Blackwell by which he paid 150 pounds for a one-third share in the five hundred plates of his wife's book, and he later acquired half of the rights (Feather 1980; Wiles 1957: 176, 183).

Blackwell describes each plant in four sections. The first is devoted to a physical description of leaves and roots; the second describes its location and growth; the third summarises its uses in the kitchen and in medicine, the latter information taken from Joseph Miller's *Botanicum Officinale*, with his consent.[11] The final section includes the plant's names in different languages. These names were provided by her husband and do not represent the author's language, so they have been omitted from the final count of words in CELiST, although they are visible in the xml file.

Other editorial changes have been made. For instance, we have edited some abbreviations such as *ye*, which have now been written on the same line rather than with the original superscript <e>. The same procedure has been followed with the abbreviations *Mr* or *Dr*, which also originally had a superscript <r>. Abbreviated *which* and *with*, in the original, have been rendered as whole words. We have also included three dashes in those places where Blackwell's work had a blank.

11. Blackwell included one of Miller's quotations of Schroder containing the word *schirrthi*, but we have not been able to check its spelling. One option is that it would have been *scirrhia*, a member of Dothideomycetes, capnodiales, mycosphaerellaceae (US National Library of Medicine 2020).

1737. John Brickell M. D.

The Natural History of North-Carolina... (Pages 107–137, 10,011 words)

The work, which extends to 408 pages, contains a folding map and four engraved plates of birds, beasts, and fishes. Printed by James Carson[12] for the author, it is another example of a discovery book with sections devoted to geographical description, religion, language, diseases, houses, history, flora, animal life, and an account of the Native American peoples of Carolina. It is classified as a Treatise in CELiST.

The book includes a dedication to Richard, Earl of Anglesey, Viscount Valentia, Lord Baron Altham, Lord Baron Mountnorris, Lord Baron of Newport Pagnel and Custos Rotulorum of the County of Wexford, and mentions the need of having a patron. In the preface Brickell presents his work as "a compendious collection of most things yet known in that part of the world; wherein I have laid down every thing with impartiality and truth, in the most plain and easie terms" (Brickell 1737: iv). After referring to the Native Americans as "these miserable people" (Brickell 1737: v) he promises to describe for the reader many secrets from that part of the world and claims exactness in his report. The work also contains a list of over two-hundred subscribers. It was republished in 1743 with a new title-page omitting the name of the printer, but including the name of the new seller, Charles Corbett.

The work is the result of Brickell's observations during his voyage to North Carolina, where he worked as a physician (1729–31), and was seen a plagiarised re-working of Lawson's *A new Voyage to Carolina* (1709), although Brickell acknowledged the use of all the material about that territory that was drawn from Lawson's work. Brickell included valuable new information of his own, such as the medical uses of plants and animals. His work is fuller, more systematic, and seems more like that of a student (Brickell 2008).[13] The style used by the author is plain, as he promises it will be, given that his descriptions and comments are directed at a wide range of readers, thus avoiding technical terms and including popular sayings and anecdotes.

---

12. An Irish publisher and printer associated in the 1720s and 1730s with the Irish newspaper press and held in high esteem as a stationer. He published the *Dublin Weekly Journal* in 1725, an idea presented to him by a group of philosophers seeking a place to publish their writings. This continued as a literary journal until 1747 when it also became involved in politics and acquired a more diverse audience. Carson is also recognised for being the first to illustrate advertisements with woodcuts in Ireland (Munter 1967: 60).

13. This is a digital version from 2008 (University of Michigan Library), at the beginning of which J. Bryan Grimes adds a Note with some explanations about the controversy.

1743. George Edwards

A Natural History of Uncommon Birds... (Pages 1–23, 10,012 words)

The book contains 52 pages and the same quantity of unnumbered plates. It was published in four parts between 1743 and 1751 under a different title, with three supplementary volumes, entitled *Gleanings of Natural History*, published in 1758, 1760 and 1764. The bird descriptions included were collected over the course of more than twenty years. In 1751, when the work was concluded and having received the Gold Medal of the Royal Society, a new title-page was added, providing a more accurate and general title. The new title-page also included the author's acknowledgement of the medal, the names of those who had been continuous subscribers to his work (as well as some who had been omitted) and some suggestions by friends. The text is organised in such a way that each page is devoted to a very detailed description of a bird, yet one which is accessible for a general reader, this because the author's "chief aim was rather to be understood than to write correctly" (Edwards 1743: xx). The book was printed for the author and has been classified as a Treatise in CELiST.

In the dedication to the Honourable President and Fellows of the Royal College of Physicians of London, Edwards mentions the favourable reception received when he presented the first essays to the College, and asks for the patronage of the President.

In the preface he writes about the differences between animals and men. He includes a section about birds and claims that drawings of nature should be promoted among young people as an important part of natural history and an essential part of descriptions here. Edwards used his own drawings as a means of reducing publication costs. He refers to the balance needed so that, on the one hand, the description of natural things "may be done without following the minute steps of some author, who have wrote large books on single birds and plants, for long descriptions lead the mind into mazes and confusion, and tired rather than instruct. On the other hand, too brief descriptions should be avoided, because they consist only of such general forms and colours that are common to many things of the same genus" (Edwards 1743: xiv). The author defends plain language as the means of making the study of nature both useful and pleasant to people. Edwards wants to convey knowledge to laymen and expresses his thanks to Dr. Mortimer, secretary of the Royal Society, for his help with the Latin names in the book.

1750. Griffith Hughes

The Natural History of Barbados (Pages 61–88, 10,005 words)

The work, which extends to 314 pages, is divided into ten parts or books, plus 29 plates, an index, and a glossary of technical terms. Each book deals with a different topic relating to natural history (people, water, climate, geography, mineral world, animals, plants, fishes, marine productions, etc.). The information was compiled by Rev. Hughes during his four-year stay in Barbados as the Rector of the parish of St. Lucy. It was described as a Treatise by the author himself and this is its classification in CELiST. It is the third voyage diary, along with those of Sloane and Brickell.

The work includes a dedication to His Grace Thomas, Lord Archbishop of Canterbury, and a list of a thousand subscribers who signed up during the promotion of the project in both London and Virginia. Nevertheless, the book was printed for the author.

In the preface Hughes reflects on Experimental Philosophy and Natural History and relevant studies that were already in existence. He also mentions that "a pompous style and language would ill fit a bare narration of facts, or the description of a grovelling plant"; thus, he decides to use a simple style (Hughes 1750: vi). His descriptions are brief and easy to understand. Those included in this sample refer to land animals, birds and flying insects.

This was a widely popular work and also constituted the first published piece of research on Barbados. It was not well received among some of his fellow scientists, who described it as lacking both accuracy and scholarship. However, Hughes had the support of Sir Hans Sloane and Martin Folkes, and his work was praised by Linnaeus himself.

1752. James Solas Dodd

An Essay towards a Natural History of the Herring (Pages 23–83, 10,006)

Dodd's work, one of the first on this topic, contains 178 pages plus plates, divided into eight chapters. Dodd, who was President of the Robin Hood Society, dedicates the work "to the Right Worshipful Slingsby Bethell, Esq. President, to the right Worshipful Stephen Theodore Janssen, Esq. Vice-president, and to the council of the free British white Herring-fishery". He describes the herring as highly nutritious and medicinal, and "good for food in the plagues." The author comments on the division of herrings and includes a description of the fish plus comments on methods of fishing and the advantages to British fisheries. He then gives several recipes for cataplasms of herrings in use at the time; and he recommends, among others, the oil of herrings as "of excellent service in cramps and convulsions" and the elixir "ossium halecum" as an effective remedy against fever. His recipe for making this elixir of herring bones is a curiosity.

The work, published for Thomas Vincent,[14] copyright-holder, is a very detailed account addressed to the men of the Herring Fishery Council, that is, to readers with experience in the field. The preface begins with an account of the situation of herring fisheries in the world and a review of the literature. Dodd notes that he has taken ideas and words from many authors, and has decided not to quote them all to avoid making the text tedious. He refers to his work as "a short account, that the curious may observe with what elegance these fish are form'd" (Dodd 1752: vii) and finally calls it an Essay (Dodd 1752: ix), which is how we have classified it in CELiST.

This sample includes material from chapters I to V, those dealing with the division and names, description and ways of fishing herrings, plus methods of salting and curing and their use as food. Since the chapters are very short, the sample included in CELiST encompasses almost half the book.

1758. William Borlase

The Natural History of Cornwall... (Pages 261–284, 10,009 words)

The book was printed by William Jackson[15] for the author, as was common in the period. Its 326 pages are divided into 28 full-page copperplates and twenty-six chapters devoted not only to the inhabitants of Cornwall and their languages but also to its animal life, plants, geology and geography. It is addressed to the nobility and gentry of Cornwall and to the "Candid Reader" and includes an explanation of the method followed, acknowledgments, and a long list of nearly four hundred subscribers.

In the dedication the author claims that Natural History is a very extensive science, dealing with both animate and inanimate substances,[16] and looking at their relations, properties and uses. In his explanations to the "Candid Reader", Borlase mentions that his situation, far from books that the "literati" had written on the same subject and also far from the assemblies of the learned, was a great disadvantage.[17] He also refers to one of the trending topics of his time: the inadequacies of the English language in naming elements of nature, and the need to use Greek

14. No information was found about this bookseller, but Timperley (1839: 600) mentions him as one of the subscribers to the William Bowyer fund after the fire.

15. Jackson was from Leeds, although he spent most of his life in Oxford. He was not only a newspaper proprietor but also the university printer and one of the founders of the Oxford's Old Bank among other activities (Headington 2014).

16. For the difference between Natural history and Life Sciences, see chapter one (Moskowich 2021).

17. It seems that Borlase's early years as a student were not very successful, as "he was put early to school at Penzance; where his master used to say, 'he could learn, but did not'" (Nichols 1812b: 291). However, he considered that to understand nature one does not need to be a specialist in the field, but a good observer.

and Latin terms here. In fact, he says that he was reluctant to use technical words, because "they are only easily understood by those who are most conversant in these studies" (Borlase 1758: viii) and because he wanted to reach a wider audience. He seems to try to gain his readers' approval by referring to the fact that he has been impartial and truthful, and also by noting that part of the work had already been published in the *Philosophical Transactions of the Royal Society*.

As editors, we have moved punctuation marks to a position before the superscript of notes, and have also changed square brackets to parentheses, because square brackets are only used for editorial marks in the CC. The sample was classified as a Treatise in CELiST not only because the author considers it so but also because it has the structure of one, with Latin terms, quotations by authorities, and very detailed descriptions.

1766. Thomas Pennant

The British Zoology (Pages 1–26, 10,010 words)

The work was printed between 1766 and 1767 in four volumes in folio edition, with one hundred and seven unnumbered copperplates of animals, presented in classes and divided according to Ray's system. The book lost money, having been printed on paper which was too large, but was so well appreciated by the public that further, smaller editions and additional volumes on all Pennant's major works were published. This work, *British Zoology*, was so popular that a fifth edition had appeared before 1812.

The book was printed by J. and J. March[18] for the Cymmrodorion Society,[19] of which Pennant was a member. The sample includes technical descriptions and a

18. John March was a printer of considerable eminence in Tower-Hill, and Master of the Stationers Company in 1790. He left a fortune to his widow and son, who continued his father's business under the name J. and J. March (Nichols 1812b: 441).

19. It was officially founded in London in 1751, emerging from an earlier society, the Most Honourable and Loyal Society of Ancient Britons, founded in 1715. This was an association of London Welshmen that sustained a charity school for their children, first in Clerkenwell, then in Gray's Inn Lane. The school, which would eventually become a girls' school, moved to Ashford in Middlesex in 1857, and closed in 2009. The aims of the society were three-fold: first, social and benevolent, holding frequent dinners where collections could be taken to help the Welsh charity school or any Welsh people in London who were in distress; second, literary and antiquarian, that is, holding discussions on Welsh history and literature, and helping to publish Welsh books; third, giving a lead to the Welsh at home, who lacked any kind of secular institutions of this kind. This book was sold for the benefit of the charity school at Clerkenwell, but since the publication in folio form lost money, the school received no benefit from it, and Pennant gave a donation of 100 pounds (The Honourable Society of Cymmrodorion 2020; Jenkins and Ramage 1951: 82–83).

classification of quadrupeds with a review of the literature and quotations by other authors, all of which led us to classify it as a Treatise in CELiST.

The book includes a dedication to the King in name of the President and Council of the Cymmrodorion Society, in which the author includes himself. In this dedication, Pennant refers to his work as *Natural History of the Quadrupeds and Birds of Great-Britain and Ireland*. Then, in the preface he mentions many well-known authors and claims that Zoology is the noblest part of Natural History, as its scope runs from man to vegetation. We can see in the work how the taxonomy of science is changing as new subfields develop. Pennant also explains that "this is a general view of the Natural History of our own country" (Pennant 1766: iii). As for his audience, he mentions the "*British* reader" for whom he wants to provide a summary of all the natural wonders in the islands and to show the usefulness of natural knowledge for all the arts. He also explains that the names used will be provided in several European languages.

We have made the following textual changes for the computerised sample. To render the information included in tables, we have repeated the headings of columns in each line. These changes do not represent any additional tokens in the word count because they have not been included in the final indexed texts. Although the work was printed in two columns, we have changed this into one, as a means of conforming to the layout common to all the CC samples. Other elements, such as the brief list of authors included at the top of each division, are only displayed in the info window of the CCT. Finally, we have also separated abbreviations from other items, as in "3000lb" which now appears as "3000 [lb]" in order to identify abbreviations when using the *CCT* software.

1769. Edward Bancroft

An Essay on the Natural History of Guiana (Pages 179–248, 10,009 words)

The book, of some 402 pages, comprises three letters and was printed in London for Thomas Becket and Peter Abraham de Hondt[20] in the Strand. These letters were written in 1766 in Guiana, where the author lived from 1763 to 1767. Our sample contains Letter II, which describes animals. In a dedication to William Pitcairn, fellow of the Royal College of Physicians in London and physician at St. Bartholomew's hospital, Bancroft refers to his work as an Essay. In the edition that

20. Previously apprenticed to British publisher Andrew Millar, Becket imported new French literature and published tracts and fashionable novels. From 1760 he was partnered by Peter Abraham de Hondt. Their partnership resulted in nearly 500 publications, and their business would ultimately be dedicated almost entirely to novels. Becket was declared bankrupt at Tully's Head in the Strand in 1779, but had reopened at Pall Mall by 1783 (Raven 2007: 160).

we have used there is also an advertisement explaining that although Bancroft had determined to conceal his name, his friends encouraged him to make it public since an anonymous publication would have suffered from a lack of credibility. The author's name was therefore added on the dedication page, since the rest of the book had been already printed. The advertisement also tells us that the first original letters dating from 1766 were not intended for publication (Bancroft 1769: ii). However, they were later edited and improved with the inclusion of the writer's recollections. A second edition appeared in Boston in 1810.

As editors, we have included the closing paragraph of this letter in the XML file in such a way that it is accessible in the info window of the CCT, although this does not alter the word count of the sample.

Regarding genre, the author refers to the whole work, both in the dedication and on the title-page, as an Essay. Nevertheless, the extract has been classified as a Letter in CELiST. Its content and vocabulary are easy to understand for the common reader. Bancroft himself (1769: 227) explains that he avoids very detailed description because of his lack of patience. He refers to the peculiar language of natural history whose main characteristic, like that of nature's operations, is simplicity (Bancroft 1769: 119).

1774. Oliver Goldsmith

An History of the Earth, and Animated Nature (Pages 111–161,10,035 words)

The work has eight volumes. We have extracted the sample in CELiST from volume VIII, which has 200 pages, plus a further 273 pages of index and plates. The volume includes Part III, "An history of insects" (Chapters I–V), Part IV "An history of the fourth order of insects" (Chapters I–VII), and Part V, "An history of the zoophytes" (Chapters VIII–XII). It was printed in London for John Nourse.[21]

The work sought to draw together virtually all that was known about the planet, its plants and animals, and even human beings, from a biological perspective. Goldsmith's intention was to write a popular scientific work, rather than a learned

21. Nourse was a bookseller to George III as well as a publisher and importer of French books and of scientific works. His dealings with a group of authors led to the opportunity to establish what would become a more familiar form of copyright agreement. Between 1735 and 1749, he acted as one of the booksellers for the Society for the Encouragement of Learning (Feather 1980). Goldsmith had signed a contract with William Griffin for an eight-volume natural history, adapting the final payment to the final number of volumes and transferring all his rights to Griffin forever. Griffin sold the rights to Goldsmith's work to Nourse in 1772, who became its sole owner and publisher (Prior 1837: 200–201, 218, 504).

or a formal one for naturalists. Therefore, this title was catalogued as a work of general and amusing information introduced with all the attraction of an elegant and perspicuous style (Prior 1837: 200). Prior describes Goldsmith's work in the following way:

> Its great charm is its style; combining that ease, freshness, and freedom which throw an irresistible attraction over his pages and render every reader of taste an admirer [...] and attracting readers who would have been repelled from the study of more elaborate works. (Prior 1837: 507)

Animated Nature went through over twenty editions, these extending into the Victorian era, and served as a popular source of information on the natural world. In the preface to the first volume, Goldsmith compares the system of natural history to a dictionary. He explains that the difference is that in the dictionary we are led from the name of a thing to its definition, whereas in the system of Natural History we are led from the definition to the name (Goldsmith 1774: iv). He also points out that books alone can never teach readers or students completely if there is no master to guide them or if they have no previous knowledge of many of the objects in nature (Goldsmith 1774: v). At one point, he admits that he has described many subjects in his own way, and that those professing themselves to be naturalists may find the work superficial (Goldsmith 1774: xiv). He also declares that he has tried to avoid difficult words or ostentatious learning, and ends the preface with the hope that the work be an innocent amusement for those who have nothing else to do, or who require some form of relaxation from work (Goldsmith 1774: xiv). Despite these considerations on the part of the author, we have classified the sample as a Treatise.

1774. William Withering

A botanical arrangement of all the vegetables, naturally growing in Great Britain... (Pages xvii–li, 32–33, 54, 93–94, 186, 193, 10,003 words)

The volume contains an initial description of its design followed by an easy introduction (page xvii to li) to Botany and directions for drying and preserving plants, plus an index. Vegetables are then arranged in orders and classes over 383 pages and many plates. The book was printed in Birmingham by Myles Swinney[22]

22. Swinney was a printer, bookseller, stationer, publisher and journalist who began printing in Birmingham around 1770. He was the printer and publisher of the Birmingham Chronicle (Equiano 2003).

for Thomas Cadell and Peter Elmsley[23] in the Strand and George Robinson[24] in Pater-noster-row, London. The sample included in CELiST is classified as a Guide because it is concise, intended for self-teaching, and written to help the non-learned reader to grasp the basics of this science, as described in the very title (*Being an attempt to render them familiar to those who are unacquainted with the learned languages*).

The first sentence in the introduction is addressed to students of Botany, explaining the main concepts and terms of this science in a simple way, and setting out the seven steps that must be followed to classify plants with the help of this book. At the end of this introduction the author concludes that "after conducting my pupils in this familiar manner through the different parts of the system, they no longer stand in need of my assistance" (Withering 1774: xlvi).

Even here, when dealing with the design of the work, Withering explains the system devised by Linnaeus and the difficulties (language and procedures) faced by people interested in Botany. This work then tries then to overcome these difficulties and to show the field "as easy as it is delightful" (Withering 1774: iv), illustrating that amateur Botany was very popular in Britain during the second half of the eighteenth century. Withering is conscious of his use of language and is concerned with conveying well-defined ideas. He tries to translate the Linnaean system into English and, like Gibson and Blair, defends the idea of using English terms whenever possible.

A second edition of this work appeared in three volumes, two in 1787 and another in 1792, produced in collaboration with the author's friend Jonathan Stokes. Withering himself produced the third edition under the title *An Arrangement of*

23. Elmsley, with Cadell, Dodsley and others, formed a literary club of booksellers who produced many important works. Elmsley was a Scot who was in trade since 1769 and succeeded Paul Vaillant in his important importing business. Cadell, nephew of a Bristol bookseller, also collaborated with William and Andrew Strahan but continued to trade under his own name until 1793, and was the richest copyright owner of his generation. In the 1760s Cadell, Robinson and Millar earned a reputation for their loyalty to writers, Cadell being part of the Royal Literary Fund. Cadell, among others, gave his name to pocketbooks (cookery, gardening, prayer or hymn books) or miscellanies, advertising their exclusive features (Cross 1985; ODNB; Raven 2007).

24. George Robinson was probably London's most prominent bookseller in the second half of the century (Nichols 1812b: 447). He had arrived in London in 1755 and established himself with John Rivington, William Johnston and John Roberts. In 1784 Robinson took into partnership his brother John and his son George. He published some of the most popular and influential periodicals of the century. His place among the country's great booksellers is unquestioned, and he gained a reputation as one of the greatest literary investors in the second half of the century and as a generous supporter of writers (Timperley 1839; Raven 2007).

British Plants, which appeared in four volumes in 1796. The work was very popular and following Withering's death further editions appeared under yet another title, *A Systematic Arrangement of British Plant*, edited first by his son and later by William MacGillivray, a Scottish ornithologist and natural historian. The final edition (the fourteenth) was published in 1877.

Withering's work is relevant because it was one of the first attempts to adapt the Linnaean system to English Botany, becoming a reference book for a wide range of readers. However, Withering's desire to protect potential women readers induced him to omit Linnaeus' mention of sexuality, dropping the sexual distinctions in the titles of Classes and Orders.

1786. William Speechly

A Treatise on the Culture of the Pine Apple and the Management of the Hot-house... (Pages 1–76, 10,017 words)

The work, 186 pages in length, is divided into three parts or books and an index. This particular edition also contains a related work, *A method to preserve peach and nectarine trees from mildew, &c.*, by a different author, Robert Browne.[25] It was printed in Dublin for Luke White.[26]

The sample in CELiST was taken from book I, dealing with the cultivation of pineapples, the description of their different parts, and how to propagate them. The book is dedicated to the Duke of Portland, for whom the author worked as head gardener. It was the Duke who first encouraged Speechly to write down and publish his knowledge of gardening. Here he thanks the Duke for his patronage and describes his work as an improvement of previous practice accompanied by a number of experiments. In the preface to the first edition of 1779 he refers to the book as a Treatise (Speechly 1786: vii, 52), the same label used in the CELiST classification. The last paragraph is addressed to his subscribers, thanking them for their support and encouragement, although no list of subscribers is included.

25. The 1779 and 1796 editions contained Speechly's work only.

26. Luke White was an Irish bookseller, operator of a lottery, and Whig politician. He started as a poor book dealer in Belfast and later in Dublin, buying and reselling books around the country. During his work as a publisher in Dawson Street, he became rich and his wealth led him to contract with the government for the supply of loans, while continuing to trade in books on a large scale (Webb 1878).

1789. James Bolton

An History of Fungusses, growing about Halifax (Pages 95–138, 10,027 words)

This work was published in three different volumes between 1788 and 1790, with copperplates drawn and etched by Bolton himself and accompanied by a separate volume with an Appendix. It was Bolton's second work and the first British work of its kind dedicated to the subject. The three volumes contain 138 beautiful hand-coloured illustrations of the specimens described by Bolton, the result of more than twenty years of observation.

Volume I contains a dedication to the Earl of Gainsborough as a promoter of the most rational of sciences, English Botany. The author also thanks the Duchess Dowager of Portland for her patronage. Bolton mentions that the cryptogamia class had been neglected in previous catalogues, thus justifying his own work.

Bolton's expected audience was of botanists as well as countrymen and "walkers spending a few hours of rural retirement". Due to this variety of readers, explanations include both a theoretical and a practical part, the latter "to divert the melancholy of an amusive walk or to pass agreeably a few hours of rural retirement" (Bolton 1789: v).

The material used for the corpus is drawn from pages 93 to 138, from volume II. It was published by Joseph Brook[27] for the author, as were most of the works we have dealt with here. It is divided into genera with unnumbered copperplates (from V to XI, 93–134), an introduction dated 1789 (xxviii–xxxii) and an Appendix with fungi (135–138). The sample was classified as a Catalogue in CELiST coinciding with the author's idea.

We have edited the text to preserve all instances of English scientific prose, whereas Latin nomenclature and descriptions of other references are restricted and can only be seen in the Info Window of the CCT so that they do not form part of the analysable material in the corpus, this because they do not represent the author's own language.

The work was so popular that it was also published in German with lower quality plates as *Geschichte der Merkwürdigsten* (1795–1820). Interestingly, in the preface to the German edition, Bolton explains that he has been cautious in making reference to other authors, since inexact references cause confusion and tend to distract, not to inform.

27. According to Law (1988) John Brook, Joseph's father, was the first printer in Huddersfield. Josiah Lancashire, with whom in 1799 Joseph registered three printing presses in accordance with the requirements of the Unlawful Societies Act of that year, was his partner until 1807. After this date Brook printed only under his own name. After his death in 1788, Joseph inherited his estate.

1794. Edward Donovan

Instructions for collecting and preserving various Subjects of Natural History... (Pages 33–84, 10,019 words)

The 86 pages of this work are divided into two parts. The first of these (1–10 and 55–86) deals with "Instructions for collecting and preserving various Subjects of Natural History; as Animals, Birds, Reptiles, Shells, Corals, Plants, &c." with two numbered plates, and the second (11–54) with a "Treatise on the Management of Insects in their several Status". The sample included in CELiST, and classified as a Guide, contains part of the instructions and the treatise. It deals with the different classes and stages of insects, plus instructions for collecting, setting and preserving them. It was printed for the author and sold by Rivingtons.[28]

The work was announced in the *Monthly Review, or Literary Journal* (1794: 237). It is also stated that a German translation was published in Zurich in 1797 by Dr. J. J. Römer (Reuss 1804), and a further edition appeared in London in 1805.

In the two-page preface, Donovan declares that the guide is aimed at the "unexperienced collector in the form of a pocket assistant". He also claims that "the treatise on insects and the instructions for the preservation of birds, appeared at the express request of many subscribers", adding extra information for early naturalists. Thus, we suppose that although a list of subscribers is not included in the book, the publication was supported by them. He also admits to have excluded some scientific data to avoid making the volume too long and expensive for a general readership. Accordingly, the style is clear and explanations easy to understand, and the overall degree of technicality is low.

28. Messieurs Rivingtons was a London-based publishing company founded by Charles Rivington (1688–1742), originally from Derbyshire. From selling books, Rivington moved into the business of publishing them. In 1741 he published the first volume of Samuel Richardson's *Pamela, or Virtue Rewarded*, persuading Richardson to write a novel in the form of correspondence. After his death in 1742, Rivington was succeeded by his two sons, John (1720–1792) and James (1724–1802). James travelled to New York and pursued his trade there. In 1760 John was appointed publisher to the Society for Promoting Christian Knowledge, and the firm maintained this relationship for over seventy years. Charles Rivington's grandsons would earn a fine reputation, publishing the works of Shakespeare, Milton, Locke, and other English classics (Lee 1896; Nichols 1812a; ODNB).

1795. Sir James Edward Smith

English Botany... (Pages 223–266, 9,999 words)

English Botany is the title of a major publication about British plants in six volumes that Smith began in 1790 and which proved to be his most enduring work. However, his name did not appear on the first three volumes, because, as he states in the preface, he was not aware of the importance that the work would assume. Hence it can be also found as *Somerby's English Botany*, in that it was Somerby who conceived, edited, published and illustrated the work with 2,592 hand-coloured plates. The volumes were issued by subscription over 23 years until the work was completed in 1814 (Lee 1898).

The descriptions given are clear as they were addressed to a wide audience. Smith wished "to stimulate the curiosity of the inexperience" (Smith 1795: i). It is in volume four, the one sampled in CELiST, that Smith acknowledges his authorship of the entire work.

Later editions, including fewer volumes, took advantage of mechanical colour printing, further increasing the affordability of the book. Phebe Lankester (an author included in the nineteenth century section of CELiST) produced a revised edition in twelve volumes (1863–1886) entitled "A Popular Portion of English Botany" published by John T. Boswell Syme. This work was based on Smith's *English Botany*, but plants were arranged according to the natural system instead of using Linnaeus' system, and it was written in a more popular and literary tone (Shteir 2003: 160).

This volume, the fourth in the series, was printed by J. Davis[29] for the author and also mentions the booksellers: Mssers. White, Johnson, Dilly and all booksellers in town and country. Pages are numbered from 217 to 288, with the same number of copperplates (as each page is devoted to one plant) and three different indexes at the end. Only those paragraphs referring to plants with abbreviations were omitted from the analysable part of CELiST, although they are available in the Info Window of the CCT. The sample was classified as a Catalogue in the corpus.

3. A note on the Appendix

The Appendix contains important details taken from our samples and provides general bibliographical information about the works sampled. Thus, when looking at the column SUBS, we can see that eight of the twenty works were published by subscription, illustrating that this was a common method of publishing in the eighteenth century. The column COPYRIGHT shows that in six of these eight cases,

29. No information has been found about this printer.

copyrights were retained by the authors, this method being one of the ways of earning money from a work without relinquishing ownership of it. Indeed, only two authors, Blackwell and Speechly, sold their copyright to the printer or seller, and in the case of Blackwell, as we have seen, this was due to financial problems. Selling one's copyright was a quick way to make money, without waiting to sell any copies, but also implied a risk of earning less if the work was very successful.

When comparing the column COPYRIGHT with the adjacent one, DESCRIPTION, we observe that these authors who kept their copyrights also had to pay for printing. There is only one author (Boreman) who shared these expenses, in his case with other two booksellers (Game and Ware), and with whom he also shared the profits. Those cases in which the authors did not need to pay for copies are the ones in which copyrights were also sold.

The table also includes the column APOLO, which indicates where an author gives some sort of apology (for not being an expert in the topic, for being a woman, for mistakes in the book, etc.). Although a common practice at the time, in fact only two of our samples, those by Sloane and by Edwards, contain such apologies.

The column DEDI identifies those works in which the author has included a dedication, also very common in light of the importance of patronage in science, and in eleven of our samples the authors have included these initial pages to praise and thank someone for their support. In this sense, the eighteenth-century section of CELiST conforms to the tendencies of the time: an author with a prestigious patron knew his or her work would be more successful and easier to sell.

Most of our texts (fifteen out of twenty) were printed in London. As mentioned in the introduction, in the eighteenth century, printing was no longer restricted to London, Oxford, Cambridge and York. Yet our texts include only two from Dublin, one from Huddersfield, and one from Birmingham. Thus, we can see that despite the Licensing Act, the power of the printing associations in London was still very strong.

Turning to the degree of technicality of the eighteenth-century CELiST samples, we can say that most of them were written not for specialists but to be understood by a wider audience, this in an attempt to bring science closer to an amateur audience. This idea may be related to the fact that all the printers involved in the process, as we have seen in this chapter, were people of considerable reputation and to a greater or lesser extent were themselves involved in the progress of science. It is also thanks to these printers, as well as the authors and others involved in the book-trade, that such important works on the development of Life Sciences, were not forgotten forever.

Acknowledgements

The complete title and printing details of each work in CELiST has also been included here, since each section contains only a short title, although other aspects of the publishing process are mentioned.

Works cited

Bancroft, Edward. 1769. *An Essay on the Natural History of Guiana, in South America. Containing a description of many curious productions in the animal and vegetable systems of that country. Together with an account of the religion, manners and customs of several tribes of its Indian inhabitants. Interspersed with a variety of literary and medical observations.* In several letters from a Gentleman of the Medical Faculty during his residence in that country. London: printed for T. Becket and P. A. Hondt.

Belanger, Terry. 1982. "Publishers and writers in eighteenth-century England". In River, Isabel (ed.), *Books and their reader in eighteenth century England*. Leicester: Leicester University Press. 2–25.

Blackwell, Elizabeth. 1737. *A Curious Herbal, containing five hundred cuts, of the most useful plants, which are now used in the practice of Physick. Engraved on folio copper plates after drawings, taken from the LIFE. To which is added a short description of ye plants and their common uses in PHYSICK.* In two volumes. Vol. I. London: Printed for Samuel Harding. https://doi.org/10.5962/bhl.title.571

Blair, Patrick. 1723. *Pharmaco-Botanologia: or, An Alphabetical and Classical dissertation on all the British Indigenous and Garden Plants of the New London Dispensatory. In which Their Genera, Species, Characteristick and Distinctive Notes are Methodically described; the Botanical Terms of art explained; their Virtues, Uses, and Shop-Preparations declared. With many Curious and Useful Remarks from proper Observation.* Decad I. London: printed for G. Strahan; W. and J. Innys, W. Mears.

Bolton, James. 1789. *An History of Fungusses, growing about Halifax. With forty-six copper-plates; or which are engraved sixty-four species of funguses, Including the Seven following GENERA, viz. CLATHRUS, HALVELLA, PEZIZA, CLAVARIA, LYCOPERDON, SPHAERIA, and MUCOR. Wherein their various appearances in the different stages of growth, are faithfully exhibited in about three hundred figures, copied with great care from the PLANTS, when newly gathered and in a state of perfection. With a particular DESCRIPTION of each SPECIES, in all its stages. From the first appearance to the utter decay of the plant; with the time when they were gathered; the soil and situation in which they grew; their duration; and the particular places mentioned, where all the new or rare species were found. The whole being a plain recital of FACTS, the result of more than twenty years observation.* In Three Volumes. Vol. III. Huddersfield: Printed by J. Brook for the Author.

Boreman, Thomas. 1730. *A description of three hundred animals; viz. beasts, birds, fishes, serpents, and insects. With a particular account of the whale-fishery. Extracted out of the best authors, and adapted to the Use of all Capacities; especially to allure Children to Read. Illustrated with copper plates, whereon is curiously Engraved every Beast, Bird, Fish, Serpent, and Insect, describ'd in the whole Book.* London: printed by J. T. for Rich. Ware, Tho. Boreman, and Tho. Game.

Borlase, William. 1758. *The Natural History of Cornwall. The Air, Climate, Waters, Rivers, Lakes, Sea and Tides; Of the Stones, Semimetals, Metals, TIN, and the Manner of Mining; The Constitution of the Stannaries; Iron, Copper, Silver, lead, and Gold, found in Cornwall. Vegetables, Rare Birds, Fishes, Shells, Reptiles, and Quadrupeds: Of the Inhabitants, Their Manners, Customs, Plays or Interludes, Exercises, and Festivals; the Cornish Language, Trade, Tenures, and Arts*. Oxford: printed for the Author by W. Jackson; *sold by W. Sandby, London; and the booksellers of Oxford.*

Briggs, Asa. 2008. *A history of Longmans and their books 1724–1990. Longevity in Publishing.* London/ New Castle: Oak Knoll Press.

Brickell, John. 1737. *The Natural History of North-Carolina. With an account of the trade, manners, and customs of Christian and Indian inhabitants. Illustrated with copper-plates, whereon are curiously engraved the map of the country, several strange beasts, birds, fishes, snakes, insects, trees, and plants, &c.* Dublin: printed by James Carson. For the Author. https://doi.org/10.5962/bhl.title.59725

Brickell, John. 1737. [2008]. *The Natural History of North-Carolina.* University of Michigan Library. Digital version.

Brown, Mary Elizabeth. 1913. *Dedications, an anthology of the forms used from the earliest days of book-making to the present time.* New York: G.P. Putman's Sons.

Cross, Nigel. 1985. *The Common Writer: Life in Nineteenth-Century Grub Street.* Cambridge: Cambridge University Press.

Derry, Margaret E. 2006. *Horses in Society: A Story of animal breeding and marketing culture, 1800–1920.* Toronto: University of Toronto Press. https://doi.org/10.3138/9781442675872

Dodd, James Solas. 1752. *An Essay towards a Natural History of the Herring.* London: printed for T. Vincent.

Donovan, Edward. 1794. *Instructions for collecting and preserving various Subjects of Natural History; as Animals, Birds, Reptiles, Shells, Corals, Plants, &c. together with a Treatise on the Management of Insects in their several Status: selected from the best Authorities.* London: Printed for the Author and sold by Messrs. Rivingtons, St. Paul's Church-Yard; and all other principal Booksellers. https://doi.org/10.5962/t.173067

Douglas, James. 1707. *Myographiæ Comparatæ Specimen: or, a comparative Description of all the Muscles in a Man and in a Quadruped. Shewing Their Discoverer, Origin, Progress, Insertion, Use, and Difference. To which is added, An Account of the Muscles peculiar to a Woman. With an Etymological Table, and several Useful Index's.* London: printed by W. B. for G. Strachan.

Downie, James Alan. 2013. "Printing for the author in the long eighteenth century". In Jung, Sandro (ed.), *British Literature and Print Culture.* Essays and Studies 2013. Cambridge: D. S. Brewer. 55–77.

Edwards, George. 1743. *A Natural History of Uncommon Birds. And of some other rare and undescribed animals, quadrupedes, fishes, reptiles, insects, &c. Exhibited in two hundred and ten copper-plates, from designs copied immediately from Nature, and curiously coloured after life. With a full and accurate description of each figure.* In Four Parts. Part I. London: printed for the author, at the College of Physicians in warwick-Lane.

Equiano, Olaudah. 2003. *The interesting narrative and other writings.* Revised edition. London: Penguin Classics.

Feather, John. 1980. John Nourse and his Authors. *Studies in Bibliography*, 34: 205–226.

Feather, John. 1985. *The provincial book trade in Eighteenth century England.* Cambridge, London, New York: Cambridge University Press. https://doi.org/10.1017/CBO9780511552892

Foxon, David Fairweather and McLaverty, James. 1991. *Pope and the early eighteenth century book trade.* Oxford: Clarendon Press.

Gibson, William. 1720. *The Farriers new Guide: containing first, the anatomy of a horse, being an exact and compendious discription of all his parts; with their actions and uses: illustrated with figures curiously engrav'd on copper plates. Secondly, an account of all the diseases incident to horses, with their signs, causes, and methods of cure; wherein many defects in the farriers practice, are now carefully supply'd, their errors expos'd and amended, and the art greatly improv'd and advanc'd, according to the latest discoveries. The whole interspers'd with many curious and useful observations concerning feeding and exercise, &c.* London: printed for William Taylor.

Goldsmith, Oliver. 1774. *An History of the Earth, and Animated Nature.* In Eight Volumes. Vol. VIII. London: printed for J. Nourse.

Griffin, Dustin. 1996. *Literary Patronage in England, 1650–1800.* Cambridge: Cambridge University Press. https://doi.org/10.1017/CBO9780511519024

Headington. Retrieved May 12, 2014, from http://www.headington.org.uk/history/famous_people/jackson.htm

Heymering, Henry. 2009. "Veterinarian or farrier?" *Veterinary Heritage*, 32/2: 42–43.

Hughes, Griffith. 1750. *The Natural History of Barbados.* By the Reverend Mr. Griffith Hughes, A. M. Rector of St. Lucy's Parish, in the said Island, and F. R. S. In Ten Books. Book III. London: Printed for the author and sold by most booksellers in Britain and Ireland.

Jenkins, Robert Thomas and Ramage, Helen. 1951. Honourable Society of Cymmrodorion. *Y Cwmrodor*, 50: 82–83.

Keill, James. 1717. *Essays on Several Parts of Animal Oeconomy.* London: Printed for George Strahan.

Lane, Joan. 1993. "Farriers in Georgian England". In Michell, A. R. (ed.), *History of the Healing Professions: parallels between veterinary and medical history.* V. 3. Wallingford: CAB International. 99–118.

Lareo, Inés; Monaco, Leida Maria; Esteve-Ramos, María José and Moskowich, Isabel (comps.). 2020. *The Corpus of English Life Sciences Texts (CELiST).* A Coruña: Universidade da Coruña. https://doi.org/10.17979/spudc.9788497497848

Law, Edward J. 1988. Huddersfield & district history. Early Huddersfield printers. Retrieved January 15, 2020, from http://homepage.eircom.net/~lawedd/PRINTERS.htm

Lee, Sidney (ed.). 1896. *Dictionary of National Biography.* Vol. XLVIII. New York: Macmillan and Co.

Lee, Sidney (ed.). 1898. *Dictionary of National Biography.* Vol. LIII. London: Smith Elder.

Monthy Review, or Literary Journal. Enlarged. 1794. vol. XV. London: Printed for R. Griffiths.

Moskowich, Isabel. 2021. "The making of CELiST, a bunch of diciplines". In Moskowich, Isabel; Lareo, Inés and Camiña, Gonzalo (eds.), *"All families and genera": Exploring the Corpus of English Life Sciences Texts.* Amsterdam/Philadelphia: John Benjamins. 1–19. https://doi.org/10.1075/z.237

Munter, Robert. 1967. *The history of the Irish newspaper, 1685–1760.* Cambridge: At the University Press.

Nichols, John. 1812a. *Literary anecdotes of the eighteenth century; comprising biographical memoirs of William Bowyer, printer, F.S.A.* Vol. III. London: printed for the author.

Nichols, John. 1812b. *Literary anecdotes of the eighteenth century; comprising biographical memoirs of William Bowyer, printer, F.S.A.* Vol. V. London: printed for the author.
Nichols, John. 1814. *Literary anecdotes of the eighteenth century; comprising biographical memoirs of William Bowyer, printer, F.S.A.* Vol. VIII. London: printed for the author.
ODNB. *Oxford Dictionary of National Biography*. Retrieved January 25, 2016, from www.oxforddnb.com
OED. *Oxford English Dictionary*. Retrieved January 26, 2016, from https://www.oed.com
Pennant, Thomas. 1766. *The British Zoology. Class I. Quadrupeds. II. Birds*. Published under the inspection of the Cymmrodorion Society, instituted for the promoting useful charities, and the knowledge of nature, among the descendants of the ancient Britons. In Four Volumes. Vol. I. London: printed by J. and J. March, for the Society: and sold for the benefit of the British Charity-School on Clerkenwell-Green.
Plomer, Henry Robert. 1900. *A Short history of English Printing 1476–1898*. London: Kegan Paul, Trench, Trübner and company, limited.
Plomer, Henry Robert. 1922. *A Dictionary of the printers and booksellers who were at work in England, Scotland, and Ireland, 1668–1725*. Oxford: Oxford University Press.
Porter, Roy. 1993. "Man, Animals and medicine at the time of the founding of the Royal Veterinary College". In Michell, A. R. (ed.), *History of the Healing Professions: parallels between veterinary and medical history*. V. 3. Wallingford: CAB International. 19–30.
Prescott, Sarah. 2003. *Women authorship and literary culture, 1690–1740*. UK: Palgrave Macmillan. https://doi.org/10.1057/9780230597082
Prior, James. 1837. *The Life of Oliver Goldsmith, M.B. from a variety of original sources*. Vol. II. London: John Murray.
Raven, James. 2005. "Publishing and bookselling 1660–17780". In Richetti, John (ed.), *The Cambridge History of English Literature, 1660–1780*. Cambridge: Cambridge University Press. 11–36. https://doi.org/10.1017/CHOL9780521781442.003
Raven, James. 2007. *The business of books. Booksellers and the English book trade 1450–1850*. New Haven and London: Yale University Press.
Reuss, Jeremias David. 1804. *Das gelehrte England, oder Lexikon der jetzlebenden Schriftsteller in Grossbritannien, Irland und Nord-Amerika nebst einem Verzeichniss ihrer Schriften, von Jahre 1770 bis 1803*. Vol. 1. Berlin und Stettin: bey Fried. Nicolai.
Ritvo, Harriet. 1985. Learning from animals: Natural history for children in the eighteenth and nineteenth centuries. *Children's Literature*, 13: 72–93. https://doi.org/10.1353/chl.0.0198
Rogers, Pat. 1993. Book dedications in Britain 1700–1799: a preliminary survey. *Journal for Eighteenth-Century Studies*, 16/2: 213–233.
Sher, Richard B. 2006. *The enlightenment and the book: Scottish authors and their publishers in the eighteenth-century Britain, Ireland and America*. Chicago: Chicago University Press.
Sloane, Hans. 1707. *A Voyage to the Islands Madera, Barbados, Nieves, S. Christophers and Jamaica, with the Natural History of the Herbs and Trees, Four-footed Beasts, Fishes, Birds, Insects, Reptiles, &c. of the last of those Islands. To which is prefix'd an introduction, wherein is an account of the inhabitants, air, waters, diseases, trade, &c. of that place, with some relations concerning the neighbouring continent and islands of America*. In two volumes. Vol. I. London: printed by B. M. for the Author.
Smith, Sir James Edward. 1795. *English Botany; or coloured Figures of British Plants with their essential Characters, Synonyms, and Places of Growth*. In Thirty Six Volumes. Vol. IV. London: printed for the Author by J. Davis.

Speechly, William. 1786. *A Treatise on the Culture of the Pine Apple and the Management of the Hot-house. Together with a Description of every Species of Insect that infest Hot-houses, with effectual Methods of destroying them by William Speechly. To which is added A method to preserve peach and nectarine trees from mildew &c.* by Robert Browne. With plates. Dublin: printed for Luke White.

Swabe, Joanna. 1999. *Animals, Disease and Human Society: Human-animal relations and the rise of veterinary medicine*. Routledge studies in Science, Technology and Society, 2. London: Routledge. https://doi.org/10.4324/9780203265116

Shteir, Ann B. 2003. "Finding Phebe: A literary History of women's science writing". In Binhammer, Katherine and Wood, Jeanne (eds.), *Women and Literary History. 'For there she was'*. Newmark: University of Delaware Press. London: Associated University Press. 152–166.

Timperley, Charles Henry. 1839. *A Dictionary of printers and printing with the progress of literature ancient and modern*. London: H. Johnson.

Webb, Alfred. 1878. *A Compendium of Irish Biography*. Dublin: MH Gill and Son.

Wigelsworth, Jeffrey R. 2010. *Selling science in the age of Newton: Advertising and the commoditization of knowledge*. Farnham: Ashgate.

The Honourable Society of Cymmdorion. Retrieved January 10, 2020, from https://www.cymmrodorion.org

US National Library of Medicine. Retrieved January 20, 2020, from https://www.ncbi.nlm.nih.gov/pmc/articles/PMC3359810

Wiles, Roy McKeen. 1957. *Serial Publications in England before 1750*. Cambridge: Cambridge University Press.

Withering, William. 1774. *A botanical arrangement of all the vegetables, naturally growing in Great Britain. With the descriptions of the Genera and species, according to the system of the celebrated Linnaeus. Being an attempt to render them familiar to those who are unacquainted with the learned languages. Under each species are added, the most remarkable varieties, the natural places of growth, the duration, the time of flowering, the peculiarities of structure, the common English names; the names of Gerard, Parkinson, Ray and Baubine. The uses as medicines, or as poisons; as food for men, for brutes, and for insects. With their applications in oeconomy and in the arts, with an easy introduction to the study of botany. Shewing the method of investigating plants, and directions how to dry and preserve specimens.* In Two Volumes. Vol. I. Birmingham: printed by M. Swinney, for T. Cadel and P. Elmsley in the Strand and G. Robinson, in Pater-noster-row, London.

Date	Author	Sex	Place Pub.	Printer/distributor	Copyright	Description	Dedi.	Apolo	Subs.	Genre
1707	Douglas	M	London	William Bowyer	George Strachan	Printed by WG for G Strachan				Guide
1707	Sloane	M	London	Benjamin Motte	The author	Printed by BM for the author	Yes	Yes		Treatise
1717	Keill	M	London	George Strachan	George Strachan	Printed for G Strachan				Essay
1720	Gibson	M	London	William Taylor	William Taylor	Printed for William Taylor				Guide
1723	Blair	M	London	G. Strahan; W and J Innys and W. Mears	G. Strahan; W and J Innys and W. Mears	Printed for G. Strahan; W and J Innys and W. Mears				Treatise
1730	Boreman	M	London	J t	R. Ware T. Boreman T. Game	Printed by JT. For Rich. Ware, Tho Boreman and Tho. Game				Textbook
1737	Blackwell	F	London	Samuel Harding	Samuel Harding	Printed for Samuel Harding	Yes		Yes	Catalogue
1737	Brickell	M	Dublin	James Carson	The author	Printed by James Carson for the Author	Yes		yes	Treatise
1743	Edwards	M	London	the author	the author	Printed for the author	yes	yes	yes	Treatise
1750	Hughes	M	London	Sold by most booksellers	the author	Printed for the author	yes		yes	Treatise
1752	Dodd	M	London	Thomas Vincent	Thomas Vincent	Printed for Thomas Vincent	yes			Essay
1758	Borlase	M	Oxford	W. Jackson	the Author	printed for the Author by W. Jackson	yes		yes	Treatise
1766	Pennat	M	London	Cymmrodorion Society	Cymmrodorion Society	Printed by J. and J. March, for the Society: and sold for the benefit of the British Charity-School	yes			Treatise
1769	Bancroft	M	London	T. Becket and P. A. Hondt	Thomas Becket and P. A. Hondt	printed for T. Becket and Peter A. Hondt	yes			Letter
1774	Goldsmith	M	London	John Nourse	John Nourse	Printed for J. Nourse				Treatise

Date	Author	Sex	Place Pub.	Printer/distributor	Copyright	Description	Dedi.	Apolo	Subs.	Genre
1774	Withering	M	Birmingham	Myles Swinney	Thomas Cadell and Peter Elmsley and George Robinson	Printed by M Swinney for T Cadell and P. Elmsley and G Robinson				Guide
1786	Speechly	M	Dublin	Luke White	Luke White	printed for Luke White	yes		yes	Treatise
1789	Bolton	M	Huddersfield	Joseph Brook	the Author	Printed by Joseph Brook for the Author	yes			Catalogue
1794	Donovan	M	London	Messrs. Rivingtons	The author	Printed for the Author and sold by Messrs. Rivingtons,			yes	Guide
1795	Smith	M	London	Printer J. Davis Sellers: White, Johnson, Dilly, etc.	The author	printed for the Author by J. Davis and sold by White, Johnson, Dilly and all booksellers &c.			yes	Catalogue

CHAPTER 4

A look beyond the texts

The samples in the nineteenth-century Corpus of English Life Sciences Texts

Inés Lareo and Isabel Moskowich
University of Vigo / University of A Coruña

1. Introduction

As in the previous chapter, we offer here relevant data for each of the samples contained in the nineteenth-century section of this corpus (Lareo *et al.* 2020). We also note conclusions drawn from the several readings and revisions that were needed to transcribe, edit and encode the samples. These tasks had an impact on genre-assignment and on establishing the degree of specialisation of the extracts, and as was the case for the previous century, these issues will also be discussed. Other features of the texts that were extracted from prefatory material, as well as the different copyright options that we have found in these works, are also dealt with.

We will not discuss matters of printing to any great extent, as these are not as varied as in the 1700s. However, it is important to note that the steam press and stereotype plates emerged in this century, reducing printing costs and thus lessening publishers' risks. Also, paper-making machines came into use (the first book made with this kind of paper was written by Freylinghausen and printed in Britain in 1804), which led to a reduction in the price of the printed copies. These changes allowed publishers to reduce expenses and to seek new ways of presenting their work to broader middle-class audiences (McKitterick 2009). We have tried to highlight the role of printers and publishers in the dissemination of knowledge by including information about them in this chapter. It is thanks to them, in addition to authors, that this knowledge has come down to us.

https://doi.org/10.1075/z.237.04lar

2. Description of the nineteenth-century life sciences texts

1804. Maria Elizabetha Jacson

Botanical Lectures. By a Lady (Pages 1–58, 10,051 words)

The work is presented in two parts, preceded by an analysis in which the author describes what she has included in each lecture, plus an advertisement (Jacson 1804: i–xxiv). Each part is divided into five lectures, the whole work extending to 225 pages. It was printed by Thomas Bensley[1] for Joseph Johnson.[2] The sample included in CELiST contains the first three lectures of the first part. It has been classified as a Lecture in CELiST, reflecting the author's intention to instruct adult pupils. Jacson's didactic aim and her intended readers are both revealed in the following passage:

> And here I recommend to my pupils, whether children or adults, to acquaint themſelves intimately with the ſeven parts of fructification, and with the various ſpecies of Calyx, Corol, Pericarp, and Seed, as deſcribed in this firſt lecture.
> (Jacson 1804: 22)

These lectures are based on her previous work *Botanical Dialogues for the Use of Schools* but now with the aim of reaching general adult readers; therefore, instead of using the maternal figure of the dialogues, the author chose a third-person account with the standard impersonal tone of a textbook. This work stressed the importance of acquiring botanical knowledge from the proper sources and using Latin terminology. The author presents her work as an introduction to Erasmus

1. Bensley was a London printer who took over the business of Edward Allen in 1783. He seems to have been a steady man of business and saw printing as a fine art. Involved in steam-powered presses, he was an innovator in the world of printing and lithography (ODNB; Plomer 1900).

2. Joseph Johnson was a notable bookseller and publisher, and was known for his business acumen and literary taste. He was the publisher of Priestley, John Newton, Wordsworth, Barlow, Beckford, Price, Coleridge, Erasmus Darwin, Cowper, Godwin, Maria Edgeworth and Mary Wollstonecraft. Not only did he discover most of these writers, but was the exclusive English publisher for all of them, except Wordsworth. His commitment to certain writers ensured publication of their works, and constituted a form of economic patronage. His backing also reflected a desire to either establish or nourish a readership for a particular kind of literature, to create or maintain a taste, but always supporting innovative writers and thinkers. Johnson also encouraged liberal notions in his periodical, familiarising the public with new ideas and literary trends. His friend Aikin praised him as the father of the book trade and promoter of young writers, including women, focusing on publishing inexpensive works addressed to a growing middle-class readership. He was against the publication of books that had a limited circulation and he also seldom published by subscription, except when a special demand was expected (Aikin 1809; Tyson 1975; ODNB).

Darwin's translation of *Linnaeus's System of Vegetables* (1783) and, as she says in the Advertisement, she offers "a complete elementary system, which may enable the student of whatever age to surmount those difficulties, which hitherto have too frequently impeded the perfect acquirement of this interesting science" (Jacson 1804: iii).

Jacson includes a note prior to the lectures with information about the pronunciation of Latin sounds. She also uses her own drawings to illustrate the development of bulbs and seeds, and shows a wide reading of the contemporary botanical literature. This work covers plant motion, propagation, and also deals with the analogies between plant and animal functions. Encouraging readers to see things for themselves, she describes her own experiments on crocus bulbs. However, she did not believe in oversimplifying the Linnaean system in order to popularise it; she was clearly more a botanist than an educator and was aware that to understand a science it is very important to know its specific language.

1808. Alexander Wilson

American Ornithology (Pages 20–52, 10,008 words)

This work was published by Bradford and Inskeep[3] and printed by Robert Carr[4] in nine volumes between 1808 and 1814 with seventy-six plates in total. It was the most extensive publication by an American author at the time. The volume selected for CELiST, number I, contains 158 text pages and it is divided into classes, orders and genera. It was illustrated with plates, engraved and coloured from original drawings from nature. It was published by subscription, with volume VIII edited by the author's friend George Ord who, after Wilson's death, completed the final volume from Wilson's notes and his own observations. The success of the work was such that printing increased to 500 copies per volume.

---

3. Samuel Fisher Bradford, who belonged to a prominent Philadelphia family of printers, worked with his father, Thomas Bradford, in 1798–99 before establishing his own firm. Bradford & Inskeep was a publishing partnership established by 1808 by Bradford (1776–1837) and his brother-in-law John Inskeep (1786–1820) in Philadelphia. Before it failed in 1815, the firm published a variety of works. After his partnership with Inskeep ended, Bradford continued printing until at least 1833. (Retrieved June 12, 2014, from http://founders.archives.gov/documents/Jefferson/03-07-02-0204-0001; Remer 1996).

4. Robert Carr was an Irish emigrant who established himself in Philadelphia in 1784. He was an apprentice to Benjamin Franklin in his printshop, later setting up on his own and earning the first gold medal for excellence in printing by the American Association of Booksellers in 1804 (Hayes and Bour 2011).

Plates were printed uncoloured and then coloured by hand, a tedious job. When colourists quit, Wilson took over the work himself. Prospectus and specimen plates were finally printed, and Wilson agreed to obtain 200 subscribers before volume I was published. Wilson's purpose was scientific rather than aesthetic, and his plates characteristically present a number of related species, accompanied by text. New editions appeared in the 1830s, 40s and 70s.[5]

The introduction to this work states that its aim is to amuse and to instruct. The author also wishes to correct the numerous errors which had prevailed in the study of Ornithology in the United States, explaining that "lucrative views have nothing to do in the business" (Wilson 1808: 1). The work is comprehensive in nature, aiming to describe and represent every species of bird, reducing the number of orders and genera to make the work easier for students (Wilson 1808: 3). Given its characteristics, it has been classified as a Treatise in CELiST.

In the XML file in the corpus all references to plates and figures appearing under the headings in square brackets (such as [Plate I- Figure 3] or [Plate 2- Figure 1]) and order identifications have been omitted from the index, but the place they occupy in the book is still visible in the Info Window in the Coruña Corpus Tool.

1816. Priscilla Wakefield

An Introduction to the Natural History and Classification of Insects
(Pages 1–51, 10,008 words)

This work contains 192 pages and is divided into seventeen letters with illustrations. It was published in a single volume printed for Darton, Harvey, and Darton.[6] The sample selected includes letters I to VI and was classified as belonging to the genre Letter in CELiST, basing our decision on both the structure of the text and the titles given to the different chapters by the author herself.

5. After Wilson's death the work appeared in different forms: (with Charles Lucien Bonaparte), American Ornithology. Edited by Robert Jameson. 4 vols. Edinburgh: Constable & Co., 1831; (with a continuation by Charles Lucien Bonaparte), American Ornithology. 3 vols. London: Whitaker, Treacher, & Arnot; Edinburgh: Stirling & Kenny, 1832 and (with Charles Lucien Bonaparte), American Ornithology. 3 vols. Philadelphia: Porter & Coates, 1871.

6. William Darton and Joseph Harvey formed a partnership in 1791, publishing a collection of Quaker works. In 1992 Joseph Harvey took his brother James on as an apprentice. Darton and Harvey decided to buy the copyright to children's books from the descendants of Newbery. From that point, they also produced specialty books for children and by the early 1800s the family firm of printers, publishers, engravers and booksellers had become an established leader in the field. In 1810 Samuel, one of Darton's sons, became the partner in the firm, now named Darton, Harvey and Darton, until William Darton's death in 1819 (ODNB).

This is a text for children, designed to "cultivate a taste in young persons for the study of nature" (Wakefield 1816: iii), written in the form of letters from a young lady, Felicia, to her sister Constance, to instruct her in the study of Entomology, or the Knowledge of Insects (Wakefield 1816: 7). Wakefield also explains in the preface that one of "the beneficial improvements of modern times is the increased attention to the education of children of all ranks" (Wakefield 1816: iii) and describes this work as "an elementary work adapted to young readers" (Wakefield 1816: iv). Each chapter is supposed to be a letter (with its formal opening and closing formulas) written in the Shrubbery.[7] The book is based on the studies of Dr. Shaw and Mr. Barbut, whom Wakefield credits in the preface.

1819. Sir William Lawrence

Lectures on Physiology, Zoology, and the Natural History of Man…
(Pages 154–188, 10,000 words)

This work is divided into four lectures (1–116), delivered in 1817, plus a treatise on the Natural History of man (117–579), which the author describes in the following way:

> The following remarks […] formed twelve lectures delivered after the three foregoing at the Royal College of Surgeons in the past summer (1818). They are here arranged according to the natural divisions of the subject, without any reference to the arbitrary distinctions of the particular Lectures, which were therefore entirely omitted. (Lawrence 1819: 117)

The treatise *Natural History of Man* is divided into two sections, with seven chapters in the first and ten in the second. For the sample in CELiST we have selected chapters III to VII from section I, and it has been classified as a Treatise. During the editing process, punctuation marks were moved and placed prior to footnote marks.

The work was published in a single volume printed for John Callow.[8] It is dedicated to I. F. Blumenbach, a Professor at the University of Göttingen and a Fellow

7. A shrubbery was a collection of hardy shrubs, quite distinct from a flower garden, which was a cutting garden to supply flowers for the house. It was a feature of English nineteenth-century gardens that had originated in the garden style of the early part the century (retrieved January 12, 2020, from https://www.synonyms.com).

8. John Callow was a bookseller and publisher specialising in medical books, and indeed was considered the father of medical bookselling. Callow was the son of a farmer from Homer, and received a limited education. His wife had a business that dealt in old books and which attracted medical practitioners and students. In a few years their shop was stocked with many books of considerable value and importance, and it was at this point that John Pearson advised Callow to establish himself as a medical bookseller and publisher (Urban 1845: 29–32).

of the Royal Society of London, recognising here Lawrence the instruction and entertainment that he has derived from Blumenbach's writings. With twelve engravings, this work was aimed at the College of Surgeons where Lawrence worked as a professor of anatomy and surgery. That is, it was directed to a professional readership and to students. The work was certainly popular and new editions appeared in London in 1822, 1823, 1834, 1838, 1840, 1844, 1848 and 1866; other editions appeared in Edinburgh and in the USA. Due to its controversial content several publishers issued their own pirated editions, keeping the work in print for several decades. It was also reprinted in 2007.

1824. Edward Jenner

Some Observations on the Migration of Birds (Pages 12–44, 9,771 words)

This sample includes an article published in the *Philosophical Transactions*. As explained in the letter by G. C. Jenner, which precedes the main content, the article was signed by the author in 1823, just before his death, and was left to his nephew G. C. Jenner. He was the one who sent it to Sir Humphry Davy, chemist and President of the Royal Society, without making any changes to the original. He says that his uncle would probably have corrected some inaccuracies in the style and order of the paper, if he had had the time to do so, since it is addressed to a learned body on the topic in question (Jenner 1824: 11–12). This sample has been classified as an Article in CELiST.

1828. John Davidson Godman

American Natural History (Pages 37–88, 10,011 words)

This sample was taken from the first edition, in which there is no preface. However, the third edition, published in 1836, adds a preface, where the author tells us that the work was begun in 1823 and that "in relation to the animals described in this work, it has been our constant aim to give none but such as certainly belong to this country, being much more desirous of presenting a faithful account of those known to inhabit it, than to produce an imposing catalogue of 'new species'" (Godman 1836: viii), thus rendering it as useful and agreeable as possible. The author states that he has used authentic sources (and references to these are made in the work itself) to facilitate the task of those who wish to extend their research further. Observations were systematised, yet narratives of first-person experiences and anecdotes were also included. The work was originally published in three volumes, from 1823 to 1828, all of these devoted to Mastology and illustrated with plates. The popularity of the work led to other editions in 1831, 1836, 1842 and 1846.

The extract we have sampled is taken from volume III, published in Philadelphia by Carey Lea and Carey.[9] It contains 264 pages divided into ten chapters, conclusions and an appendix, and chapters II to VIII, divided into species, genera and orders, are the ones included in CELiST. Descriptions on the aspect, habits and manners of Cetacea are easy to understand, and the author seeks to involve the reader into his own experiences, to the extent that Appendix B mentions the expected reader: "considering that this work is destined for general Readers, we have deemed it advisable to omit what we have prepared on this subject, and to make the facts we have been enabled to collect and observe, the subject of a paper to be published in a work exclusively devoted to students of natural science" (Godman 1828: 241).

Godman devotes a significant portion of the book to dolphins, stories about whaling and captains' stories, using a first-person narrator here to describe his own experiences. This section was also published independently as *American Whale Fishery* in 1828, and is considered to be one of the first texts in the field of American maritime literature.

Some changes have been made to the file so that references to previous literature on the topic and Latin nomenclature are not included in the index. That is, they cannot be searched by using software as they do not represent the author's own language. However, for the sake of textual integrity, they can be seen in the Info Window of the CCT. This sample has been classified as a Treatise in CELiST.

1832. Almira Hart Lincoln (Phelps)

Familiar Lectures on Botany (Pages 278–305, 10,013 words)

Lincoln will be remembered in the history of Nature writing as a woman of firsts. She was the first woman in the United States to write and publish a textbook on a Natural Science, the first innovator of a textual, hands-on method of nature study, and the first American female populariser of science. She presented the beauty and wonder of nature in an orderly, multidisciplinary manner that generated the kind

9. Carey & Lea, Philadelphia, was one of the most significant publishers in America in the 1820s and 1830s, and can be considered as the first publisher in the modern sense. With the appointment of William Blanchard as a new partner, the firm was formed in 1833. The company, founded by Matthew Carey in 1785, had made publishing history. His son Henry worked in the business from a very early age. In 1817 Henry became a junior partner and the company's name was changed to Carey & Son. His son-in-law, Isaac Lea, was accepted as a junior partner. He was a member of the Academy of Natural Sciences of Philadelphia. The firm name was changed to Matthew Carey & Sons. When Matthew Carey retired, his son Henry bought out his share in 1824. Henry's younger brother, Edward Carey, joined the firm in 1827 and the company name changed to Carey, Lea & Carey. When he departed in 1929, Carey & Lea was formed (Gassan 2014).

of interest in the sciences among female students as had never previously been felt. With a motherly tone and great rhetorical skill, she describes a system for encouraging and enabling young minds to explore and nurture all that nature has to offer. Her textbooks served as a doorway to nature writing for the masses. Because no earlier work had made so much knowledge so accessible to students, her textbooks became a very powerful educational tool. Lincoln faced a challenge in altering the long-established belief that women's intellectual pursuits should include all those areas that would best serve the needs of their future husbands and children. She was aware of the role of women as children's instructors and husbands' companions, yet she believed in female intellect. Prior to her textbooks, formal training in Botany was limited to male medical students, but she managed to present this type of education for women as a natural extension of femininity (Patterson, Thompson and Bryson 2008).

This work was originally published in 1829 and was used widely, going through nine editions in a decade and including, after the fifth edition, a preface for teachers. For CELiST we have used the third edition, published in one volume with four parts, fifty-eight lectures, and 440 pages. It contains a dedication to the author's mother and an address to teachers. For this edition the author included corrections originating from the remarks of teachers and others who had used her work previously. The sample selected here comprises text from lectures XXXIX to XLIII. The work was published by F. J. Huntington[10] in Hartford as well as by other publishers in different cities.

The author herself designed the textbook, which is not simply a series of monotone lectures, but rather encourages students to work with real plants. In the preface she points out that she wrote the work because her pupils needed a suitable book for beginners in the study of Botany. In that her purpose is a didactic one, she says that "I have no claim to any discoveries, neither have I ventured to make any innovations upon the science itself" (Lincoln 1832: iv). This is also why, when explaining the language and the structure of the book, she says that "t is desirable that school books should be easy to teach, and easy to learn", going on to list the three essentials for such purposes: "1st. A clear and methodical arrangement of subjects, 2d. perspicuity of language and 3d. a pleasing style, and interesting illustrations" (Lincoln 1832: v). When criticised for the low level of its scientific content, the author answers that "from the first time, the book was only intended as a popular introduction to the science" (Lincoln 1832: v).

10. Francis Junius Huntington was born in Suffield, Connecticut, the son of Hezekiah Huntington. He was active as a publisher, printer and bookseller, first in Hartford and later in New York City. He published many schoolbooks, including books on geography and atlases (retrieved January 12, 2020, from https://goodwingenealogy.fandom.com).

> a small text book, in a dry, concise style, may answer very well where a teacher has leisure and ability to amplify and explain; […] But many teachers have neither the time, nor the confidence in themselves to attempt to enlarge or illustrate: […] It is desirable that school books should be easy to *teach* and easy to *learn*.
>
> (Lincoln 1832: v)

Although divided into lectures, the sample has been classified as a Textbook in CELiST due to its character. As regards the editorial process, two main changes have been made. First, punctuation marks after footnote marks were placed before these. Second, and because of the special nature of the encoding conventions adopted, the list of topics included by the author at the bottom of each page has been omitted from the XML files.

1835. Sir William Jardine

The Naturalist's Library (Pages 83–141, 10,007 words)

This series encompasses forty volumes, although it was divided into four main sections: ornithology (14 volumes), mammalia (13 volumes), entomology (7 volumes) and ichthyology (6 volumes). Each of these was prepared by a leading naturalist. The sample in CELiST is from volume III, which contains a memoir of Camper (1–82) and a part devoted to ruminating animals. It also contains thirty-five plates.

The book contains also an advertisement in which Jardine refers to his expected readership when he says: "Every one is aware of the importance and interest of this subject to the zoologists, and we hope to render it equally so to the general reader […] and in a pleasing way introducing the female and the young into the groupes [sic.]" (Jardine 1835: ix, x).

In terms of its price and design *The Naturalist's Library* appealed to all but the lowest levels of the Victorian middle classes; at six shillings, each volume was a fraction of the price of most books on Natural History. Hand-coloured plates were included in higher-priced luxury editions. The series was described as "another spirited attempt to render the study of natural science popular, by placing before the public treatises from the pens of the ablest writers at a cheap rate" (Albion 1833: 8,[11] quoted in Sheets-Pyenson 1981: 56). The work was so successful that by the time the eleventh volume appeared, 60,000 copies of the whole work had been sold. Longman & Co. withdrew *The Naturalist's Library* from sale when the Natural History section of their own Lardener's *Cabinet Cyclopaedia* was published.

11. "The Naturalist's Library". *Albion. A weekly chronicle of literature, science and the fine arts.* 1833. 1, April.

The sample selected here was published in Edinburgh by W. H. Lizards.[12] Some changes have been made during preparation: simple inverted commas in the original text (pages 84 and 107) have been replaced by double inverted commas, this because the CCT takes the former as part of its syntax. Paragraphs with identification nomenclature and abbreviations for books have been omitted from the index but can be seen in the Info Window. This sample has been classified as a Treatise in CELiST.

1840. Anne Pratt

Flowers and their Associations (Pages 161–232, 10,021 words)

This work was published in one volume by Charles Knight[13] & Co. and numbers some 409 pages divided into thirty chapters. The work includes Pratt's own drawings. The preface says that the book is addressed to those fond of flowers but not to those who have made them the object of their study, thus it was not addressed to an expert readership. In fact, the work is presented as an elementary work, with contents expressed in a manner which, given the general avoidance of scientific terms, she hopes will render her work pleasing to the general reader (Pratt 1840: iii)

12. William Home Lizars was born at Edinburgh in 1788. He was the eldest son of the engraver Daniel Lizars, to whom he was apprenticed in 1802. His father, however, then acquired a place for him as a student with the famous John Graham, Master of the Trustees' Academy, where for a time Wilkie was a fellow pupil. William, who already had extensive experience co-producing plates with his father, first made an independent reputation by producing banknotes for local banks. On his father's death, he was compelled to take over the family printing and engraving business.

Lizars perfected a method of etching which performed the functions of wood-engraving and which was used for book illustrations. He was a prolific engraver and an energetic presence on the Edinburgh publishing scene (retrieved June 24, 2014, from http://ctgpublishing.com).

13. Charles Knight, born in 1791, was an English publisher, editor, and author. The son of a bookseller and printer in Windsor, and was apprenticed to his own father. He promoted early periodicals and magazines such as *Knight's Quarterly Magazine*, which with only six issues made Knight's name as publisher and author, launching a career which lasted over forty years.

In 1827 Knight was forced to give up publishing and become the superintendent of the publications of the *Society for the Diffusion of Useful Knowledge*, for which he conceived of and edited *The British Almanack and Companion* in 1828. In 1829 he resumed business on his own account with the publication of *The Library of Entertaining Knowledge*, writing several volumes of the series himself. In 1832 and 1833 he established *The Penny Magazine* and *The Penny Cyclopaedia*, both of which had a large circulation. He edited many well-known books and in 1853 became the editor of *The English Cyclopaedia* and of the *Popular History of England (Encyclopaedia Britannica.* Retrieved January 10, 2020, from https://www.britannica.com).

and also "to those Readers who have just commenced the study of botany, or who may Intend. To pursue it at some future time, the author hopes these pages may not be unacceptable" (Pratt 1840: iv).

Pratt was to some extent responsible for the popularisation of Botany, but she never achieved critical acclaim due to the stigma of having been self-taught. The sample selected is taken from chapters XIII to XVII. Each chapter is devoted to a flower and begins with a poem. During editing and encoding these poems have been deleted from the xml file, since the corpus represents scientific prose only. The extract has been classified as a Guide in CELiST.

1848. Sir John Graham Dalyell

Rare and remarkable Animals of Scotland (Pages 138–166, 10,010 words)

This work was published in two volumes by John van Voorst.[14] The sample in the corpus, chapters VII and VIII, has been taken from volume II, the twelve chapters of which extend to 322 pages and fifty-six coloured plates, and are devoted to rarities in Scotland. In the preface the author refers to his work as a treatise based on a series of observations over many years (Dalyell 1848: iii).

The work is addressed to his countrymen, those inspired by a love of science, and to eager students, with the following phrase: "But nothing can be more appalling to the eager student, in quest of knowledge, than a repulsive array of unintelligible phrases, intercepting the objects of curiosity from his understanding" (Dalyell 1848: x). Despite these comments, the degree of technicality and difficulty of the text is medium, perhaps due to the nature of the topic itself, with which readers would typically not have been familiar.

The sample has been classified as a Treatise in CELiST and during the encoding we have removed the captions of plates, which are not included in the CCT index.

14. John Van Voorst, born in 1804, was an English publisher. From the age of 16 he served a six-year apprenticeship in Wakefield before returning to London to work for publishers Longman, Green, Orme, Hurst & Co. until he set up his own business in Paternoster Row in 1833. He soon began to specialise in Natural History books, these often illustrated, and was appointed bookseller to the Zoological Society in 1837. Some of his most noted publications were *British Fishes* and *British Birds* written by Yarrel in 1835 and 1837, and *British Quadrupeds* (by Bell 1836). Voorst worked with most of the notable naturalists of his day including Wallace, but not Darwin. He was a founder member of the Royal Microscopical Society in 1839, and a Fellow of the Linnean Society in 1853 (ODNB).

1859. Elizabeth (Cabot Cary) Agassiz

A First Lesson in Natural History (Pages 7–66, 10,035 words)

The entire book contains just 82 pages and is divided into four chapters. For CELiST we have used a second edition that appeared the same year as the first. A revised edition was printed twenty years later. It was published in Boston by Little, Brown and Company[15] and by Sampson Low,[16] Son and Company in London. The material that is searchable in the CCT includes 10,035 words from chapters I to chapter IV. However, we have reproduced (although not indexed) the whole work up to page 82 so that researchers can read the entire text in the Info Window.

Agassiz was married with no children of her own, but organised the household and took care of the finances and her three stepchildren, Alexander, Ida, and Pauline, and her grandchildren. She worked closely with her husband in his scientific research. In 1856 in their home in Cambridge, Agassiz founded a school for girls from Boston, which closed in 1863. During this time Elizabeth, with no previous scientific knowledge and without teaching any classes herself, began to take notes on her husband's lectures. She played an important role in preserving, elucidating and popularising his ideas. Her husband is said to have despised her scientific abilities. In fact, in Paton's words, when looking at his wife's notes, Agassiz said: "My dear, these are most gracefully expressed, but from the point of view of science they are such nonsense as I never uttered" (Paton 1919: 50).

A true educator, Elizabeth learned not to sacrifice scientific truth for the sake of a happily turned phrase and became a remarkably proficient assistant to her husband (James, James and Boyer 1971; Oakes 2007). Her first book, the one selected here, was prepared under his husband's direction, which she humbly thanks in the unnumbered preface, mentioning also her expected readership:

> This little book, which is hoped may be interesting for children, and perhaps of some use to parents […] has been prepared under the direction of Professor Agassiz and owes any little merit it may possess to his advice and assistance.
> (Agassiz 1859)

15. The Massachusetts publisher James Brown worked for many booksellers including, latterly, Charles C. Little & Co. Indeed, in 1837 the firm became Charles C. Little and James Brown. After Brown's death, Augustus Flagg joined the firm and its name was changed to Little, Brown and Company in 1847. They published fiction as well as non-fiction books (Wilson and Fiske 1900).

16. Sampson Low was a London publisher and bookseller. He was an apprentice to Lionel Booth, the proprietor of a circulating library, and spent a few years in the house of Longman & Co. He began his own business in 1819, and his reading room was the meeting point of many literary men, lawyers and politicians. In 1848 Low and his son Sampson Jr. Opened a publishing office, continuing the business until 1964 (Brake and Demoor 2009).

Paton (1919: 64) explains that Agassiz's book was written in the form of letters to her niece and nephew, Lisa and Connie Felton. They were written in a clear and simple language, talking about sea-anemones and corals, hydroids and jellyfishes, star-fishes and sea-urchins, and the text succeeds in relating the basic scientific facts concerning these whilst also conveying something of the charm of their life cycles. These first modest pages anticipate the style that would make her subsequent writing on scientific subjects so agreeable and successful. As already noted, to appreciate any of her books, it should be born in mind that she had had no technical scientific training and that practically all the information conveyed in her published works had been acquired not by study of her own but by association with her husband. Her writings are admirable expressions of her peculiar ability to present second-hand knowledge accurately and with as much animation and authority as if it were the result of her own scientific observations.

According to Paton's description, and taking into account the initial and final paragraphs of the book, which have a typical letter form, this sample has been classified as a Letter in CELiST, although many libraries catalogue it as a guide for the teaching of science. The closing words of the work are "The End", usually found in tales. This may be due to the fact that the addressees of the assumed letters were children, and the book itself was probably also intended for parents to read aloud to their children. This also explains its low degree of difficulty and technicality.

1859. Charles Robert Darwin

On the Origin of Species by means of Natural Selection, or the Preservation of favoured Races in the Struggle for Life (Pages 422–454, 10,014 words)

One of Darwin's most influential scientific books considered the foundation of evolutionary biology, this work contains 490 pages plus an index, divided into fifteen chapters. The sample in CELiST includes part of chapter XIII, being classified as a Treatise in this corpus. It was published by John Murray[17] in London.

Darwin's book, which introduced the theory of evolution and natural selection, presented a body of evidence in support of the claim that the diversity of life arose

17. John Murray was the grandson of a publisher known for publishing books of well-known authors as well as the *English Review* and *The Star*. His father continued the business and made the publishing house one of the most important and influential in Britain, thus contributing significantly to the spread of knowledge. Murray's father was a friend of many leading writers of the day and launched the *Quaterly Review* in 1809. Murray continued the family business and published Eastlakes's first English translation of Goethe's *Theory of Colours* (1840), Livingstone's *Missionary Travels* (1857), and Darwin's *Origin of Species* (1859). He also established the *Murray Handbooks* in 1836, a series of travel guides from which modern-day guides are directly descended (Carpenter 2009).

by common descent through a branching pattern of evolution, a model which was also applied to other fields of knowledge, such as historical linguistics.

The author includes evidence gathered whilst on the *Beagle* expedition in the 1830s and his subsequent findings from research, correspondence, and experimentation, all expressed in his own perspicuous language. The matter was of the deepest interest, not only to naturalists and men of science. Thus, the book was written for non-specialist readers and attracted widespread interest. Given its popularity, it had (and continues to have) many editions. For the sixth edition in 1872, the title was shortened to *The Origin of Species*. Given that Darwin was an eminent scientist, his findings were taken seriously and the evidence he presented generated scientific, philosophical, and religious discussion. The debate over the book contributed to the campaign by T. H. Huxley and his fellow members of the *X Club* to secularise science by promoting scientific naturalism (Wilberforce 1860).

1863. Thomas Henry Huxley

On the Origin of Species, or, the Causes of the Phenomena of organic Nature. A Course of six Lectures to working Men (Pages 80–113, 10,048 words)

The book contains 150 pages divided into six lectures. We have selected lectures IV and V as our sample in CELiST, which we have classified as a Lecture. This book was the first American edition, published by D. Appleton and Company[18] in New York (although the work had been already published in London the previous year). The second British edition came out in London in the same year as the one selected here, 1863, with an additional title: *on our knowledge of the causes of the phenomena of organic nature*. Several other editions were published in New York over the century, in 1869, 1870, 1871, 1872, 1873, 1878, 1880, 1883 and 1886.

In the preface to this first American edition that we reproduce in CELiST, the editor states that these lectures have clarity, simplicity and directness and are presented in so lucid and attractive a manner as to impress the commonest imagination of "working men" (the audience Huxley mentions in a note prefixed to the English edition, as well as in the title of this edition), although they deal with the most abstruse and fundamental questions of the mind and its organisation (Huxley 1863: 3–4).

18. This was the American company founded by Daniel Appleton, publishing its first book in 1831. The company issued the works of contemporary scientists at moderate prices, thus helping to make science more popular. Among other writers published we find Herbert Spencer, John Tyndall, Thomas Huxley and Charles Darwin. Medical books and books in Spanish, for the South American market, also formed special departments (Rines 1920).

Huxley was regarded as one of the leading scientists in England by the age of 26, gaining fame primarily from his support for Darwin's theory of evolution. He wrote extensively, and in the field of Biology he wrote both from a scientific perspective and also to popularise the subject. The editor of the book we have sampled here praises the popular tone of these lectures in the preface, thus:

> The gift of translating the high questions of science into popular forms of expression, without sacrificing accuracy and introducing error, is a very rare one among scientific men, but Professor Huxley possesses it in an eminent degree: his lectures are models of their class. (Huxley 1863: 4)

1867. Herbert Spencer

The Principles of Biology (Pages 419–452, 10,001 words)

The work was published in two volumes, the first of these in 1864, and the second in 1867. It is the latter which is sampled in CELiST; this book comprises 574 pages and includes Part IV (divided into seventeen chapters), Part V (ten chapters) and Part VI (thirteen chapters). This edition appeared in London and Edinburgh and was published by Williams and Norgate.[19] The popularity of the work resulted in other editions appearing in America in 1881–1884, 1896–1897 and 1898–1900. For the corpus we have reproduced chapters V to VIII and have classified the sample as a Treatise.

Under the influence of Darwin, Spencer says in volume I that his aim is "to set forth the general truths of Biology, as illustrative of, and as interpreted by, the laws of Evolution" (Spencer 1864: v). He also believes that the only way to gain knowledge is through a scientific approach to learning. He thus sought to replace the theological systems of the Middle Ages with his own philosophical system, which claimed that all knowledge could be placed within the framework of modern science. Science was the only way to gain "useful" knowledge. It was through this "scientific" knowledge that people learned to live in society. However, this work is clearly addressed to a minority of specialists in the field and its degree of technicality is high.

19. Edmund Sydney Williams and Frederick Norgate were booksellers and publishers specialised in foreign scientific literature, with premises at Covent Garden, London, and South Frederick Street, Edinburgh. They were the publishers of the *Natural History Review* journal. Frederick Norgate was born at Hethersett. Initially destined for the church, in 1843 he became a bookseller and a partner in the publishing firm Williams and Norgate. After two decades he retired to devote himself to his scholarly interests, especially the poetry of William Caxton (Burkhart *et al.* 2010; ODNB).

1876. Alexander Macallister

An Introduction to Animal Morphology and Systematic Zoology
(Pages 1–35, 10,023 words)

This work was published in one volume with forty-eight chapters, extending to 421 pages plus an index. For the corpus we have selected chapters I to VI of the first part, and it has been classified as a Textbook in CELiST. It was published in London by Longmans, Green and Co.[20]

Macallister mentions in the preface that he was encouraged to write this work because he understood that "students desire to have a text-book in their hands to enable them to learn the terminology of the science, [...] to assist them in remembering the practical instruction of the class-room" (Macallister 1876: iii); hence the aim of this manual was not to supplant existing works in the field, but rather to prepare students for the profitable perusal of such works (Macallister 1876: iii). He also points out that he wrote the book with junior students in mind, organising the work in the clearest possible way, printing the paragraphs in larger type and the details in smaller, so that the general principles could be more easily grasped (Macallister 1876: vi). Although the expected readers are undergraduates, the degree of technicality is quite high.

As with other samples, some editorial changes have been made in the XML file. Square brackets have been replaced by parentheses, since square brackets are only used in the XML files for editorial marks. In general, formulas have not been reproduced in CELiST, and the editorial mark [formula] has been used instead. When the inclusion of these was not necessary for the understanding of the sentence, they were omitted from the index, but remain visible in the Info Window in the CCT to provide a wider context for research.

20. Thomas Norton Longman died in 1842, leaving his business to his sons, Thomas and William, who were well known for their literary talent. Thomas Longman edited a beautifully illustrated edition of the New Testament, and William Longman was the author of several important books, among them a *History of the Three Cathedrals dedicated to St Paul* and a work on the *History of the Life and Times of Edward III*. In 1863 the firm took over the business of John William Parker, and with it *Frazer's Magazine*, and the publication of the works of John Stuart Mill and J. Froude. In 1890 they incorporated into their own company all the publications of the old firm of Rivington, established in 1711. The family control of the firm (later Longmans, Green & Co.) was continued by Thomas Norton Longman, son of Thomas Longman (Briggs 2008; *Encyclopedia Britannica* retrieved January 25, 2020, from http://www.britannica.com).

1879. Phebe Lankester

Wild Flowers worth Notice (Pages 66–117, 10,020 words)

First published in 1861, this is the second edition of the work, with one hundred coloured figures from drawings by J. E. Sowerby. Other editions appeared in 1885, 1896, 1900, 1903, 1900, 1905 (simultaneously in London and New York), and 1923.[21] The edition we have used for CELiST was published in London by David Bogue.[22] It contains 150 pages plus an index and is divided into sections devoted to different plants. The work is aimed at young and amateur people, and the sample is classified as a Guide in CELiST.

In the preface to the second edition Lankester refers to the success of the previous one and thanks the publisher for producing this corrected, improved and updated version, which also manages to retain its original character. She recognises that hers is not an exhaustive treatise on British Flora, such as "exists in many forms and in large ponderous volumes compiled by profound and learned botanists" (Lankester 1879: iv). When referring to the growing importance of Botany and her expected readers, she says:

> Facilities for the study of Botany, both in the metropolis and in the provinces, are much greater now than when I first wrote this little book, and urged it upon young people of both sexes. The Government recognize Botany as one of the subjects for science teaching; [...] But these efforts of the Government are simply in the way of examinations to test the knowledge of teachers who are supposed to have qualified themselves elsewhere, and to submit to this test of their fitness for teaching the subject in the board and other schools of the country. About 1500 candidates present themselves annually for examination, and botany is now recognized as one of the subjects to be taught in our national schools. (Lankester 1879: viii–ix)

21. Different editions were given slightly different titles. It seems that the 1861 and 1872 editions appeared as *Wild Flowers Worth Notice. Being a Selection from the British Flora.* It was also published under the title *Wild Flowers Worth Notice for Their Beauty, Associations, Or Uses.*

22. David Bogue was a Scottish publisher and bookseller who worked as an assistant to the Edinburgh bookseller Thomas Ireland before moving to London in 1836. He then entered the shop of Charles Tilt, publisher and bookseller, and later joined him as a partner. He imported French lithographic sheets that yielded over £ 100 a week and also published and sold children's books, reprints of novels, and scrapbooks and annuals embellished with steel engravings. Some years later, Bogue was able to purchase the firm. He experimented with multi-volume compendia (*Comic Natural History*, 17 vols., *Manuals of Utility*, 8 vols., and *Books for the Race, Road and Fireside*, all 1847), and Traveller's Guides (to Belgium and the Rhine, Paris and its environs, Switzerland and Savoy, as well as the Traveller's Dictionary, 1852–3). Bogue, who wrote children's stories anonymously, expanded the list of publications for the nursery (ODNB).

When the author explains the divisions and subdivisions of Botany, she compares this system in a popular way to the classification used in libraries (Lankester 1879: xv).

Since we use single inverted commas to indicate the genitive case in the XML files, any such commas appearing in the original work (often used on titles) have been changed into double inverted commas, other than those indicating the genitive case.

1880. Francis Maitland Balfour

A Treatise on Comparative Embryology (Pages 44–67, 10,004 words)

This work was published in two volumes, one devoted to vertebrata and the other to invertebrata. The sample in CELiST, classified as a Treatise, reproduces chapters I and II of the first volume, which is divided into twenty-one chapters and contains 485 pages with figures plus an index and appendix. It was published in London by Macmillan & Co.[23]

In the preface the author states that his aim in writing this work was to give the kind of account of the development of animal forms that might prove useful to both students and those engaged in embryological research (Balfour 1880: v). As with the work by Macallister (1876), different font sizes are used in this edition for different purposes, which makes the text more useful for students. The printer has used two different types, a larger one for general parts and a smaller one for theoretical matter. As compilers, we have nor retained this font-size difference and have not transcribed captions in the XML file.

As it is addressed to students with a good knowledge of Embryology, professionals, physicians or research in general the degree of technicality of the text is quite high.

1889. Sir Francis Galton

Natural Inheritance (Pages 4–49, 10,019 words)

The work was published in one volume by Macmillan & Co. It contains 198 plus an index (259 pages) and is divided into twelve chapters. Other editions of the work appeared in 1894, 1968, 1973, 1996 and 1997.

23. Macmillan was founded in 1843 by Scottish brothers Daniel and Alexander Macmillan. Daniel took care of business issues, while Alexander laid the literary foundations of the firm, publishing notable authors. Beyond literature, the company created such enduring titles as *Nature* (1869), the *Grove Dictionary of Music and Musicians* (1877) and Palgrave's *Dictionary of Political Economy* (1894–99). Macmillan established an office in New York City, which was sold in 1896 (*Encyclopedia*, retrieved January 25, 2014, from https://www.britannica.com; James 2002).

The sample selected covers chapters II to V and is from the first edition of the work. In the introductory chapter the author states: "I have a great subject to write upon, but feel keenly my literary incapacity to make it easily intelligible without sacrificing accuracy and thoroughness" (Galton 1889: 3). In this book Galton presents a summary of his work on correlation and regression. He also provides an account of the concepts that he introduces as well as the techniques which he has discovered. It is the first statistical study of biological variation and inheritance. Galton uses statistical methods and proposes a "law of filial regression", and thus he has a double interest in mathematics and nature (Dewey 1889). Given that the topic is quite new at the time, and that, as he says, he is not able to make it easily intelligible (in fact, the degree of technicality of the sample is quite high), we consider that the expected audience would be a professional one. It is classified as a Treatise in CELiST.

1895. Emily Lovira Gregory

Elements of Plant Anatomy (Pages 85–118, 10,020 words)

This work was published in one volume by Ginn and Company.[24] It contains 144 pages plus an index and figures. Its two parts are, in turn, divided into seven chapters. Chapter VI in Part II, devoted to the anatomy of cormophytes, is included in the corpus.

Gregory's was one of the first American botanical works devoted solely to plant anatomy. In the preface the author explains that this "textbook" is based on lectures delivered during the second half of the year's course in Botany at Barnard College. Given that the subject was quite new in the country, and that she was familiar with the most advanced German research and theories in the field, Gregory tries to take advantage of her immersion in the European academic system. Accordingly, she applies her experience to the organisation of the subject, dividing the work into two parts: the first devoted to the anatomy of the cell occupying about one third of the book, and the second to the anatomy of tissues, occupying roughly two thirds. Prior to these two parts is an introduction to plant physiology. This little textbook

---

24. Edwin Ginn was born in Maine into a farming family who were descendants of early settlers of Maryland, Virginia, and Salem, Massachusetts. He attended a rural school, an academy and later the Westbrook Seminary, a Universalist preparatory school. Forgoing the ministry, he enrolled at Tufts University and, after graduation, Ginn had a successful career selling schoolbooks. In 1867 he founded Ginn & Co., which became one of the leading textbook publishers in America. Later, Ginn became a noted philanthropist and peace advocate. In his late 50s, Ginn turned his focus to philanthropy being the American peace movement his primary concern (Sauer *et al.* 2000).

demonstrates how Gregory had thoroughly mastered a wide range of subjects and adapted them for the use of her readers, "all students of botany who wish to use this science in any direction" and students of pharmacy (Gregory 1895: iv). The sample has been classified as a Textbook in CELiST, and has a medium degree of technical difficulty.

1898. Alpheus Spring Packard

A Text-Book of Entomology... (Pages 1–35, 10,001 words)

This work (708 pages with plates plus an index) was published in one volume by Macmillan &Co.[25] It is divided into three parts: morphology and physiology, embryology, and the metamorphoses of insects. Further editions appeared in 1903 and 1909. The sample selected here includes pages 1 to 35 and has been classified as a Textbook in CELiST.

The author had previously written books for beginners, such as *Guide to Study of Insects* and *Entomology for Beginners*, which he recommends to his readers in the preface as a good preparatory reading. This work is more technical and specialised in nature, with a medium degree of difficulty. Readers are also informed that the author, when writing this general work on structural and developmental entomology, had in mind both students and teachers in agricultural and technical schools and colleges (Packard 1898: iii) as well as working entomologists. As Packard, Professor of Zoology and Geology at Brown University, makes clear on the very title page.

As Wheeler (1998) predicted in his review of this work, Packard's book has attracted many American students to the study of the intricate organisation and development of insects in the course of time. He made the following statement:

> With the books of Professor Packard, Comstock and Sharp on his shelves, the beginning entomologist of to-day will find before him a short and pleasant path to a knowledge of his subject instead of the long tortuous course which many American entomologists have had to pursue. (Wheeler 1998: 836)

During the editing of this sample we have omitted captions, figures and tables from the XML file.

25. See note 23.

Works cited

Agassiz, Elizabeth. 1859. *A First Lesson in Natural History*. Boston: Little, Brown and Company. London: Sampson Low, Son and Company.

Aikin, John. 1809. "Biographical account of the late Mr. Joseph Johnson". *The Gentleman's Magazine*, LXXXIX: 1167–68.

Balfour, Francis Maitland. 1880. *A Treatise on comparative Embryology*. In Two Volumes. Vol. I. London: Macmillan &Co. https://doi.org/10.5962/bhl.title.3756

Brake, Laurel and Demoor, Marysa. 2009. *Dictionary of Nineteenth-century Journalism in Great Britain and Ireland*. Gent: Academia Press.

Briggs, Asa. 2008. *A history of Longmans and their books 1724–1990. Longevity in Publishing*. London: British Library Publishing Division.

Burkhart, Frederick et al. (eds.). 2010. The Correspondence of Charles Darwin. Vol. 18. Cambridge: Cambridge University Press.

Carpenter, Humphrey. 2009. *The Seven Lives of John Murray: The Story of a Publishing Dynasty, 1768–2002*. London: John Murray.

Dalyell, Sir John Graham. 1848. *Rare and remarkable Animals of Scotland, represented from living Subjects: with practical Observations on their Nature*. In Two Volumes. Vol. II. London: John van Voorst.

Darwin, Charles Robert. 1859. *On the Origin of Species by means of Natural Selection, or the Preservation of favoured Races in the Struggle for Life*. London: John Murray.

Dewey, John. 1889. "Galton's Statistical Methods". *American Statistical Association*, 331–334. https://doi.org/10.2307/2276500

Encyclopaedia Britannica. Digital version. Retrieved January 16, 2020, from https://www.britannica.com/topic/publishing/The-19th-century-and-the-start-of-mass-circulation#ref398335

Freylinghausen, Johann A. 1804. *An abstract of the whole Doctrine of the Christian Religion*. London: Edward Harding. Printed by A. Wilson.

Galton, Sir Francis. 1889. *Natural Inheritance*. London/New York: Macmillan &Co. https://doi.org/10.5962/bhl.title.32181

Gassan, Richard H. 2014. *Carey & Lea, Printer and Publisher – Seasonal Variations in its Business Cycle 1833–1836*. Retrieved June 21, 2014, from http://www.let.rug.nl/usa/essays/1801-1900/carey-lea-printer-and-publisher/publishers-of-philadelphia.php

Godman, John Davidson. 1828. *American Natural History*. In Three Volumes. Vol. III. Philadelphia: Carey Lea and Carey.

Godman, John Davidson. 1836. *American Natural History*. In Two Volumes. Vol. I. Philadelphia: Hogan & Thompson.

Gregory, Emily Lovira. 1895. *Elements of Plant Anatomy*. Boston/London: published by Ginn and Company.

Hayes, Kevin J. and Bour, Isabelle (eds.). 2011. *Franklin in His Own Time: A Biographical Chronicle of His Life, Drawn from Recollections, Interviews, and Memoirs by Family, Friends, and Associates*. Iowa: University of Iowa Press.

Huxley, Thomas Henry. 1863. *On the Origin of Species: Or, The Causes of the Phenomena of Organic Nature. A Course of Six Lectures to Working Men*. New York: D. Appleton and Company.

Jacson, Maria Elizabetha. 1804. *Botanical Lectures. By a Lady. Altered from "Botanical Dialogues for the use of schools", and adapted to the use of persons of all ages, by the same author*. London: printed for J. Johnson by T. Bensley.

James, Edward T.; James, Janet Wilson and Boyer, Paul S. (eds.). 1971. *Notable American Women, 1607–1950: A Biographical Dictionary*, Volume 1. Harvard: Harvard University Press.

James, Elizabeth (ed.). 2002. *Macmillan a publishing tradition*. London: Palgrave.

Jardine, Sir William. 1835. *The Naturalist's Library. In Forty Volumes. Mammalia Vol. III. Ruminantia. Part I. The Natural History of the Ruminating Animals containing Deer, Antelopes, Camels, &c. Illustrated by thirty-five Plates; with memoir and portrait of Camper*. Edinburgh: W. H. Lizards, London: Samuel Highley, Dublin: W. Curry.

Jenner, Edward. 1824. "Some Observations on the Migration of Birds. By the late Edward Jenner, M.D. F.R.S.; with an Introductory Letter to Sir Humphry Davy, Bart. Pres. R.S. By the Rev. G. C. Jenner". Philosophical Transactions of the Royal Society of London.

Lankester, Phebe. 1879. *Wild Flowers worth Notice: A Selection of some of our Native Plants which are most attractive from their Beauty, Uses, or Associations*. London: David Bogue.

Lareo, Inés; Monaco, Leida Maria; Esteve-Ramos, María José and Moskowich, Isabel (comps.). 2020. *The Corpus of English Life Sciences Texts (CELiST)*. A Coruña: Universidade da Coruña. https://doi.org/10.17979/spudc.9788497497848

Lawrence, William. 1819. *Lectures on Physiology, Zoology, and the Natural History of Man, delivered at the Royal College of Surgeons*. London: printed for J. Callow, medical bookseller.

Lincoln, Almira Hart. 1832. *Familiar Lectures on Botany, including practical and elementary Botany, with generic and specific Descriptions of the most common native and foreign Plants and a Vocabulary of botanical Terms. For the use of Higher schools and academies*. Hartford: F. J. Huntington.

Macallister, Alexander. 1876. *An Introduction to Animal Morphology and Systematic Zoology*. London: Longmans, Green, and Co. https://doi.org/10.5962/bhl.title.23419

McKitterick, David. 2009. Changes in the look of the book. In McKitterick, David (ed.). *The Cambridge History of the book in Britain. 1830–1914*. Vol. 6. Cambridge: Cambridge University Press. 75–116. https://doi.org/10.1017/CHOL9780521866248.003

Oakes, Elizabeth H. 2007. *Encyclopedia of World Scientists*. New York: Infobase Publishing.

ODNB. *Oxford Dictionary of National Biography*. Retrieved June 12, 2014, from http://www.oxforddnb.com

Packard, Alpheus Spring Packard. 1898. *A Text–Book of Entomology including the Anatomy, Physiology, Embryology and Metamorphosis of Insects. For Use in agricultural and technical Schools and Colleges as well as by the working Entomologist*. New York/London: Macmillan &Co.

Patterson, Daniel; Thompson, Roger and Bryson, Scott J. (eds). 2008. *Early American Nature Writers: A Biographical Encyclopedia*. Westport: Greenwood Publishing Group.

Paton, Lucy Allen. 1919. *Elizabeth Cary Agassiz: A Biography*. Boston and New York: Houghton Mifflin Company. https://doi.org/10.5962/bhl.title.6551

Plomer, Henry R. 1900. *A short history of English printing. 1476–1898*. London: Kegan Paul, Trench, Trübner and company, limited.

Pratt, Anne. 1840. *Flowers and their Associations*. London: Charles Knight and Co.

Remer, Rosalind. 1996. *Printers and men of capital. Philadelphia book publishers in the new republic*. Philadelphia: University of Pennsylvania Press.

Rines, George Edwin (ed.) 1920. Appleton, Daniel. *Encyclopedia Americana*. Vol. II. Danbury: Grolier, Inc.

Sauer, Anne; Branco, Jessica; Bennet, John and Crowley, Zachary (eds.). 2000. *Concise Encyclopedia of Tufts History*. Digital Collection.

Sheets-Pyenson, Susan. 1981. "War and Peace in Natural History Publishing: the Naturalist's Library, 1833–1843". Isis, 72/1: 50–72. Retrieved January 4, 2020, from www.jstor.org/stable/231177
Spencer, Herbert. 1864. *The Principles of Biology*. In Two Volumes. Vol. I. London/Edinburgh: Williams and Norgate.
Spencer, Herbert. 1867. *The Principles of Biology*. In Two Volumes. Vol. II. London/Edinburgh: Williams and Norgate.
Tyson, Gerad P. 1975. "Joseph Johnson, an Eighteenth-century Bookseller". *Studies in Bibliography*, 28: 1–16.
Urban, Mr. B. S. G. S. 1845. "John Callow the medical bookseller". *The Gentleman's Magazine*, XXIII: 29–32.
Wakefield, Priscilla. 1816. *An Introduction to the Natural History and Classification of Insects. In a Series of familiar Letters. With illustrative Engravings*. London: printed for Darton, Harvey, and Darton. https://doi.org/10.5962/bhl.title.157681
Wheeler, William Morton. 1898. "Review: A Text-book of Entomology, including the Anatomy, Physiology, Embryology and Metamorphoses of Insects by A. S. Packard". *Science* 17, 7/181: 836–837.
Wilbeforce, Samuel. 1860. "Review of On the origin of species". *Quaterly Review*, 108: 225–264.
Wilson, Alexander. 1808. *American Ornithology; or, The Natural History of the Birds of the United States. Illustrated with plates and colored from the original drawings taken from Nature. In Nine Volumes*. Vol. I. Philadelphia: published by Bradford and Inskeep and printed by Robert Carr. https://doi.org/10.5962/bhl.title.97204
Wilson, James Grant and John Fiske (eds.) 1900. *Appleton's cyclopaedia of American Biography*. Vol. 1 (revised edition). New York: D. Appleton and Co.

CHAPTER 5

The Corpus of English Life Sciences Texts and representativeness

An information and documentation analysis of Late Modern English scientific Texts

Elena Alfaya-Lamas
University of A Coruña

> [...] science is the interpretation of the world of a particular individual (the scientist) and not an independent entity or an absolute truth. This being so, language is a central element both for the interpretation of facts and for their transmission.
>
> (Moskowich 2012: 48)

1. Introduction

It is my aim in this chapter to address the broad and complex issue of representativeness with respect to the Corpus of English Life Sciences Texts (CELiST) (Lareo *et al.* 2020). To this end, empirical quantitative data will be analysed to assess the variability and reliability of the corpus. The qualitative principles governing CELiST will also be examined.

Although the compilers of the Coruña Corpus of English Scientific Writing (CC) have repeatedly claimed that their overall corpus is representative of the Late Modern English scientific register, my research question here is to ask specifically whether CELiST is indeed representative. The primary objective, then, is to study CELiST systematically by means of ReCor, a tool used to analyse quantitative data.

Since CELiST has been compiled following strict qualitative criteria, we might expect that it would also be quantitatively representative. Here, I will address issues of quantitative representativeness by focusing exclusively on the lexicon, leaving aside syntax, rhetorical patterns and any other linguistic aspects.

The chapter is structured as follows. First, I will discuss corpora in general, and CELiST in particular, in terms of representativeness as a distinctive attribute of corpora. Next, my methodological approach will be described, followed by a presentation of findings, a discussion of these, and some concluding remarks.

https://doi.org/10.1075/z.237.05alf

2. The corpus and representativeness

There are many definitions of corpora, and these tend to agree on certain key characteristics. McEnery, Xiao and Tono (2013: 5) observe that "there is an increasing consensus that a corpus is a collection of (1) machine-readable (2) authentic texts (including transcripts of spoken data) which is (3) sampled to be (4) representative of a particular language or language variety". Taking this definition as a point of departure, I intend to describe how these CELiST authentic, machine-readable-sampled texts are quantitatively and qualitatively representative of Life Sciences Scientific Writing, between 1700 and 1900. As Leech (1992: 116) says: "[...] corpora are rarely haphazard collections of textual material: They are generally assembled with particular purposes in mind, and are often assembled to be *representative* of some language or text type". That said, representativeness is an absolutely necessary distinctive attribute of a corpus and is related to the extent to which samples in a corpus comprise a complete range of variability in a given specific language (Atkins 1992). Representativeness in a corpus is related to the idea of accurately representing the language of which the corpus takes samples. It is also "typically achieved by balancing the corpus through sampling a wide range of text categories which are defined primarily in terms of external criteria" (McEnery, Xiao and Tono 2013: 125). In this sense, I have seen that the texts included in CELiST differ in nature a great deal. The corpus, devised as a highly specialised corpus, contains different types of texts that cover two centuries of academic writing in the Life Sciences. Also, the number of texts selected, and the samples included have been carefully considered before inclusion in CELiST. As Biber notes:

> Some of the first considerations in constructing a corpus concern the overall design: for example, the kinds of texts included, the number of texts, the selection of particular texts, the selection of text samples from within texts, and the length of text samples. Each of these involves a sampling decision, either conscious or not. (Biber 1993: 243)

CELiST is the fourth subcorpus of the CC, following Astronomy, Philosophy and History. As with its predecessors, the selection of sampled texts includes two ca. 10,000-word text extracts for every decade. This is a general principle of the CC towards its aim of achieving representativeness for the two-hundred-year period.

CELiST contains samples from the fields of Biology, Zoology, Entomology and Botany, plus related fields of knowledge and expertise. As with all other subcorpora in the CC, each text file is paired with a corresponding metadata file that includes information not only on the text sampled but also on the social, educational and linguistic background of the author. Thus, CELiST is meant to represent the kind of language used in scientific discourse by native speakers in the Late Modern English period when writing on Life Sciences.

Whether or not a sample is "representative" depends on the extent to which it is selected from the range of text types in the target population (Atkins 1992). In this respect, the compilers of CELiST have sought to define the language that the corpus is intended to represent. They have also made explicit the techniques used to select samples for inclusion. Chapter one in this volume complies with Atkins (1992) claims when he states, as McEnery, Xiao and Tono (2006: 126) put it, that "Definition of the target population has at least two aspects: (1) the boundaries of the population – what texts are included and excluded from the population; (2) hierarchical organization within the population – what text categories are included in the population, and what are their definitions". Descriptions of the CC and its different subcorpora tell us these conditions are met. However, to achieve corpus representativeness adequate internal linguistic criteria and external non-linguistic criteria are required. These criteria are interrelated and mutually dependent. Looking further than compilers' descriptions, an analysis of these criteria can be carried out following the methodology presented in the next sections.

3. Methodology

A mixed approach has been used in the assessment of the texts that make up CELiST. On the one hand, the approach is quantitative, positivist and experimental in the sense that data are processed, and results are generalisable. On the other hand, it is qualitative and interpretative, in that the object of study is analysed deepening in its unique nature.

Quantitative data analysis has been carried out using the ReCor tool. This was devised by Gloria Corpas Pastor and Miriam Seghiri Domínguez at the research group "LEXYTRAD. Lexicografía y Traducción".[1] This computer application is based on the N-Cor algorithm of lexical density analysis and allows researchers to confirm whether a corpus is statistically representative.

> By using the N-Cor algorithm it is possible to quantify a posteriori the minimum number of documents and words that should be included in a specialised language corpus, in order that it may be considered representative. This method has been implemented by means of a computer program (ReCor).
>
> (Corpas and Seghiri 2007a: 165)

The base equation of the N-Cor algorithm is Cn = d1+d2+d3+...+dn, where *d* is the lexical density and *C* the corpus growth. The present algorithm calculates the

1. "LEXYTRAD. Lexicography and Translation" research group. University of Málaga, Spain. ReCor is patented (ref. P200695657) by the "Oficina española de patentes y marcas" of the Ministry of Industry, Tourism and Commerce.

minimum size needed for a corpus to be representative by means of the relation between *d* and *C*.

Three versions have been implemented thus far: 1.0, 2.0 and 2.1. The version I am using is 2.1 and presents data in plain text format (.txt) and simultaneously in Excel tables. The three versions function in a similar way. This software application makes it possible to graphically illustrate the point from which a corpus that is compiled following qualitative criteria is also representative from a quantitative point of view. As Corpas Pastor & Seghiri Domínguez (2007b: 171) note: "se ha implementado un programa (ReCor), que permite ilustrar gráficamente el punto a partir del cual un corpus que ha sido compilado según criterios cualitativos comienza a ser representativo en términos cuantitativos".[2]

All the samples in the corpus are included in the data, from which ReCor uses types and tokens as basic indicators. In ReCor, a corpus begins to be representative when the lexical density, expressed as the type/token ratio (Ty/To) continues to decrease when more new texts are added to the corpus (Corpas and Seghiri 2010).

The selection and analysis of examples is done according to separate two criteria: alphabetically, which in my study is also chronological (as explained below), and randomly. This is done so as to avoid the order in which the data are processed having an influence on the final results. ReCor processes one sample after the other, either alphabetically or randomly, and calculates types and tokens for all of them, providing overall totals, and also calculates lexical density. This algorithm is used in both the alphabetical and the random analysis of n-grams to analyse the point at which almost no new types are found, this point being an indication that the size of the corpus has reached adequacy. In a graph, this is represented by the lines for types and tokens being close together. In practical terms, it would be almost impossible to reach the point of zero new types in a corpus, so what we try to find is a constantly low incorporation of new lexical types.

CELiST samples are analysed in several ways:

a. The closure or saturation degree at a lexical level of the corpus as a whole.
b. Ty/To in relation to the total number of documents (40) in CELiST, analysed both chronologically-alphabetically and randomly.
c. Ty/To in relation to the total number of words analysed in CELiST samples, both chronologically-alphabetically and randomly.
d. Ty/To in relation to the total number of words of the corpus from the eighteenth century, both alphabetically-chronologically and randomly.

2. A program (ReCor) has been implemented. It lets us graphically illustrate the point from which a corpus compiled following qualitative criteria begins to be quantitatively representative.

e. The ratio Ty/To in relation to the total number of words of the corpus from the nineteenth century, both alphabetically-chronologically and randomly.
f. Ty/To in relation to the number of documents in the corpus (20) from the eighteenth century, both alphabetically-chronologically and randomly.
g. Ty/To in relation to the number of documents in the corpus (20) from the nineteenth century, both alphabetically-chronologically and randomly.
h. Each sample is divided in two halves as if CELiST contained 80 rather than 40 samples. This enables us to check whether the results obtained for the 80 extracts are the same as those obtained for the real 40 samples in the corpus.
i. Comparison of the evolution of Ty/To in full, first and second halves of the documents, and the eighteenth century and nineteenth centuries.

The findings of the analyses are discussed in what follows.

4. Findings and discussion

Having seen the importance of the representativeness of corpora as a means of obtaining reliable results in the analyses of these, I will here try to answer my research question about the extent to which CELiST is representative and thus is valid as a source of data on the register it seeks to represent.

I will use a quantitative approach here, assessing the lexical density of the texts that comprise CELiST, in a variety of ways. Also, I will consider the representativeness quality criteria described by the compilers of CELiST as a basis and a starting point for the corpus.

Qualitative representativeness

The corpus has been conceived of and compiled following a variety of relevant quality criteria that are designed to entail not only balance but also to lead to a corpus that is a reflection of Late Modern English scientific discourse. Compilers have aimed at carefully choosing samples to represent English in Life Sciences in such a way that is a valid tool for research.

Towards achieving qualitative representativeness, a number of steps were taken:

a. samples of all possible genres representing the textual reality for Life Sciences in the eighteenth and nineteenth centuries in English-speaking countries are provided in CELiST. These text-types reflect the different categories classified and characterised in the corpus that were in use at the time the texts were written, which provides balance. The compilers have tried to make the text chunks sampled in CELiST proportional in number to their presence in the population from which they are taken. Thus, not all genres are equally represented, which is a reflection of the realities of academic writing during the timespan covered. Of all the genres, there is a higher number of treatises, that is, informative academic writing, in that these were common in the dissemination of Life Sciences knowledge in the past.

b. Two Late Modern English texts were sampled for every ten-year period so as to accurately represent the language in use across the whole timespan covered by the corpus.
c. Only samples of texts written directly in English are included in the corpus, which thus excludes translations, this in order to avoid linguistic interference from other languages. The corpus intends to represent one specific language in a specific time frame.
d. Different parts of texts are included, and hence represented, in the corpus. In this way, samples are taken from the beginning, middle or final sections of the original pieces, irrespective of genre.
e. The same number of samples and a very similar number of words have been collected for both centuries with the intention of achieving not only representativeness but also balance.
f. Both female and male authors are sampled, the former to a lesser extent. This is the case not only to reflect the social invisibility of women in the period, but also because women's writing is more difficult to find precisely due to the social conditions in which they worked at the time.

These qualitative criteria, as described by the compilers of the CC, will be used as a complement to the quantitative analysis undertaken with ReCor. Hence, a comparative study of the samples from the eighteenth to the nineteenth centuries can be carried out, the population size here also being comparable. The qualitative homogeneity of same-size samples makes a quantitative lexical analysis possible, avoiding biases of this kind.

Quantitative representativeness

Table 1 below sets out the results of the first analysis carried out with ReCor. The samples have been organised alphabetically because this is one of the ways in which the application processes the data. The order of rows is also chronological since the names of the files in CELiST begin with a year, which ReCor takes as alphabetical, thus alphabetical and chronological ordering coincide here. Table 1, then, provides an overall picture of the data generated by ReCor, and these will be the basis for Figures 1 and 2, below.

Column 1, types, accounts for the number of different words processed in each CELiST text. The results shown here are cumulative. Thus, in the first row, 1,013 types is the number of different words in the first document included in the corpus; in the second row, 2,646 types is the number of different words in both the first and second documents included in CELiST; and so on. At the end of the process, then, the forty samples of CELiST texts have 21,041 types.

Column 2, tokens, sets out the total number of words in the texts included in the CELiST corpus. The first document processed has 10,005, as seen in the first row in Table 1 show. Every row then sees the cumulative total of new words corresponding to the texts added thus far. For all 40 texts, this total is 400,305 tokens.

Column 3, Ty/To, sets out the type-token ratio. This ratio is the main variable in my analysis and expresses the lexical density of the texts, that is, the number of different words in a set of texts in relation to the total number of words of the texts under analysis. As new texts are processed, the Ty/To should decrease because the number of new words found in the texts should be less. Thus, the first row, showing data about the first processed text, is 0.09, and the final one is 0.05.

Column 4, V1, includes the cumulative number of *hapax legomena* in the texts, that is, words with only one occurrence in every processed text. The first processed text has 530 such words, and over the forty texts of CELiST there are 10,866.

Column 5, V2, shows the number of words with only two occurrences, once more cumulatively. There are 152 in the first processed text, and a total of 3,609 overall.

Table 1. Types, Tokens, Ty/To, V1 and V2 in chronological and alphabetical order

Types	Tokens	Ty/To	V1	V2
1,013	10,005	0.09389	530	152
2,646	20,007	0.11808	1,544	462
3,230	30,021	0.09850	1,795	548
3,788	40,021	0.08757	2,048	651
4,723	50,039	0.08792	2,587	826
5,339	60,056	0.08422	2,895	968
6,651	70,075	0.08931	3,759	1,258
7,073	80,086	0.08397	3,931	1,320
7,332	90,098	0.07810	4,026	1,340
7,836	100,103	0.07543	4,294	1,433
8,399	110,109	0.07362	4,600	1,523
9,196	120,107	0.07355	5,105	1,631
10,195	130,111	0.07456	5,778	1,798
10,581	140,120	0.07217	5,956	1,866
10,898	150,155	0.06946	6,050	1,961

(continued)

Table 1. (*continued*)

Types	Tokens	Ty/To	V1	V2
11,249	160,158	0.06735	6,219	1,997
11,396	170,175	0.06440	6,215	2,028
11,971	180,202	0.06353	6,550	2,142
12,390	190,221	0.06131	6,759	2,173
12,872	200,220	0.06038	6,926	2,269
13,210	210,271	0.05912	7,079	2,309
13,928	220,279	0.05991	7,547	2,420
14,478	230,287	0.06002	7,837	2,524
15,047	240,287	0.06010	8,113	2,627
15,465	250,058	0.05970	8,290	2,708
16,004	260,069	0.05957	8,552	2,797
16,520	270,082	0.05942	8,797	2,930
17,034	280,089	0.05882	9,027	3,000
17,504	290,110	0.05858	9,239	3,092
17,821	300,120	0.05798	9,304	3,170
17,993	310,155	0.05645	9,351	3,184
18,230	320,169	0.05568	9,439	3,221
18,398	330,217	0.05473	9,482	3,240
18,735	340,218	0.05426	9,646	3,305
19,425	350,241	0.05486	10,068	3,412
19,900	360,261	0.05469	10,345	3,452
20,141	370,265	0.05404	10,445	3,491
20,464	380,284	0.05355	10,582	3,534
20,631	390,304	0.05279	10,655	3,548
21,041	400,305	0.05256	10,866	3,609

Table 2 sets out the same data as in Table 1, but rather than ordered chronologically/alphabetically, the texts are in random order. This is useful to find out if the order in which I incorporate the documents has an influence on the final results. My intention is to find out if there is a moment at which the way the data are included, alphabetically/chronologically or randomly, has no effect on the results obtained. If we find a point at which both lexical density variables have nearly the same value, whatever way the data are processed, alphabetically/chronologically or randomly, then we can say that lexical representativeness begins to be present in the corpus. The data shown in the last row in both Table 1 and Table 2 are similar, since the same number of documents, all the ones processed in CELiST, have been included.

Table 2. Types, Tokens, Ty/To, V1 and V2 randomly

Types	Tokens	Ty/To	V1	V2
1,126	10,010	0.10841	565	202
2,713	20,012	0.11266	1,516	489
3,446	30,026	0.10524	1,835	622
4,367	40,026	0.10437	2,297	811
5,510	50,044	0.10759	3,056	1,005
6,308	60,061	0.10526	3,487	1,159
6,896	70,080	0.10051	3,742	1,299
7,389	80,091	0.09383	3,969	1,354
8,239	90,098	0.09124	4,355	1,532
8,684	100,103	0.08785	4,488	1,617
9,177	110,109	0.08441	4,732	1,688
9,819	120,107	0.08299	5,076	1,801
10,156	130,111	0.07887	5,204	1,857
10,598	140,120	0.07691	5,380	1,933
10,834	150,155	0.07263	5,449	1,960
11,471	160,158	0.07158	5,722	2,091
11,793	170,175	0.06940	5,859	2,137
12,079	180,202	0.06703	5,987	2,172
12,534	190,221	0.06620	6,219	2,246
12,850	200,220	0.06446	6,368	2,318
13,569	210,271	0.06432	6,826	2,424
13,906	220,279	0.06285	7,004	2,450
14,396	230,287	0.06246	7,244	2,527
15,116	240,287	0.06317	7,685	2,629
15,601	250,058	0.06270	7,939	2,721
16,371	260,069	0.06284	8,413	2,884
17,373	270,082	0.06396	9,097	3,084
17,609	280,089	0.06279	9,195	3,114
18,031	290,110	0.06199	9,405	3,200
18,526	300,120	0.06161	9,682	3,270
18,863	310,155	0.06083	9,816	3,318
19,020	320,169	0.05966	9,865	3,337
19,120	330,217	0.05821	9,892	3,356
19,459	340,218	0.05746	10,084	3,373
19,545	350,241	0.05607	10,072	3,408
19,735	360,261	0.05507	10,162	3,399
20,183	370,265	0.05458	10,422	3,490
20,389	380,284	0.05357	10,507	3,512
20,920	390,304	0.05343	10,844	3,592
21,041	400,305	0.05256	10,866	3,609

The representativeness of a specialised corpus, if measured at the lexical level, is related to the closure (McEnery and Wilson 2001: 166) or saturation degree (Belica 1996: 61–74) and not so much to the broad range of text types sampled, as is the case with general corpora. Saturation is related to not finding new lexemes in new samples, and when a saturation point is reached compilers can stop including new samples. An almost constant decrease in the number of new lexemes relates to an increase in the degree of saturation. This closure-saturation degree is measured by tokens, column 2, where we can see a constant lexical growth towards the mid-end, an indication that CELiST has become saturated at the lexical level. This data is useful to assess the degree of lexical saturation of the texts sampled with respect to the whole lexicon. Every sample added either chronologically/alphabetically or randomly, as Tables 1 and 2 show, adds approximately the same number of new words as the previous one towards the mid-end of the table, and considerably less than at the beginning. Thus, the corpus becomes more and more complete as new samples are added. Also, since the lexical field of Life Sciences is so wide, we will always find new words, these decreasing in number, as new samples are added and growing in a constant manner. This being the case, after these partial results I can say that the corpus becomes more and more complete with every new sample added, and reaches its saturation point towards its mid-end.

As a means of assessing the Ty/To ratio in relation to the total number of documents processed in CELiST, Figure 1 sets this out alphabetically and randomly. In this way I can check whether enough documents have been included in the corpus, or if more new samples might be necessary to achieve representativeness. Figure 1 illustrates how this ratio evolves when more samples are added to the corpus.

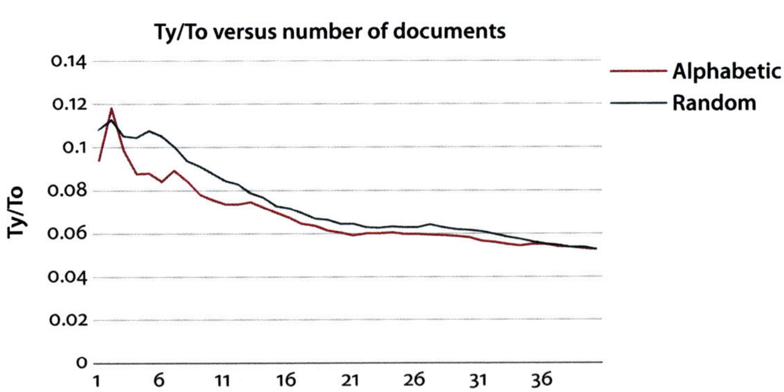

Figure 1. Metric evolution Types/Tokens according to the number of documents processed in CELiST

Figure 1 above presents the metrics of the Ty/To on the vertical axis, and the number of documents analysed, that is, the number of documents processed in CELiST, on the horizontal axis. The Ty/To metrics here are the type/token ratio for the whole processed corpus, ranging from 1 document up to 40 documents.

As Figure 1 shows, the Ty/To metrics are inversely proportional to the number of documents processed, which makes sense because as more documents are incorporated fewer new words (types) appear. Also, the frequency of occurrence of each word increases, which reduces the relation between types and tokens. As each new document is processed, there are more common words that have already been used in previously incorporated documents, so fewer new words are found. The corpus, and the set of types it contains, is thus seen to become increasingly complete with every document processed.

This Ty/To metric has been calculated by organising the texts in two different ways: (a) alphabetically-chronologically, represented in Figure 2 with a blue line, and (b) randomly, represented with a red line. In the former case, documents are processed one after the other following a chronological order; in the second case, documents are processed in random order according to a sequence generated by the software. In the same way as the other values calculated and described above, this Ty/To metric is performed in both ways to see if the variable Ty/To behaves in a similar way independently of the order of the documents being analysed; we need both alphabetical and random measurements to check whether they behave similarly when new words are added and thus avoid bias related to the organisation of the calculation. These two lines overlap towards the end of the graph. They depict not only documents alphabetically processed but also those processed in random order. When they become close and touch each other, and maintain proximate, without separating, it is considered that the corpus has begun to be balanced and that evidence of the representativeness of a corpus has been shown (Seghiri 2011: 26). Figure 1 above plots data illustrating that after document 35 the blue and the red lines overlap. Hence, from this document onwards the corpus begins to be representative. This means that to get a representative corpus of these 40 documents we need 35. After document 35, whether texts are added to the analysis alphabetically-chronologically or randomly, the variation in new lexicon decreases to such a point that the two lines overlap.

Figure 2 below shows the same variables, Ty/To alphabetically and Ty/To randomly, but here they are depicted not in relation to the number of documents in the corpus but according to the number of tokens in all the documents processed in CELiST. Both variables overlap, as the graph shows, when the number of tokens reaches 372,694 tokens. The total number of tokens of CELiST is 400,305. This is the point from which the corpus begins to be representative according to the guidelines of the creators of ReCor.

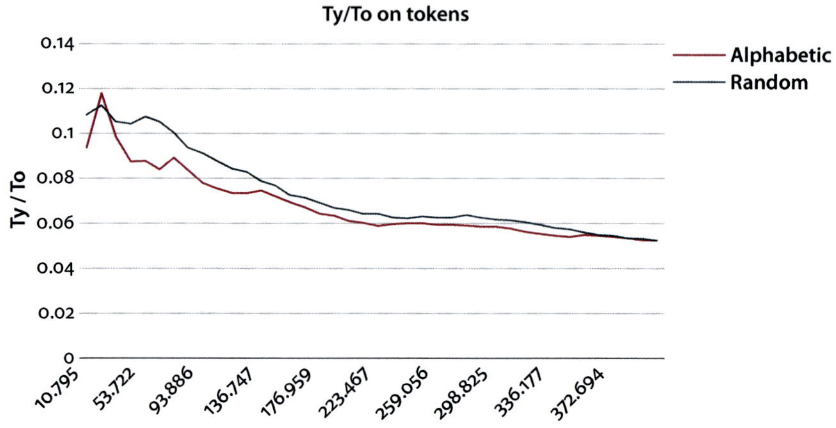

Figure 2. Metric evolution of the ratio Types/Tokens according to the number of words of the documents processed in CELiST

Since some scholars will clearly use only one of the sections of the corpus in their studies, it is also worth conducting a comparative analysis of the texts from the eighteenth century and the nineteenth century, so as to verifying that the data obtained in this way are similar. Figure 3 below shows the evolution of Ty/To when analysing the eighteenth-century and the nineteenth-century texts separately.

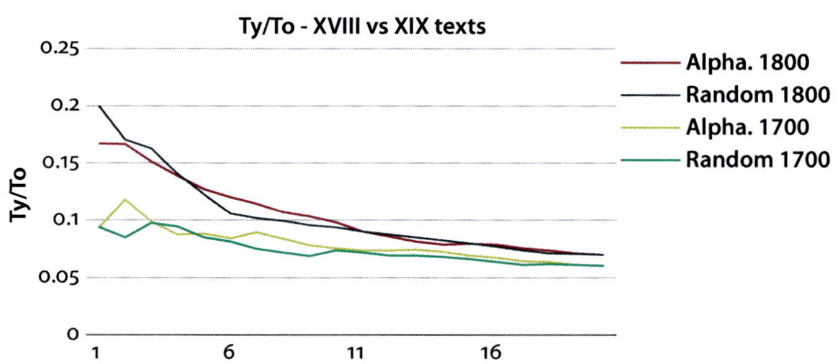

Figure 3. Metric evolution Types/Tokens according to the number of documents processed in each century

The orange line represents the Ty/To ratio in relation to the number of documents for the nineteenth century analysed randomly. It overlaps with the blue line depicting Ty/To in the same century, analysed alphabetically-chronologically. In document 15 both lines begin to come together, in parallel or overlapping each other, and never separate thereafter, which is a sign of balance in the nineteenth-century section of the corpus.

The grey and the yellow lines represent the same information in the eighteenth-century samples. When looking at the horizontal axis, we can see that the lines begin to show a parallel pattern at document 10. This pattern becomes more stable towards the seventeenth document processed, either randomly or chronologically.

It is noteworthy that the four lines depicted overlap and run together towards the right of the graph, which means a minimum of seventeen documents are needed in each century to reach the point from which some balance begins to be shown. It is also worth noting that when comparing both centuries, the graph shows that variable Ty/To is constantly higher in the nineteenth century. This is an indication of the higher lexical variation in the nineteenth century in comparison to the eighteenth century. The reason for this might be that in the nineteenth century scientific topics might be more varied than in the preceding one, thus having more terms (hence Types) reflecting new areas of research.

Figure 3 also shows that the variable Ty/To decreases in every case when the number of documents processed from CELiST increases, which is normal. The more tokens we have, the more likely they are to be used repeatedly, thus making the Ty/To ratio decrease. The tendency is the same in the results obtained when both centuries are analysed together, that is, for the whole corpus.

Figures 4 and 5 illustrate the diachronic evolution of variable Ty/To in CELiST in the eighteenth and nineteenth centuries. On this occasion, the number of tokens in both centuries is different and the data cannot be plotted on a single graph as we did previously when analysing the number of documents for this variable. Figure 4 illustrates the evolution of Ty/To in relation to the number of words included in CELiST to represent the eighteenth century.

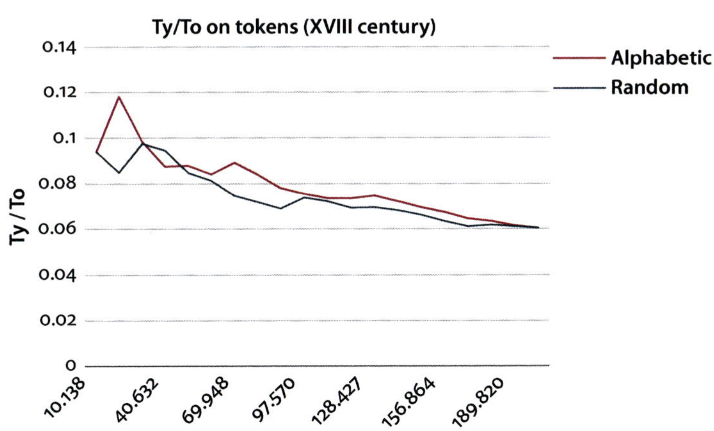

Figure 4. Evolution Types/Tokens according to the number of words processed in the eighteenth century

Figure 5, in turn, illustrates the evolution of Ty/To in relation to the number of words included in the nineteenth-century section of CELiST.

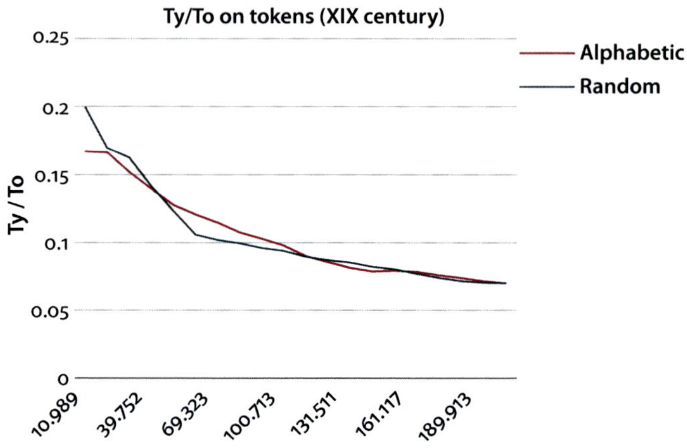

Figure 5. Evolution Types/Tokens according to the number of words processed in the nineteenth century

Figures 4 and 5 above show that eighteenth-century texts begin to show representativeness from 189,820 tokens onwards, whereas in the nineteenth century this is from 189,913 tokens onwards. These partial results show that the Ty/To ratio is 0.06 in the eighteenth century and 0.07 in the nineteenth. This means that we find differences in the analysis of the two centuries. The ratio is 16% higher in the nineteenth century, possibly because the texts in CELiST published from 1800 to 1900 exhibit a richer vocabulary. Table 3 below shows this data.

Table 3. Types and Tokens in the eighteenth and nineteenth centuries

Century	Types	Tokens	Ty/To	V1	V2
18th	12,089	200,220	0.06037893	6,926	2,269
19th	14,013	200,085	0.07003623	6,659	2,311

Although Biber (1993) claims that 1,000-word samples should be representative of variation in scientific writing, the compilers of the CC resorted to extracts of 10,000 words per decade, and CELiST is no exception to this principle. These 10,000 words samples are what we have scrutinised so far. It is now my purpose to see what happens when smaller samples are analysed with ReCor, that is, whether they are equally representative from a lexical point of view. To this end, we now use the texts in CELiST in halves. To do this we imagine that we have 80 samples instead of the 40 in the corpus, in order to see whether the results we obtain are the same as the ones above, when analysing the samples as they are. First, we study the first halves of the texts sampled; then, we study the second halves of the text-types sampled; finally, we study the corpus as if first and second halves were different text samples.

In Table 4 we provide the results for the comparative analysis of the two halves of the samples in CELiST. The data are alphabetically-chronologically organised. Column 1 shows the number of types in the first halves of the samples; the second column shows the number of tokens in the first halves; the ratio between types and tokens in these texts is shown in column 3; columns 4, 5 and 6 show also types, tokens and Ty/To ratio for the second halves of the documents.

Table 4. Types and Tokens in the first and second halves of the documents

Types 1st	Tokens 1st	Ty/To 1st	Types 2nd	Tokens 2nd	Ty/To 2nd
710	5,406	0.13141228	1,027	4,088	0.2513469
1,719	11,009	0.1562087	1,535	9,475	0.1620172
2,076	15,674	0.13249591	2,527	15,484	0.1632106
2,488	20,855	0.11932518	2,912	21,200	0.1373936
3,276	27,098	0.12089717	3,292	26,481	0.1243242
3,705	31,399	0.11802394	3,736	30,704	0.1216854
4,377	36,248	0.12075196	4,184	36,067	0.1160179
4,672	40,485	0.1154082	5,093	42,296	0.1204255
4,874	45,535	0.10704617	5,413	47,801	0.1132557
5,171	49,415	0.10465318	5,617	52,406	0.1071827
5,477	53,302	0.10277042	6,002	58,517	0.1025771
5,953	58,812	0.10123579	6,408	64,820	0.0988642
6,627	64,888	0.1021399	6,965	70,258	0.0991463
6,872	69,191	0.0993257	7,559	75,885	0.0996215
7,150	74,441	0.09605787	7,884	81,444	0.0968054
7,379	80,388	0.09179588	8,090	86,470	0.0935599
7,523	85,741	0.08774137	8,357	90,662	0.0921774
7,875	91,015	0.08653285	8,490	95,961	0.0884739
8,211	97,880	0.08389235	8,893	102,137	0.0870723
8,551	102,865	0.08313014	9,172	108,971	0.0841698
8,823	108,155	0.08158144	9,543	115,050	0.0829547
9,340	113,249	0.08247726	9,767	120,037	0.0813718
9,696	117,174	0.08275373	10,168	123,978	0.0820173
10,134	122,043	0.08303697	10,571	128,763	0.0821013
10,424	126,062	0.0826916	10,918	133,047	0.0820624
10,709	129,367	0.08278244	11,239	137,713	0.0816138
11,117	134,483	0.08266972	11,683	143,980	0.0811449
11,516	140,529	0.08195204	12,004	148,241	0.0809782
11,958	146,062	0.08187152	12,361	153,748	0.0803976
12,279	151,174	0.08122678	12,615	157,475	0.0801089

(continued)

Table 4. *(continued)*

Types 1st	Tokens 1st	Ty/To 1st	Types 2nd	Tokens 2nd	Ty/To 2nd
12,451	158,502	0.07855707	12,815	160,921	0.0796361
12,607	162,038	0.07780413	12,919	164,978	0.0783092
12,788	167,578	0.07631517	13,101	170,107	0.0770189
13,054	172,646	0.07561213	13,177	173,333	0.0760252
13,409	175,820	0.07626764	13,395	177,365	0.0755250
13,791	181,838	0.07584317	13,959	182,978	0.0762916
13,998	186,940	0.07488174	14,217	186,759	0.0761264
14,269	192,297	0.07420681	14,374	190,493	0.0754568
14,424	197,708	0.07295951	14,469	193,749	0.0746829
14,738	202,155	0.0729089	14,762	198,150	0.0744994

Clearly, the data in this table show that the two halves of each sample present nearly the same lexical distribution. The first half of the samples contains 50.25% of the tokens in the full collection and has a lexical density of 0.0729. The second half of the samples contains 49.75% of all the tokens and has a lexical density of 0.0744 which is only 2% higher than the first half of the documents and, therefore, very similar.

A comparison of the metrics of the first halves, the second halves, and the complete samples is graphically presented in Figure 6. We intend to find out which is the tendency in each case and we see that the two lines corresponding to the first halves of the sampled texts, grey and yellow, respectively, overlap when they reach text 26, from the total of 40, and maintain the same tendency after that. The second halves of the documents represented with the blue and orange lines begin to overlap when they reach texts 36–37. In turn, the navy blue and green lines represent the complete samples and depend on the patterns previously discussed. They follow an intermediate pattern. This means that representativeness is reached towards the right side of the graph, independently of the fact that we study halves or the complete samples.

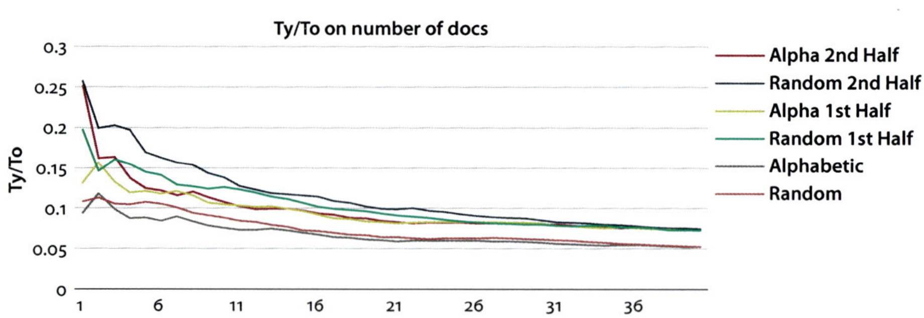

Figure 6. Comparison of the metric evolution Types/Tokens in full/1st/2nd halves of the documents according to the number of documents sampled in CELiST

Figure 6 shows graphically that the lexical density of this corpus is inversely proportional to its size, that is to say, the bigger the corpus is, the lower lexical density it presents. In order to facilitate comparisons, the following table shows the data for the different sections analysed in this corpus: the corpus as a whole, the eighteenth century, the nineteenth century, the first halves of sampled texts, and the second halves of sampled texts.

Table 5. Sampled data for comparative analysis

Set of texts	No. of documents	Types	Tokens	Ty/To	V1	V2
Complete	40	21,041	400,305	0.052564	10,866	3,609
18th Century	20	12,089	200,220	0.060379	6,926	2,269
19th Century	20	14,013	200,085	0.070036	6,659	2,311
1st halves	40	14,738	202,155	0.072909	7,927	2,563
2nd halves	40	14,762	198,150	0.074499	8,021	2,647

As a general observation, we can see that the data obtained is quite similar independently of the section analysed. The main difference is to be found when comparing the eighteenth and the nineteenth centuries, a fact that has already been commented on.

My findings show that the number of types of the first half of the corpus added to the number of types of the second half of the corpus amount to 29,500, which is considerably higher than the 21,041 types found for the corpus as a whole. This is an indicator that there are many types or terms present in only one of the two halves of each text sampled in CELiST. This is in line with the quality criteria used to select the texts in the CC.

Let us turn now to the list of terms that are present in the first half of the texts sampled in CELiST but not in the second half.

Table 6 shows that 7,038 types, from a total of 14,738 in the first halves of the texts sampled, that is 48%, are not present in the second halves of these same texts. This is an indication that the content of the documents varies quite a lot. We can see in Table 7 the terms that appear in the second half of the texts sampled in CELiST but not in the first half.

The results are similar again, 7,244 types out of a total of 14,762 appearing in the second halves of the samples in CELiST are not present in the first halves of texts, an indication of the rich variety of vocabulary that the corpus comprises.

The quantitative analysis described above indicates that the samples included in the Corpus of English Life Sciences Texts are quantitatively representative of the register and period sampled. When studying the complete samples, we can see that CELiST begins to be representative from samples 36–37 onwards. As it contains a total of 40 samples, we can conclude that it is representative in this respect.

Table 6. Number of occurrences in the first halves of the texts

No. of occurrences in first half	No. of types
1	5,216
2	1,059
3	369
4	175
5	87
6	46
7	33
8	14
9	10
10	5
11	5
12	5
14	2
15	1
16	2
17	1
20	2
21	1
22	2
23	1
29	1
31	1
Total general	**7,038**

The texts sampled show that V1 is very high, 10,866, which means there are lots of terms that we can only find once. This is an indication of the British and American variation present in the samples. There are words that ReCor only finds once because the spelling varies in the text, and in this sense these words are unique and representative of a geographical area. It also indicates the specificity of the scientific register. The comparative analysis above summarises these findings.

5. Concluding remarks

In this chapter I have looked at representativeness in CELiST. My quantitative analysis of the texts has revealed that the relevant quality criteria governing the corpus and followed by the compilers entail balance and reflect the Late Modern English scientific discourse satisfactorily. The compilers have made a careful selection of samples that represent Late Modern English in Life Sciences in such a way that the corpus is a valid tool for research.

Table 7. Number of occurrences in the second halves of the texts

No. of occurrences in second halves	No. of types
1	5,339
2	1,090
3	433
4	175
5	78
6	40
7	24
8	23
9	9
10	6
11	6
12	6
13	2
14	2
15	4
17	3
19	1
21	1
22	1
36	1
Total general	**7,244**

The research problem that I have framed (Is CELiST representative?) has been answered. Quantitative analysis has shown that whatever the samples, centuries or parts of texts we study, the corpus has proved to be balanced and representative. Also, different comparisons of the texts in CELiST have been outlined and relevant quantitative results obtained. The closure-saturation degree of the lexicon in CELiST shows a constant lexical growth, indicating that the corpus is saturated at the lexical level. Every sample added provides approximately the same number of new words as the one before.

Not only has ReCor shown itself a useful instrument for quantitative representativeness analysis, but my hypothesis has been confirmed: CELiST has been compiled following strict qualitative criteria, and the corpus is also quantitatively representative.

This representativeness analysis of the samples collected for CELiST from a quantitative and qualitative perspective allowed me to ascertain its validity as a tool for the study of the evolution of Life Sciences writing in English, along with the study of variation. CELiST reflects in a balanced way the singular and precise language variety that its sampled texts are intended to represent.

Works cited

Atkins, Sue; Clear, Jeremy and Ostler, Nicholas. 1992. "Corpus design criteria". *Literary and Linguistic Computing*, 7/1: 1–16. https://doi.org/10.1093/llc/7.1.1

Belica, Cyril. 1996. *Analysis of Temporal Changes in Corpora*. Amsterdam/Philadelphia: John Benjamins Publishing Company. https://doi.org/10.1075/ijcl.1.1.05bel

Biber, Douglas. 1993. "Representativeness in corpus design". *Literary and Linguistic Computing*, 8/4: 243–57. https://doi.org/10.1093/llc/8.4.243

Corpas Pastor, Gloria and Seghiri Domínguez, Miriam. 2007a. *El concepto de representatividad en lingüística de corpus: aproximaciones teóricas y consecuencias para la traducción*. Málaga: Servicio de Publicaciones de la Universidad.

Corpas Pastor, Gloria and Seghiri Domínguez, Miriam. 2007b. "Determinación del umbral de representatividad de un corpus mediante el algoritmo N-Cor". *Procesamiento del Lenguaje Natural*, 39: 165–172.

Corpas Pastor, Gloria and Seghiri Domínguez, Miriam. 2010. "Size Matters: A Quantitative Approach to Corpus Representativeness". In Rabadán, Rosa (ed.). *Lengua, traducción, recepción. En honor de Julio César Santoyo/ Language, translation, reception. To honor Julio César Santoyo*. León: Secretar. 112–146.

Lareo, Inés; Monaco, Leida Maria; Esteve-Ramos, María José and Moskowich, Isabel (comps.). 2020. *The Corpus of English Life Sciences Texts (CELiST)*. A Coruña: Universidade da Coruña. https://doi.org/10.17979/spudc.9788497497848

Leech, Geoffrey. 1992. "Corpora and Theories of Linguistics Performance". In Svartvik, Jan (ed.), *Directions in Corpus Linguistics*. Proceedings of Nobel Symposium 82, Stockholm, 4–8 August 1991. Berlin/New York: Mouton de Gruyter. 105–134.

McEnery, Tony and Wilson, Andrew. 2001. *Corpus Linguistics*. Edinburgh: Edinburgh University Press.

McEnery, Tony; Xiao, Richard and Tono, Yukio. 2006. *Corpus-based Language Studies. An Advanced Resource Book*. London and New York: Routledge.

McEnery, Tony; Viao, Richard amd Tono, Yukio. 2013. *Corpus-based Language Studies. An Advanced Resource Book*. London and New York: Routledge. (Retrieved January 20, 2020, from http://index-of.co.uk/Computational-Linguistics/Unit%201%20Corpus%20linguistics-%20the%20basics.pdf

Moskowich, Isabel. 2012. "CETA as a tool for the study of modern astronomy in English." In Moskowich, Isabel and Crespo, Begoña (eds.), *Astronomy "playne and simple". The Writing of science between 1700 and 1900*. Amsterdam: Benjamins. 35–56. https://doi.org/10.1075/z.173.03mos

Seghiri Domínguez, Miriam. 2011. "Metodología protocolizada de compilación de un corpus de seguros de viajes: aspectos de diseño y representatividad". *RLA. Revista de Lingüística Teórica y Aplicada*, 49/2: 13–30. https://doi.org/10.4067/S0718-48832011000200002

CHAPTER 6

Lexical fixedness within the field of Life Sciences in Late Modern English

Evidence from the Corpus of English Life Sciences Texts

Magdalena Bator
WSB University in Poznań

1. Introduction

There is a substantial body of research discussing conjoined pairs (or binomials) in terms of their structure and ordering, based on both contemporary and historical corpora. A wide range of topics related to binomials have been recently discussed in the volume edited by Kopaczyk and Sauer (2017b), and thus it is not my aim to repeat or summarise these studies. Instead, the discussion will be limited to the issues necessary for a better understanding of the analysis that follows.

The line between binomials and simple coordinated phrases is by no means clear-cut. Thus, a definition of binomials which will be followed in the present study is a good starting point. Greenough and Kittredge (1920: 133–14) describe what has later been called *binomials* as "a peculiar trick of repetition. Again and again he [Bede] uses two synonymous nouns or verbs or adjectives, where one would suffice to convey his whole meaning." Their definition, however, refers to a single type, that is, synonymous binomials. A more general explanation was offered by Malkiel, who coined the term 'binomial'. He interprets such phrases as "the sequence of two words pertaining to the same form-class, placed on an identical level of syntactic hierarchy, and ordinarily connected by some kind of lexical link" (Malkiel 1959: 113). His approach, based on the structural characteristics, required neither for such a word pair to be fixed or formulaic, nor the presence of any coordinating conjunction. These elements were gradually added by later scholars (cf. for instance Bhatia 1993). In the present study, I will follow the definition promoted by Bhatia (1993), who apart from the structural characteristics pays attention to the semantic relations between the two elements:

https://doi.org/10.1075/z.237.06bat

> What I mean by a binomial or multinomial expression is a sequence of two or more words or phrases belonging to the same grammatical category having some semantic relationship and joined by some syntactic device such as 'and' or 'or' (see Bhatia 1984: 90). Typical examples include 'signed and delivered', 'in whole or in part', 'to affirm or set aside', 'act or omission', 'advice and consent', 'by or on behalf of', 'under or in accordance with', 'unless and until' (…). (Bhatia 1993: 108)

The above definition is also favoured in the contributions edited by Kopaczyk and Sauer (2017b). Additionally, the editors in their introductory chapter make a distinction between fixed and flexible binomials and write about fixedness and formulaicity as "good criteria for a binomial" (Kopaczyk and Sauer 2017a: 3), but not obligatory ones (as advocated by Mollin (2014)). Malkiel (1959) and Gustafsson (1976: 624) place binomials on "a continuum with varying degrees of irreversibility and 'formulaicness'". Moon (1998: 152) allows for "compositional binomials" which (rather than being fixed) "demonstrate clear tendencies for preferred ordering". In the present chapter, the approach that word pairs have to be neither fixed nor formulaic to be categorised as binomials will be adopted. Instead, their frequency will be used as the criterion to distinguish possible binomials from "accidental pairings", as Chapman (2017: 42ff) calls them.

Semantically, binomials are divided into categories based on the relationship between the meanings of the two components. For instance, Gustafsson (1975: 85–87) distinguishes four major types of such relations, which result in four categories of binomials:

a. semantic homeosemy = (near-)synonymous binomials, such as *aches and pains*;
b. semantic opposition = antonymous binomials, such as *hot and cold*;
c. hyponymy = one element of the binomial is the hypernym of the other, e.g. *birds and animals*;
d. semantic complementation = both constituents share some characteristics, other than (a)–(c), as in *hand and foot*, *fair and reasonable*.

Kopaczyk (2009: 92), in her analysis of multi-word units in sixteenth-century legal Scots, is more meticulous in her division of legal binomials, and she distinguishes seven possible types of relationship: (a) binomials proper, which she defines as "the group involving semantic repetition", such as *clenge and make clean*; (b) complementation; (c) cause and effect; (d) contiguity (metonymy); (e) hyponymy; (f) antonymy (contrast); and (g) metaphor.

Different types of binomials are determined by diverse motivation for their use, or the register in which they occur. For instance, in literary texts binomials are introduced for stylistic purpose, as rhetorical figures, or for emphasis (Koskenniemi 1968); in sermons, in which antonymous binary concepts are frequently evoked, binomials may result from "a Latin-like rhetorical style" of a text, or a specific subject matter (Mollin 2017: 291, 292); legal texts gain precision and all-inclusiveness (Bhatia 1993; Kopaczyk 2009). Kopaczyk and Sauer (2017a: 11–15) distinguish three major sources of motivation for using binomials: (a) semantic – in which case the same meaning or a shade of meaning is added; (b) phonological – "to achieve a sound effect"; and (c) etymological – as an explanation or translation of one element with the other.

Despite the claim of early scholars, such as Greenough and Kittredge (1920), concerning the decrease in the use of binomials (or even their disappearance from the language),[1] recent studies leave the issue of their frequency open. Mollin (2017), for instance, asserts that their frequency is highly register-dependent, and she advocates the idea that "the use of specific types of binomials (…) is due to the contents of a text" (2017: 294).

In light of the above, it is my aim to scrutinise Late Modern English texts on Life Sciences, representing different genres, in search for possible binomials. The study shall illustrate their frequency, syntactic structure and fixedness. Moreover, diachronic and genre variables, as well as the semantic relations between the components, will be discussed.

Being in agreement with Mollin (2017), who concluded that the choice of binomials is highly related to the type and contents of a text, I intend to demonstrate that the language used in texts on Life Sciences is not void of binomials. A chronological illustration of the selected binomials will hopefully prove that they are not in decline. The investigation of a single-register corpus will illustrate differences in the use of binomials between genres.[2] It is expected that educational texts directed at a lay audience (such as lectures, guides or textbooks) should include a higher number of binomials, especially the synonymous ones used for clarification of technical or scientific terms; whereas in the more formal texts (such as essays and treatises) aimed at a learned audience, this type of binomials should be less frequent than complementary or antonymous word pairs.

1. Greenough and Kittredge (1920: 114) claimed that binomials were "an English literary habit of the ninth century. (…) the habit survived in English prose until the end of the eighteenth century."

2. The terms *register* and *genre* are often confused. A detailed explanation of differences between the two and various approaches to their analysis would constitute a discussion on its own rights. Suffice it to say that in this chapter the two notions are understood after Biber, Connor and Upton (2007: 8), according to whom *register* refers to "a general kind of language associated with a domain of use, such as a 'legal register', 'scientific register'. (…) *genre* has been used to refer to a culturally recognised 'message type' with a conventional internal structure, such as an affidavit, a biology research article, or a business memo."

2. Corpus and methodology

The primary source for the collection of data was the electronic version of the Corpus of English Life Sciences Texts (hereafter referred to as CELiST), which was compiled at the University of A Coruña as a part of the Coruña Corpus of English Scientific Writing (hereafter: CC).[3] The CELiST is a compilation of forty text samples on Life Sciences (such as Biology, Entomology, Botany and Zoology, among others) written and published between 1700 and 1900 (Lareo *et al.* 2020). The corpus is well-balanced and illustrative of the period. It consists of fragments of equal length (of approximately ten thousand words), and evenly represents each decade of the two centuries (with two texts per decade). The texts represent eight different genres and together render a sample of over 400,000 words:

- article (1 text; 9,771 words; ART)
- catalogue (3 texts; 30,045 words; CAT)
- essay (2 texts; 20,020 words; ESS)
- guide (6 texts; 60,068 words; GUI)
- lecture (2 texts; 20,099 words; LEC)
- letter (3 texts; 30,052 words; LET)
- textbook (5 texts; 50,074 words; TXTB)
- treatise (18 texts; 180,176 words; TREA)

As far as the particular text samples are of approximately the same length, the particular genres are not equally represented, ranging from a single fragment (Article) to eighteen samples (Treatise). Such an unequal rendering has been accounted for by the pursuit of the CC compilers to represent "as closely as possible the reality of text production" in the Late Modern English period (Crespo 2016: 35). Therefore, in the discussion that follows, the respective results referring to particular textual categories will be given as normalised values.

The searches were conducted with the use of the Coruña Corpus Tool (cf. Parapar and Moskowich 2007; Barsaglini-Castro and Valcarce 2020). First, all the occurrences of the elements *and, &* and *or* were automatically extracted, giving the total number of 16,700 units. At this stage, the texts in the corpus have not been tagged for the part of speech category, thus the next step was to scan the initial results manually in order to select those combinations which agreed in the grammatical (part of speech) category from other concordances of *and, &* and *or*. Cases which did not agree in terms of the part of speech or represented a complex type of coordination (e.g. a phrase on either side of the conjunction, as *at the contact or*

3. http://www.udc.es/grupos/muste

extremely near) were removed. Additionally, at this stage, numerals (e.g. *the fourth and fifth*), days of the week and months (*iune an iuly*), coordinated pronouns (*you and I*), and proper nouns (personal and place names) were discarded from the analysis. Unfortunately, such a search was still far from ideal and the results contained "possible false positives", as Mollin (2017: 283) calls tokens in which a conjunction coordinates two items representing the same part of speech found, for instance, across clause boundaries, as opposed to "true binomials". Her viewpoint, however, is that these are rather infrequent, among others due to the use of punctuation. The results were thus scanned, and all the coordinated phrases broken by a punctuation mark were deleted (e.g., *never found in woods, or in company*). Finally, the study has been limited to conjoined pairs, excluding longer strings, such as *the fans of the clappers or forceps, or the fowling-net; The whiting pollack – Rawlin pollack, bib or blind-hadoc – Whiting, Ray*.[4] This limited the primary search to 4,473 coordinated phrases (tokens) which were potential binomials.

The vast majority of the selected coordinated phrases were *hapax legomena* or unique types (using Chapman's (2017) terminology) – that is, pairs used by a single author or in a single genre, and as such they were excluded from further analysis, even though some of them are recognised as binomials in the available publications, for instance *cities and towns, night and day, cry or scream, father and mother, disk or discus, girdles and garters, able and willing, bow and arrow, rise or fall* (see Gustafsson 1976; Mollin 2014; Kopaczyk and Sauer 2017b, among others). Looking at the frequencies with which the word pairs surfaced in the CELiST, only 391 types occurred more than once throughout the corpus (this amounts to 1,300 tokens). Among these, 191 types were found in a single text. Only 49 binomials (types) occurred in more than 5 texts (altogether 475 hits), and only 14 pairs were found in more than ten texts (altogether 248 tokens).

All in all, for the present study I have decided to include only pairs which were found at least three times and in more than one text. This resulted in a corpus of 108 phrases which were used 625 times in the analysed texts. They were grouped into (i) parts of speech, and (ii) semantic categories by establishing the meaning of each constituent and the relationship between the two elements. All the meanings were established with the help of the *Oxford English Dictionary*.

4. They were discarded since (a) the chapter concentrates on word pairs only, and (b) they are perceived as lists rather than multinomials. However, some scholars do take such formations into account when discussing multinomials; see for instance Sauer (2017).

3. The analysis

The analysis below is to verify whether the assumptions and hypotheses mentioned above hold true when it comes to a corpus of material relatively homogenous in terms of its content. The search of the CELiST corpus allowed for the extraction of 108 coordinated word pairs which represent the properties of a binomial, as defined by Bhatia (1993). They were repeated in the corpus at least three times and occurred in two or more texts. The most frequent of the phrases are enumerated in Table 1.

Table 1. The most frequent binomials found in the CELiST

more or less	*here and there*	*wings and tail*	*breast and belly*	*simple and/or compound*
male and female	*figure and/or describe*	*stem and leaf*	*stamens and pistils*	*arteries and veins*
legs and feet	*figure and/or description*	*black and white*	*to and fro(m)*	*leaves and flowers*
greater or less(er)	*fingers and toes*	*large and small*	*up and down*	*plant(s) and animal(s)*
now and then	*stalk and branches*	*upper and lower*	*stem(s) and branch(es)*	*earlier or later*

The majority of the selected pairs were rather infrequent; only single types occurred more than twenty times within the corpus. These were binomials such as *more or less, legs and feet, male or female*; thus, structures which (apart from a scientific context, as in the case of the latter two) are frequently used in the general lexicon. As expected, most of the phrases were joined by *and*, with *or* being used mostly with the antonymous binomials.

In the sections to follow, the results of my analysis will be discussed with reference to the structure of binomials, time of occurrence, genre, semantic relations between the two components, and reversibility of the respective elements.

3.1 Structure variable

It is of no surprise that nominal pairs are the most frequent in the analysed corpus, constituting over half of the phrases (both in terms of the types and tokens). As evident from Table 2 and Figure 1, the other parts of speech lag behind, with 29 adjectival, 10 adverbial and 6 verbal types. Only two prepositional binomials were selected from the corpus.

Table 2. Number of binomials with reference to the part of speech variable (minimum of three occurrences in at least two texts)

Part of speech	No. of types	No. of tokens
nouns	61	327
adjectives	29	134
adverbs	10	120
verbs	6	28
prepositions	2	16
Total:	**108**	**625**

As one might expect, binomials joined by *and* predominate, constituting 65.5% of all the types. The second most frequent group in terms of the conjunction used are binomials which occur with either *and* or *or* (that is 28% of types), and the rest (6.5%) are *or*-binomials. When we look at the particular genres, it turns out that the *and*-binomials are the least common in Essay (only 33% of all the types). On the other end of the scale is Article, in which 80% of all the phrases are joined with the conjunction *and*. One has to bear in mind, however, that Article is poorly represented in the corpus (with only one sample). Thus, the results may not be fully credible and should not be generalised to the genre as such. Similarly, 75% of the binomials found in letters were joined by *and*. The remaining 20% and 25% of phrases in Article and Letter respectively allow for variation between *and* and *or*. The high percentage of phrases which allow for either of the conjunctions (almost one third of all the combinations found in the corpus) proves that the structure of the phrases used in the analysed texts was relatively free, irrespective of the genre.

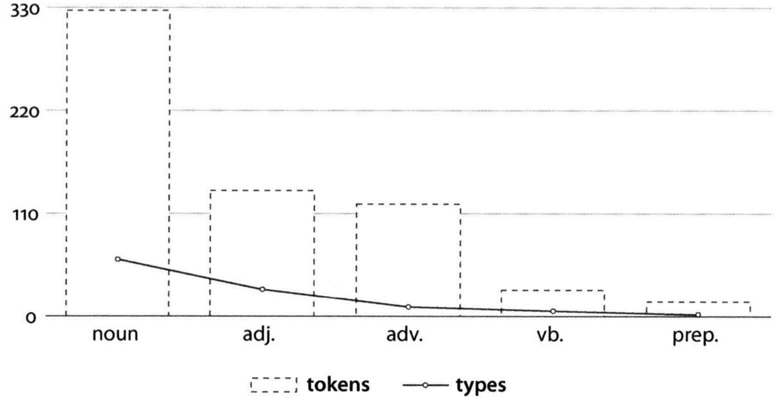

Figure 1. The number of binomials according to the part of speech variable

3.2 Chronological distribution

The CELiST ensures an equal representation of texts per decade. Figure 2 presents the chronological distribution of the binomials singled out from the corpus. With the exception of the first half of the eighteenth century, when a peak in the number of word pairs has been noted, the numbers seem to be relatively stable, not exceeding 40 occurrences per decade.

The small ratio of types to tokens suggests that individual binomials are rather infrequent and rarely repeat across the analysed text samples. This does not attest to their being unpopular in the particular period but rather indicates that larger text samples with a more homogenous range of topics would be needed. The period between 1720 and 1750 reflects a higher number of both types and tokens (although it seems that the two are not proportional). The peak prior to the 1750's may have resulted from the genres chosen to represent this period, that is, Catalogue, Textbook and Treatise – two of which, Catalogue and Textbook, contain most of the binomial phrases found in the CELiST (see Section 3.3).

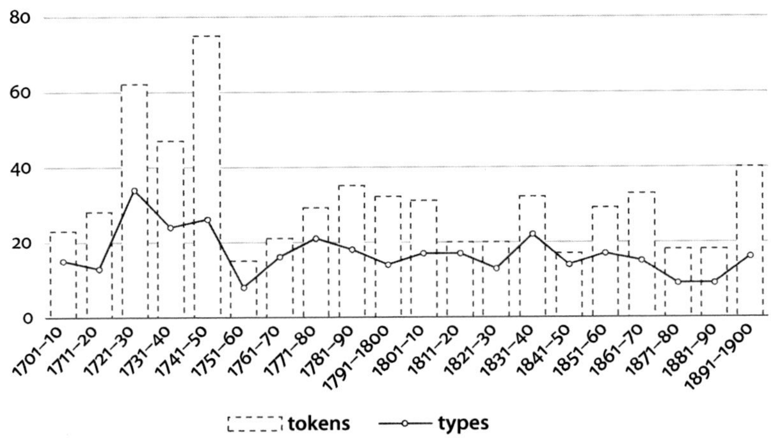

Figure 2. Chronological distribution of binomials in the CELiST (minimum of three occurrences in at least two texts)

3.3 Genre variable

The corpus comprises texts belonging to eight different genres: Article (ART), Catalogue (CAT), Essay (ESS), Guide (GUI), Lecture (LEC), Letter (LET), Treatise (TREA) and Textbook (TXTB). They are, to a certain extent, related to one another. For instance, Swales (1990) claims that the research article has been shaped by letters and treatises. It emerged only after the establishment of the *Philosophical Transactions of the Royal Society*, and developed from informative letters exchanged among scholars long before the seventeenth century. "Another powerful force that shaped the early scientific article came from the existing tradition of published scientific treatises" (Swales 2004: 203).

The genres may be divided into two groups with respect to their target audience: (a) those aimed at learned professionals, for instance Article, Essay, and Treatise; and (b) those aimed at non-professionals, as for instance Lecture, Textbook or Guide. The respective genres are not equally represented in the CELiST (with, for instance, one single article, but eighteen treatises; see Section 2). Thus, the number of binomials found in each of the genres is presented as a normalised value (see Figure 3). The best represented as to the number of binomials are Catalogue, Textbook and Lecture–with an almost equal number of types. In terms of tokens, Catalogue contains most of the repetitions of phrases in the analysed fragments. It is of no surprise that the texts which abound in binomials represent the non-formal, educational or instructive genres. This idea finds support in the studies of Tani (2008) and Sprau (2017), among others.

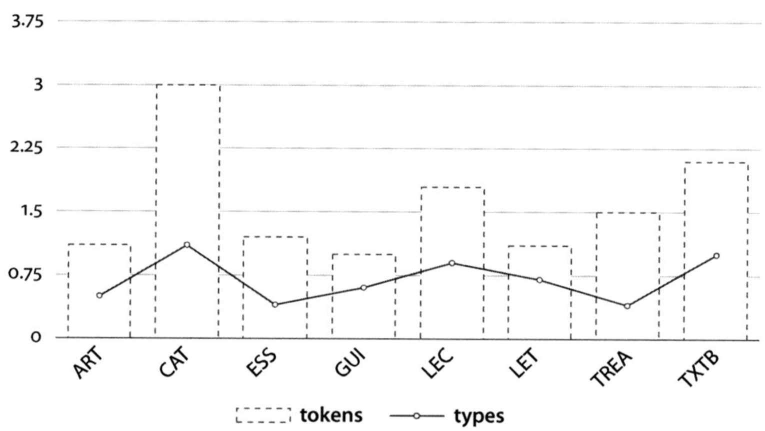

Figure 3. Distribution of binomials per genres, nf/1,000 words (minimum of three occurrences in at least two texts)

When we juxtapose the use of binomials in the respective genres with their chronological distribution, as illustrated in Figure 4, it turns out that the use of binomials in the genres with the highest number of occurrences in the CELiST (Catalogue and Textbook) is high disregarding the period. On the other hand, Treatise, which was highly represented in terms of the number of samples included in the corpus, reveals the largest number of binomials in the twenties and forties of the eighteenth century, that is, in the peak period. It had relatively fewer cases in the other decades. However, the unequal representation of the text categories throughout the analysed period does not allow to formulate any tenable conclusions as to the rise or decrease in the use of binomials across genres.

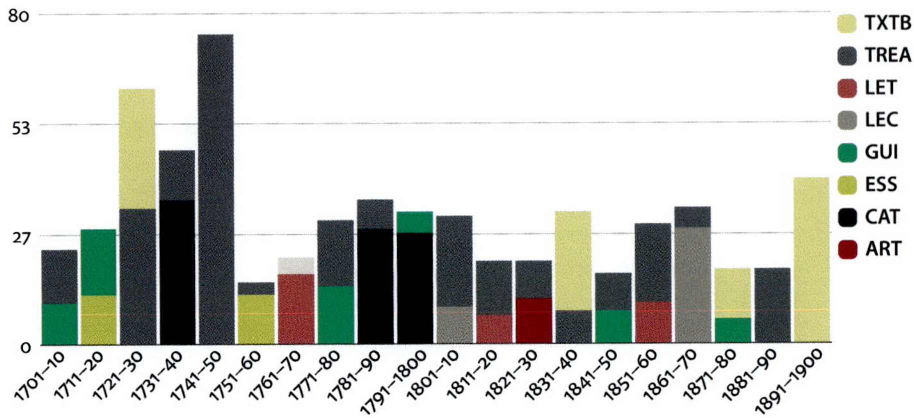

Figure 4. Chronological distribution of binomials juxtaposed against the genre variable

3.4 Semantic categorisation

Having extracted the binomial phrases from the CELiST, I decided, on the one hand, to follow Kopaczyk and Sauer's suggestion (2017a: 13) and "single out complementation as a separate category". But, on the other hand, the degree of variety of the material did not require as detailed a division of the binomials as that applied by Kopaczyk (2009). Therefore, four semantic categories have been distinguished: (a) synonymy – which comprises words of close meaning (identical meaning, if possible at all, is rare), e.g. *bifid or forked, concoct and digest*; (b) antonymy – which includes words of opposite meanings, e.g. *black and white, contract and dilate*; (c) hyponymy – which refers to word pairs in which some hierarchical relation is noticeable, e.g., *class or order* (the former being superior to the latter); and (d) complementation – in which the two constituents share some common property (but do not belong to any of the above categories (a)–(c)), e.g., *leaves and flowers, legs and feet*. Figure 5 illustrates the ratio of the particular semantic relations as found in the corpus.

Complementation is by far the most prevalent semantic relation among the binomials analysed. This is in agreement with Gustafsson's study (1975), which shows that complementation may account for most binomials. Most of these cases are nominal phrases, for instance *animals or plants, arteries and veins, ale or wine, fields and meadows, form and colour, legs and feet, leaves and branches*, etc. The first two expressions were found in a reversed order as well: *plants and animals*, and *veins and arteries* (the latter, additionally, formed a frequently used multinomial: *veins, arteries and nerves*). When we compare the use of complementation across the genres, it revolves around 50% of all the semantic relations in each of the genres but two: Article and Essay, in which it comes to one fourth of the relations.

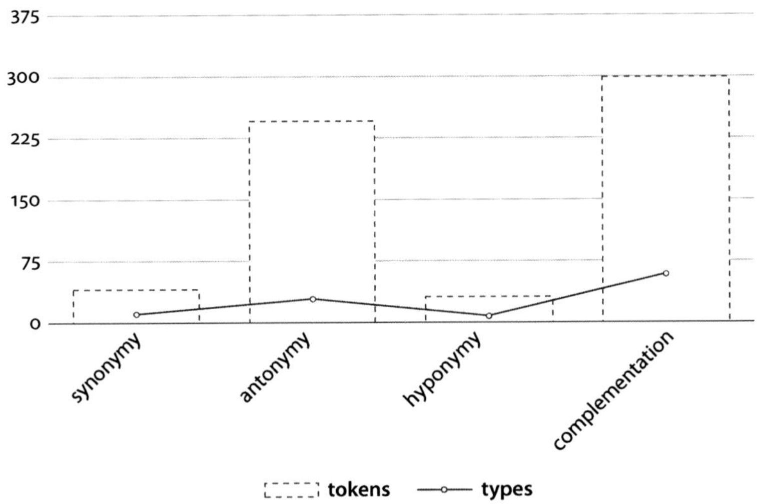

Figure 5. Semantic categorisation of binomials in CELiST

Adjectival and adverbial binomials are mostly based on contrasts, as in *right and left, earlier or later, black and white, upper and lower*. The fact that these parts of speech are the second and third most frequent in the corpus analysed contributed to antonymy being the second most common semantic category in the corpus (see Figure 5). Most cases of antonymous binomials were recorded in Essay and Article (71% and 55% of all the relations, respectively). The other two semantic relations, synonymy and hyponymy, lag behind with fewer than fifty occurrences each, that is, 11 and 8 types respectively. When considering the particular genres, most cases of the former were found in Article and, of the latter–in Catalogue (18% and 14% of all the semantic relations, respectively).

Against initial expectations, synonymous binomials are relatively insignificant in number. This is especially striking if one bears in mind that (a) they reflect a Latin-like style (Mollin 2017); (b) this was the time when English was ousting Latin from the scientific registers (Webster 1975; Taavitsainen 2001); and (c) at least some of the texts included in the corpus are of an educational character and, as texts directed at non-professional readers, one would expect that the explanatory function of binomials should be induced. Thus, one would also expect certain specialised terms to be explained by the repetition of the meaning, among others by means of (near)-synonymous equivalents.

Although the data discussed so far do not show such structures, they were present in the corpus analysed. However, most of them were found in single texts or with single occurrences only; and therefore, they were excluded from the discussion. Such an occasional distribution of these terms may have been caused by the fact that the field of Life Sciences comprises a wide range of topics and the texts which build up the corpus are not homogenous in terms of their subject matter. Therefore, the selection of data on the basis of repetition and frequency triggered a higher number of common than technical vocabulary.

Amidst the less popular binomials (that is, either those with fewer than three occurrences within the corpus analysed or those which were found in a single text), manifold synonymous repetitions were found. Next to phrases which belong to general vocabulary, such as: *esteemed and valued, jump and spring, repair and mend, shrubs and bushes, storms and tempests, mane and hair, name and epithet, cleansing and purifying*, there are numerous synonymous pairs which can be classified as belonging to one of the registers representing Life Sciences (usually medical, botanical or zoological), see Table 3.

Table 3. Examples of synonymous binomials found in the CELiST less frequently

Binomial	Register	Semantic type	Binomial	Register	Semantic type
apozems and decoctions	medical	synonymous	*calyx or empalement*	botany	synonymous
cramp and convulsions	medical	synonymous	*obtuse or blunt*	botany / zoology	synonymous
epilepsy or falling-sickness	medical	synonymous	*ruminated and chewed*	botany / zoology	synonymous
infectious and pestilential	medical	synonymous	*stings or darts*	zoology	synonymous
haemorrhoids or piles	medical	synonymous	*exuviae or skin*	zoology	synonymous
pains and aches	medical	synonymous	*talons or claws*	zoology	synonymous

3.5 Reversibility of components

The majority of the selected word pairs seem not to be arbitrary in terms of element ordering. They tend to have a fixed order, with the exception of a few binomials which were recorded in a reversed form next to the more frequent pattern. However, the latter had only single occurrences, for instance, *legs and feet* was five times more frequent than *feet and legs; stem and leaf*–twice as common as *leaf and stem; arteries and veins*–similar frequency to *veins and arteries*, but the latter was found exclusively in a single text; *male and female* (the second most frequent phrase in the corpus) was reflected as *female and male* only once. Only a few binomials seem not to have any preferred ordering (with similar frequencies of both patterns), e.g. *colour and form / form and colour; shape and size / size and shape; twigs and branches / branches and twigs*.

Following the available publications (cf. for instance Cooper and Ross 1975; Landsberg 1995; Benor and Levy 2006; Mollin 2014; Renner 2014), semantic and phonological constraints prevail among the principles which govern the ordering of elements within a binomial. Renner (2014: 441), drawing from earlier studies, enumerates "experiential closeness, temporal iconicity, syllable number, vowel length, initial consonant obstruency, final consonant obstruency, stress alternation and lexical frequency" as the "statistically significant" principles decisive of the order within word pairs; Mollin (2014: 87), who tested twenty principles put forward by Benor and Levy (2006), adds: power, formal markedness, vowel backness and height, syllable openness and weight, initial and final sonority, and alphabetic order. It has to be pointed out that there is a certain degree of overlap between the constraints proposed by different scholars. For instance, the two constraints under (c) and (d) below have been grouped together by Benor and Levy (2006), and later by Mollin (2014), as the "perceptual markedness" constraint; whilst their "power constraint" partly correlates with Cooper and Ross's "me-first principle".

In the present study I have concentrated on the semantic and (as called by Mollin) "other" constraints. The former, following Cooper and Ross (1975), are of a higher rank than the phonological criteria. Among the "other constraints", Mollin categorises lexical frequency and alphabetical order. The major principles which may be responsible for the ordering of elements in the material analysed are:

a. male before female referents (Landsberg 1995: 71ff), e.g. *male and female, boys and girls, monks and nuns*;
b. human before animate, and animate before inanimate, e.g. *man and beast, animal and plant* (Landsberg 1995: 71ff);
c. the "more positive or dominant, or logically prior" item first (Moon 1998: 152), also referred to by Renner (2014: 443) as the "superior first" constraint, e.g. *light and dark, joys and sorrows, class or order, languages and dialects*;
d. the "me-first principle" (Cooper and Ross 1975: 67) or the "closeness hierarchy" (van Langendonck 1995: 83ff). In the CELiST this constraint was found in such binomials as, for instance *here and there, now and then*;
e. the "temporal iconicity", which refers to chronology as the ordering factor (Landsberg 1995: 72), e.g. *cause and effect, to kill and devour, arrivals and departures*;
f. the food and drink hierarchy (Cooper and Ross 1975: 66), e.g. *fruits and spices, fruit(s) and vegetables*;
g. lexical frequency, on the basis of which the more frequent word precedes the other component of a word pair (Sobkowiak 1993), e.g. *liver and spleen*;
h. alphabetical order (Sullivan and Casagrande 1997; Mollin 2014: 70), e.g. *bifid and forked, black and white*. This constraint, however, has been disparaged by Benor and Levy (2006: 259).

The phonological constraints seem to be poorly represented in the CELiST. For instance, the principle that morphologically shorter words should be placed before the longer ones (Pinker and Birdsong 1979: 499), as in *care and protection, heat and sharpness, form and structure*, seems not to be very strong in the corpus, as the opposite structures were also frequent.

4. Conclusions

The aim of the present chapter was to scrutinise a corpus of Late Modern English texts on Life Sciences in order to examine the use and nature of word-pairs which could be classified as binomials. I have adopted a considerably free definition of the term binomial, thus avoiding too many restrictions, such as formulaicity or fixed word order. Discarding the former of the features has been dictated by the fact that the texts analysed cover a wide range of themes, which results in a low ratio of repetition of the selected phrases–and frequency is one of the defining features of formulaic patterns (cf. for instance Wray 2009; Filatkina 2018). As a result, the relatively rare repetition of the selected binomials was an impediment in answering how formulaic the Late Modern English language of Life Sciences was. Similar reasoning led me not to discard word pairs which exhibit a free word order. After all, these turned out not to be numerous, as discussed in Section 3.5 above. The choice of possible binomials was based on their frequency in the corpus. All in all, one hundred and eight types which were found at least three times in more than one text were extracted.

On the one hand, as scholarly texts on Life Sciences tend to be down-to-earth and substantive, they do not require any stylistic devices, thus the use of binomials might seem superfluous. But on the other hand, similarly to legal or medical texts, these writings should be precise and coherent, which could be rendered by making the texts understandable by means of clarification and explanation of technical vocabulary with the help of binomials. Against initial expectations to find numerous cases of synonymous binomials, these phrases were in the minority. They were by far outnumbered by complementary and antonymous word pairs, which correlates with the frequencies related to the part of speech variable: complementary binomials being typical of nominal phrases, antonymous binomials–of adjectival and adverbial phrases.

In terms of the genre variable, educational or instructive genres, such as Lecture or Textbook, benefit from the use of binomials much more than the ones aimed at learned professionals, such as Letter or Essay.

Chronologically, most of the binomials were found in the 1st half of the eighteenth century. This, on the one hand, might have been caused by the compilers' choice of genres to represent this period. But on the other hand, it would be interesting to further examine the fluctuation in frequency with respect to the peak period (that is, 1720–1750), as it substantially deviates from the general trend in the use of binomials. One option would be to investigate whether similar tendencies are observed in the other scientific registers of the period.

Even though a few reversible structures (with similar frequency of both patterns) were found, it seems that the majority of the selected phrases followed one preferred order. There was, however, a certain degree of freedom in the choice of the conjunction: almost one third of all the types allowed for either of the conjunctions (even if one of them was used sporadically).

To sum up, Life Sciences turns out to be an extensive area which encompasses such fields as Biology, Botany, and Zoology, to name but a few. Each of them represents a vast area on its own and thus could be studied in isolation. The extent of topics covered in the texts analysed was partly responsible for the rare repetition of the word pairs, and in turn for their exclusion from the analysis. A number of steps are required to further investigate the use of binomials in scientific texts of the Late Modern English period. The differences between individual authors and their preferences as to their use and choice of binomials are one of such issues.

Works cited

Barsaglini-Castro, Anabella and Valcarce, Daniel. 2020. The Coruña Corpus Tool: Ten Years On. *Procesamiento del Lenguaje Natural*, 64: 13–19.

Benor, Sarah Bunin and Levy, Roger. 2006. The Chicken or the Egg? A Probabilistic Analysis of English Binomials. *Language*, 82/2: 233–278. https://doi.org/10.1353/lan.2006.0077

Bhatia, Vijay. 1984. "Syntactic Discontinuity in Legislative Writing and its Implications for Academic Legal Purposes". In Pugh, A. K. and Ulijn, J. M. (eds.), *Reading for Professional Purposes – Studies and Practices in Native and Foreign Languages*. London: Heinemann Educational Books. 90–96.

Bhatia, Vijay. 1993. *Analysing Genre: Language Use in Professional Settings*. London: Routledge.

Biber, Douglas, Ulla Connor and Upton, Thomas. 2007. *Discourse on the Move. Using Corpus Analysis to Describe Discourse Structure*. [Studies in Corpus Linguistics 28] Amsterdam: John Benjamins. https://doi.org/10.1075/scl.28

Chapman, Don. 2017. "Fixity and Flexibility in Wulfstan's Binomials". In Kopaczyk, Joanna and Sauer, Hans (eds.), *Binomials in the History of English*. Cambridge: CUP. 41–62. https://doi.org/10.1017/9781316339770.003

Cooper, William and Ross, John Robert. 1975. "Word Order". In Grossman, Robin E.; San, L. James and Vance, Timothy J. (eds.), *Papers from the Parasession on Functionalism*. Chicago: Chicago Linguistic Society. 63–111.

Crespo, Begoña. 2016. "Genre Categorisation in CEPhiT". In Moskowich, Isabel; Camiña, Gonzalo; Lareo, Inés and Crespo, Begoña (eds.), *The Conditioned and the Unconditioned: Late Modern English Texts on Philosophy*. Amsterdam: John Benjamins. 25–44. https://doi.org/10.1075/z.198.02cre

Filatkina, Natalia. 2018. "Expanding the Lexicon through Formulaic Patterns". In Arndt-Lappe, Sabine et al. (eds), *Expanding the Lexicon*. Berlin: Mouton de Gruyter. 15–41.
Greenough, James Bradstreet and Kittredge, George Lyman. 1920. *Words and Their Ways in English Speech*. London: MacMillan.
Gustafsson, Marita. 1975. Binomial Expressions in Present-Day English: A Syntactic and Semantic Study. [Ph.D. dissertation: Annales Universitatis Turkuensis]. Turku: Turun Yliopisto.
Gustafsson, Marita. 1976. The Frequency and 'Frozenness' of Some English Binomials. *Neuphilologische Mitteilungen*, 77: 623–637.
Kopaczyk, Joanna. 2009. "Multi-Word Units of Meaning in 16th-Century Legal Scots." In McConchie, Roderick; Honkapohja, Alpo and Tyrkkö, Jukka (eds.). *Selected proceedings of the 2008 Symposium on New Approaches in English Historical Lexis (HEL-LEX 2)*. Somerville, MA: Cascadilla Proceedings Project. 88–95.
Kopaczyk, Joanna and Sauer, Hans. 2017a. "Defining and Explaining Binomials." In Kopaczyk, Joanna and Sauer, Hans (eds.), *Binomials in the History of English*. Cambridge: CUP. 1–23.
Kopaczyk, Joanna and Sauer, Hans (eds.). 2017b. *Binomials in the History of English*. Cambridge: CUP. https://doi.org/10.1017/9781316339770
Koskenniemi, Inna. 1968. *Repetitive Word Pairs in Old and Early Middle English Prose*. Turku: Turun Yliopisto.
Landsberg, Marge. 1995. "Semantic Constraints on Phonologically Independent Freezes." In Landsberg, Marge (ed.), *Syntactic Iconicity and Linguistic Freezes: The Human Dimension*, Berlin: Mouton de Gruyter. 65–78. https://doi.org/10.1515/9783110882926.65
Langendonck, Willy van. 1995. "Categories of Word Order Iconicity." In Landsberg, Marge (ed.), *Syntactic Iconicity and Linguistic Freezes: The Human Dimension*. Berlin: Mouton de Gruyter. 79–90. https://doi.org/10.1515/9783110882926.79
Lareo, Inés; Monaco, Leida Maria; Esteve-Ramos, María José and Moskowich, Isabel (comps.). 2020. *The Corpus of English Life Sciences Texts (CELiST)*. A Coruña: Universidade da Coruña. https://doi.org/10.17979/spudc.9788497497848
Malkiel, Yakov. 1959. Studies in Irreversible Binomials. *Lingua*, 8: 113–160. https://doi.org/10.1016/0024-3841(59)90018-X
Mollin, Sandra. 2014. *The (Ir)reversibility of English Binomials. Corpus, Constraints, Development*. Amsterdam: John Benjamins. https://doi.org/10.1075/scl.64
Mollin, Sandra. 2017. "Developments in the Frequency of English Binomials, 1600–2000." In Kopaczyk, Joanna and Sauer, Hans (eds.). *Binomials in the History of English*. Cambridge: CUP. 281–295. https://doi.org/10.1017/9781316339770.015
Moon, Rosamund. 1998. *Fixed Expressions and Idioms in English*. Oxford: Clarendon Press.
Oxford English Dictionary. Retrieved January 20, 2020, from http://www.oed.com
Parapar, Javier, and Moskowich, Isabel. 2007. The Coruña Corpus Tool. *Revista del Procesamiento de Lenguaje Natural*, 39: 289–290.
Pinker, Steven and Birdsong, David. 1979. Speakers' Sensitivity to Rules of Frozen Word Order. *Journal of Verbal Learning and Verbal Behavior*, 18/4: 497–508. https://doi.org/10.1016/S0022-5371(79)90273-1
Renner, Vincent. 2014. A Study of Element Ordering in English Coordinate Lexical Items. *English Studies*, 95/4: 441–458. https://doi.org/10.1080/0013838X.2014.908039
Sauer, Hans. 2017. "Milton's Binomials and Multinomials in *Samson Agosistes*." In Kijak, Artur et al. (eds.), *Current Developments in English Historical Linguistics. Studies in Honour of Rafał Molencki*. Katowice: Wydawnictwo Uniwersytetu Śląskiego. 275–303.

Sobkowiak, Włodzimierz. 1993. Unmarked-Before-Marked as a Freezing Principle. *Language and Speech*, 36/4: 393–414. https://doi.org/10.1177/002383099303600403

Sprau, Melanie. 2017. "Binomials and Multinomials in Sir Thomas Elyot's *The Boke Named The Gouernour*." In Kopaczyk, Joanna and Sauer, Hans (eds.), *Binomials in the History of English*. Cambridge: CUP. 201–221. https://doi.org/10.1017/9781316339770.011

Sullivan, William, and Casagrande, Jean. 1997. "Irreversible Binomials: Preliminary Considerations and Initial Findings." In Hammond, Robert (ed.), *Linguistic Studies in Honour of Bohdan Saciuk*. West Lafayette: Learning Systems. 341–352.

Swales, John. 1990. *Genre Analysis: English in Academic and Research Settings*. Cambridge: CUP.

Swales, John. 2004. "Episodes in the History of the Research Article." In Seale, Clive (ed.), *Social Research Methods. A Reader*. London: Routledge. 203–206.

Taavitsainen, Irma. 2001. "Language History and the Scientific Register." In Diller, Hans-Jürgen and Görlach, Manfred (eds.), *Towards a History of English as a History of Genres*. Heidelberg: Winter. 186–202.

Tani, Akinobu. 2008. "The Word Pairs in *The Paston Letters and Papers* with Special Reference to Text Type, Gender and Generation". In Amano, Masachiyo et al. (eds.), *Historical Englishes in Varieties of Texts and Contexts*. Frankfurt a/M.: Peter Lang. 217–231.

Webster, Charles. 1975. *The Great Instauration: Science, Medicine and Reform 1626–1660*. London: Duckworth.

Wray, Alison. 2009. "Identifying Formulaic Language: Persistent Challenges and New Opportunities." In Corrigan, Roberta; Moravcsik, Edith; Ouali, Hamid and Wheatley, Kathleen (eds.), *Formulaic Language*. Amsterdam: John Benjamins. 27–51. https://doi.org/10.1075/tsl.82.02ide

CHAPTER 7

Engagement in the botanists of the Corpus of English Life Sciences Texts

Flourishing female scientific writing

Margarita Mele-Marrero
University of La Laguna

1. Introduction

The earlier CETA (Moskowich *et al.* 2012), CEPhiT (Moskowich *et al.* 2016) and CHET (Moskowich *et al.* 2019), subcorpora of the Coruña Corpus of English Scientific Writing (CC) have all included women writers in their sources but, even if they worked in the same field in each case, their specific themes presented some divergence. There is an important group of women in CELiST (Lareo *et al.* 2020) who write on the same topic, Botany, and most of them had their works published in London. This implies they probably found fewer impediments to research and write on this subject, which has a large female tradition, although it is also populated by relevant male figures. The fact that in the works compiled on Botany most writers are women allows us to consider the confidence they may have had in their knowledge on this field and their capacity to convey this authority to their readers.

Focus can be laid on different linguistic elements to determine the interaction of an author with his/her readers. Key resources to mark stance and engagement in terms of Hyland (2005: 176) provide us with information about the authority transmitted by the writer. On the part of *stance*, hedges, boosters, attitude markers and self-mentions might be easier to measure, whereas, comparatively and on the whole, *engagement markers* like reader pronouns, directives, questions, shared knowledge and personal asides may not be so easily quantifiable. This might have caused the study of engagement to be more neglected. But even if stance seems to correlate more with evidentiality, affect and presence – the position to the assertions made, attitude, emotions and direct manifestation of his/her person – and engagement more with how the reader becomes involved in the discourse (Hyland 2005: 178, 182), both are relevant to measure the authoritative voice of the writer. Engagement encompasses forms of persuasion and as such it can also demonstrate the capacity a writer has to make his/her text be read as an authoritative discourse. Among these

https://doi.org/10.1075/z.237.07mel

latter forms, directives mark the interaction between writer and reader. Following and elaborating Searle's definition (1976: 11), directives are understood as attempts made, in our case by the writer, to obtain a specific response or action, physical or cognitive, from the reader. Different approaches towards directives offer classifications that vary mainly in those that are more difficult to apprehend (Hyland 2002, 2005; Pérez Hernández and Ruiz de Mendoza 2002; Kohnen 2008; Van Olmen and Heinold 2017). For my purpose I will basically rely on Hyland (2002); Hyland and Jiang (2016) and Zou and Hyland (2019) who consider three categories depending on the type of request: *textual acts, cognitive acts* and *physical acts*; these would be "expressed through imperatives and obligation modals" (Zou and Hyland 2019: 4). Textual acts correspond to the indications the author gives to the reader in order to follow his/her discourse (e.g.: *and now let's move on to…*); cognitive acts request the mental processing of the content (e.g.: *compare this with previous results*); finally, physical acts require more tangible actions on the part of the reader (e.g.: *use the scalpel to cut…*).[1]

The previous categorisations might be complemented with the "manifestations of directives" suggested by Kohnen (2008: 298). His study starts from a search by hand to delimit the most common forms of directives in a historical corpus, with the aim that his results could also be applied to other research, even though, as he reckons, some of the manifestations he elicited as samples might not be easy to detect. Kohnen is, therefore, more concerned with the types of clauses or constructions through which directives can be expressed. Thus, he speaks of *performatives, imperatives, modal expressions* and *indirect manifestations*. For the performatives, he indicates they usually contain "a directive speech act verb in the first person singular or plural indicative active, an object referring to the addressee and the requested act" (2008: 298); within the imperatives he distinguishes: periphrastic forms with first person plural, that is, with *let us* or *let's*; subjunctive forms with inversion; strict imperatives with the explicit presence or not of the second person pronoun; third person imperatives involving *let*; modal expressions which imply an obligation, permission or possibility; finally, the indirect manifestations go from indirect questions and conditionals to the expression of wishes (Kohnen 2008: 298–300). It must be mentioned that Kohnen was working with a corpus that included very early texts (tenth/eleventh-, fifteenth-, seventeenth-century) before he deals with Early Modern English texts; this means that not all the manifestations he encountered will be present in our corpus.

I intend to combine, as far as possible, the perspectives of these authors, Kohnen (2008); Hyland and Jiang (2016) and Zou and Hyland (2019), to elicit the directives

1. My examples based on typical cases.

which botanists in CELiST may have used. The extent to which our authors use these markers may indicate the extent to which they consider themselves capable of providing instruction. This might be even more excelling in a discipline like Botany which is mainly descriptive.

In the following sections I will first contextualise the situation of female botanists in the eighteenth and nineteenth centuries and provide the sources from CELiST that constitute the subcorpus of analysis; three male authors present in the corpus will also be included in my subcorpus to serve as counterpoints. The findings on directives obtained through the Coruña Corpus Tool (CCT) will be presented, classified, examined and compared, where possible, with the results obtained by Kohnen (2008), especially those concerning his conclusions about directives in Early Modern English in the Helsinki Corpus.

2. Eighteenth- and nineteenth-century English botany

2.1 Eighteenth- and nineteenth-century English botanists

Botany has not been unfamiliar to women, although their access to it as a science has had restrictions in history. The knowledge of healing herbs and gardening in their domestic environment was to a certain extent acceptable; writing on Botany as scientific lecturers was something different. For Easterby-Smith (2018: 101) during the eighteenth century, Botany was a non-elite science with a wide number of publications that, within Natural History, made of it the most consumed genre by the nineteenth century in Britain. She (Easterby-Smith 2017: 282) considers that the translation from Latin of the works of Carl Linnaeus helped women to have more access to this field. The attractiveness that Botany exerted on "middle-and-upper-ranking women" was eased by the simpler terminology and classification used by Linnaeus. Sam George (2005) had previously stated that:

> Botanists wooed female readers, drawing on familiar analogies between women and flowers to celebrate the virtues of the 'British fair'. That traditional Pastoralism, looking nostalgically to some lost Eden, employed flowers as symbols of innocence; this was dramatically disturbed when the Swedish botanist and taxonomist, Carl Linnaeus, focussed on the flower in order to detail the sexuality of plants by offering precise descriptions of their organs of generation. In the *Systema Naturae* of 1735 Linnaeus abandoned previous formal systems of classification and founded the 'sexual system'. (George 2005: 2)

In fact, Linnaeus is mentioned more than one hundred times in CELiST, and specially by the male and female authors in our subcorpus. Elizabeth Jacson (1804: 31) urges her students to use Linnaeus original nomenclature, so that they can understand botanical descriptions in depth and converse with other colleagues all over the world.

But not everyone approved of the Swedish botanist's system. On the contrary, some considered it offensive to female's ears and his anatomisation of plants inadequate. George (2014: 681) remarks that one of our male authors, William Withering (1776), who served as a main source for other female writers, "masked" Linnaeus' classes and orders by eliminating sexual distinctions.[2] It appears that further poetic elaborations of Linnaeus's supposedly immodest plant descriptions may have added to Botany appeal among women (George 2005: 3–4). By the nineteenth century, Almira Lincoln (also in CELiST) will, in a note, describe Linnaeus's language as "bold and graphic" (1834: 294), although to what extent she despised it is not clear. As Shteir (1990) observes, even if there was a consensus about the benefits of female study of plants, historical perspective can be diverse:

> One can argue, for example, that botany served as an inoffensive hobby for women excluded from more significant work by changing social and economic norms. One also can argue, by contrast, that botany was a cultural, intellectual, and even an economic resource for women. (Shteir 1990: 304)

For Shteir (1990: 304, 313) some women came to acquire "scientific expertise" through Botany, and such evolution might be seen in their publications.

This whole scenery makes it interesting to know if in pragmatic terms there is variation from male to female authors, and to what extent women were capable of rendering their discourse authoritative.

2.2 Sources

The following table presents the nine authors (three men and six women), abbreviated titles and number of words in each case – a total of 90192–, that constituted the CELiST subcorpus, analysed in this chapter. According to the metadata provided by the corpus, all the authors are of English origin except Lincoln and Gregory, who were American. Most of their works had been published in London, except for Lincoln's which was published in Hartford and Gregory's whose publisher note includes both Boston and London. Our male authors received education in their place of origin, in the case of our female writers they either received some education or were surrounded by culturally relevant figures, usually men. In the case of Emily. L. Gregory she even received a doctorate in Botany and was the first female lecturer and professor at Columbia University.

2. He appears to have distinguished flowers depending on how their sexual organs, pistils and stamens, were distributed in it.

The samples have been selected taking into account they deal with Botany but also considering the type of discourse used, thus, the work of Smith (1795), even though directly related with Botany, has not been included since it is limited to the description of plates, allowing for little interaction with the reader. In the case of Agassiz (1859), a female writer, she is also excluded. Her text, addressed to "my dear Lisa and Connie", is a continued monologue or letter which, although divided in chapters, seems to refer to a supposed previous experience these children may have had with nature and specially with sea animals often confused with plants like sea anemones and corals. Although it includes scientific data and references, these do not concern Botany.

Table 1. Authors and data

Year	Author	Title	Words
1723	Blair, Patrick	Pharmaco-botanologia: or, an alphabetical and classical dissertation on all the British indigenous and garden plants of the new London Dispensatory. [...]	10018
1737	Blackwell, Elizabeth.	A Curious Herbal, containing five hundred cuts, of the most useful plants, which are now used in the practice of Physick [...]	10019
1776	Withering, William	A botanical arrangement of all the vegetables, naturally growing in Great Britain. With the descriptions of the Genera and species, according to the system of the celebrated Linnaeus. Being an attempt to render them familiar to those who are unacquainted with the learned languages. [...]	10003
1789	Bolton, James.	An History of Fungusses, growing about Halifax. [...], the result of more than twenty years observation.	10027
1804	Jacson, Maria Elizabeth	Botanical Lectures. By a Lady. Altered from "Botanical Dialogues for the use of schools", and adapted to the use of persons of all ages, by the same author.	10051
1832	Lincoln, Almira Hart Phelps	Familiar Lectures on Botany, including practical and elementary Botany [...] For the use of Higher schools and academies	10013
1840	Pratt, Anne	Flowers and their Associations.	10021
1879	Lankester, Phebe	Wild Flowers worth Notice: A Selection of some of our Native Plants which are most attractive from their Beauty, Uses, or Associations	10020
1895	Gregory, Emily Lovira	Elements of Plant Anatomy.	10020
			Total: 90192

3. Directives in female and male botanists

I have sustained before (Mele-Marrero 2019: 233) that a qualitative analysis of data is essential for reliable results, but previous to that analysis we need a considerable quantity of examples. The above-mentioned difficulties in compiling data for engagement (see Section 1) lie in locating, or rather, devising a form to locate imperatives and other forms of directives using corpora. Reading all the texts and marking them does not seem to be the most effective use of a corpus. A method to find the different "manifestations" in terms of Kohnen (2008) or "directives" in the form of imperatives and modals as indicated by Hyland (2002), Hyland and Jiang (2016) and Zou and Hyland (2019), is the first necessary step.

English adds to the difficulty posed by any other language since "plain" imperative is not morphologically marked. Something different is the use of modals and imperative constructions with *let*. Additionally, our particular topic, Botany, is a quite descriptive science, it is not gardening where more directives, particularly from the physical acts category, are expected when receiving instructions on how to start cultivating plants. Abounding on this, in his *Botanical Latin*, Stearn (1966) commented:

> Since a major function of verbs is to express action and since botanical descriptions usually state the characters of plants as observed in the most inactive of all states [...]. Botanists manage verbs best by avoiding them altogether.
>
> (Stearn 1966: 130)

This seems to apply to English too. As a first approach for my purpose I tried searches with the CCT which included the few verbs compiled in Botany glossaries[3] to obtain plain imperative forms, this with little (one) or no results. Although predictable, after having read Stern's consideration, I think this was a necessary caveat to focus the research in other common verbal forms which could also justify more findings related to textual and cognitive acts. Thus, I considered the list of "engagement features: directives" used by Hyland and Jiang (2016) when searching for these in a collection of academic articles within a frame of fifty years. Their list includes sixty-two verbs, from which I excluded some that would give duplicated results[4] and modals. Of these, only sixteen rendered more than one positive result when applying the CCT. That is, only a 2.5% of these verbs was found used in a directive

3. Collins Dictionary and the Canadian University of Saskatchewan offer glossaries mainly populated with nouns and adjectives.

4. Hyland and Jiang's list includes *let, let us* and *let's* or *think* and *think of*. Using CCT the results when searching for the first form will include the others which were afterwards discriminated by hand.

structure and others did not appear at all; some very common verbs like *arrange, choose* or *classify*, which one would consider viable when talking about collecting or classifying plants, are absent from our subcorpus. If we consider only imperatives, the usage of these verbs is quite limited. Figure 1 presents total numbers and number of imperative forms. Since *let* was considered apart, to avoid duplications a verb that appears combined with it was only counted in this structure unless it also appears as a plain imperative form.

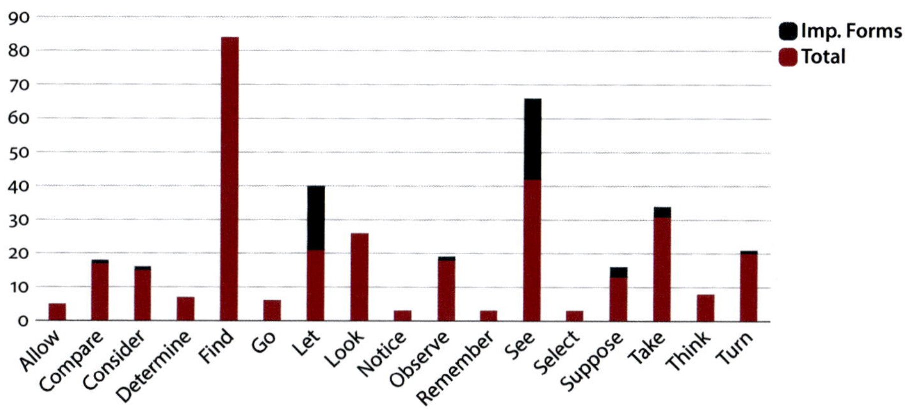

Figure 1. Verbs and their imperative forms

Of the total of verb occurrences (322 items), only 16% (53 items) corresponds to imperatives (Examples (1) and (2)) or imperatives with *let* (Examples (3) and (4)). This latter construction renders the highest proportion, out of 25 findings, 23 are imperatives and the other two are used to develop a scientific argument (see Example (5)). None of the constructions corresponded to third person constructions:[5]

(1) *Consider* then which of these subdivisions it best agrees with (Withering 1776: xxx)

(2) for an explanation of these *see* the glossary (Withering 1776: xviii)

(3) if that will not do *let* it be frequently plow'd up and fallow'd (Blair, 1723: 7)

(4) *let us select* a few instances of well known plants (Withering 1776: xxix)

(5) *let* the shaded bands represent the vascular tissue and the white (Gregory 1895: 100)

5. Kohnen exemplifies this use in the sentence: "let no man think the worse of Religion, because some are so bold as to despise and deride it" (2008: 299)

The most frequent verb is *see*, used by most of the authors to indicate textual acts, directing the reader to glossaries, plates, or even other books by the authors or other external references (Example (6)). Apart from imperatives and *let* constructions, there are constructions equivalent to them such as (7), which can also be a textual act to bring back the reader to a more specific field, parallel to structures with *let*, so as to say: *let's now consider the woodland flowers*. Examples like this or (8), "he is not desired to examine", similar to *there's no need to examine*, that is, *don't examine*, are not so easy to detect but should be pondered. Kohnen considers these cases as indirect directives, "highly variable and extremely difficult to classify" (Kohnen 2008: 300).

(6) see so well as the elastic spines of the teazle [note] *see* [dr] Lankester's lectures on the uses of animals (Lankester 1879: 78)

(7) But it's time to turn to the woodland flowers (Pratt 1840: 182)

(8) He is not desired to examine and compare all the examples: perhaps it will be better he should sometimes try his strength at unknown flowers that he chances to pick up in his walks. (Withering 1776: xxx)

Of all the authors, the one with the highest number of imperative forms is Withering and the ones with the lowest, Bolton, followed by Blackwell and Blair.

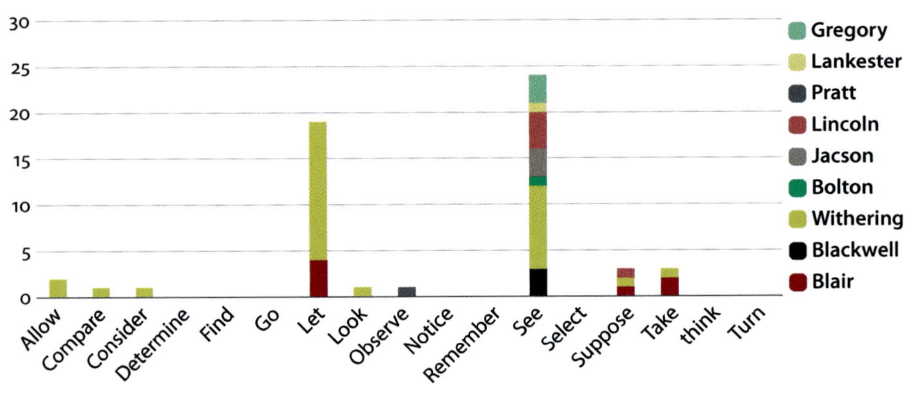

Figure 2. Verbs and their imperative forms per author

In order to obtain a more complete profile of the use of directives I also ran a search for personal pronouns *you* and *we*. This search helped not only to review cases with a second person explicit imperative, but also to discern modal uses that signified a directive towards the reader. In the case of *we*, a much closer selection by hand was necessary to distinguish self-inclusive *we*, which can act as an engagement marker, from its exclusive form, a majestic plural substituting the singular form *I* (see Example (18)). Thus, modals and other directive structures could be detected.

Figure 3 presents findings, totals of the mentioned pronouns used per author, as well as those items that actually represent a directive.

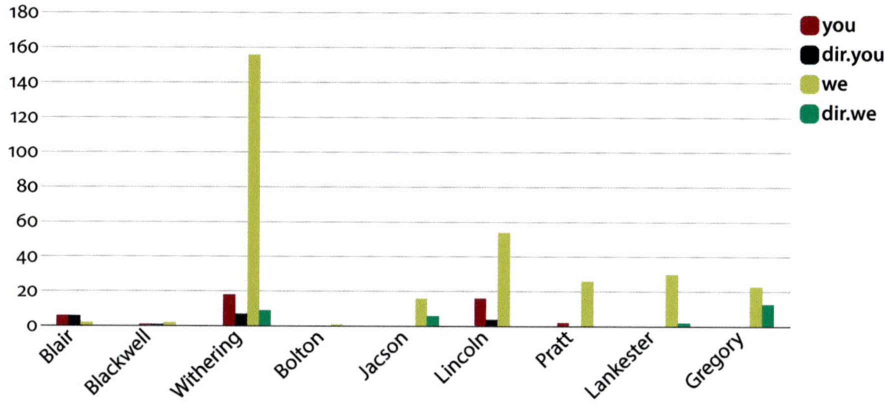

Figure 3. *You* and *we* directives

From the figure, it can be seen how directives with *you*, which would correspond to forms of imperatives with a specified subject (9), modals (Examples (10) and (11)) and other indirect manifestations (Examples (12) and (13)) are quite low. This search also allowed me to find verbs not included in Zou and Hyland's (2019) list, like *dig* (14). Nevertheless, *dig* appears only once in my material, the unique imperative form of the three examples of this verb located in the whole CELiST.

(9) and at [fig] 8 *you see* a particle of dust greatly magnified (Withering 1776: xxii)

(10) in any of these distant places *you may thin* them by pulling and transplanting the tenderest (Blair 1723: 7)

(11) but *you must recollect* that they are the workmanship of his hand (Lincoln 1832: 282)

(12) an invariable rule *not to pass over a term you do not thoroughly understand* without consulting the glossary (Withering 1776: xxxi)

(13) that is interesting to those who understand their structure but *you will not be called on* to examine the mosses (Lincoln 1832: 281)

(14) *dig* several places at convenient distances if you please regularly by applying the line about a foot square (Blair 1723: 7)

The total number of *wes* is higher, but the ones that can be considered directives grow in our final author surpassing those of Withering (1776). Most of the first-person plural pronouns appear to be either exclusive or inclusive forms of engagement that intend to involve the reader in the general procedures presented by the authors, but without requiring specific actions from the readers ((19)–(21)). These examples cannot be considered performatives as such because they are not asking for any concrete action, they simply describe facts. That is the case of most of the uses of the verb *find* recorded in Withering. Conversely, the total number of *we*s used duplicates that of *I*s. Withering (1776) prefers the self-mention plural in spite of having the greater number of imperatives, opposite to Blair (1723) and Bolton (1789) who use more often the first-person singular. Most women prefer *we*, although Lankester (1879) combines this with the use of *I* when referring to personal opinions or experiences. When directives are given these are mainly, in Zou and Hyland's terms (2019), cognitive acts (15) or textual acts (17), physical acts are the least (18).

(15) *we may infer* this from seeing the different positions of several flowers (Pratt 1840: 200)

(16) upon opening the blossom *we find* four chives (Withering 1776: xxxix)

(17) we *shall now have to return* to the figure (Gregory 1895: 91)

(18) but *we should with a good magnifying-glass endeavour to show* that each one of the little white strap-shaped petals (Lankester 1879: 79)

(19) *we trust you have learned to* consider as presenting much (Lincoln 1832: 281)

(20) it had not even obtained a name and *we have yet to acquire* the certain knowledge of its use (Jacson 1804: 12)

(21) that we are led to this remark for *we instinctively compare him* with the poor child of the town (Pratt 1840: 205)

It is worthy of mention that although combined with *you* and *we*, the modal *must* is relatively frequent, most female authors prefer *may* in any type of request (22). This hedging or attenuation is also perceived when all authors, men and women, include conditionals in their directives, as in (23):

(22) we may perceive with the naked eye (Jacson 1804: 32)

(23) *if you please you may plant* ashes conveniently tall from a nursery (Blair 1723: 7)

In their Grammar edition of 1999, Biber *et al.* (1999: 222) state that in imperative clauses "the specification of the addressee and the softening devices […] are rare in news and academic prose". Whereas it is true that directives with *you* seem to decrease in the nineteenth century, female authors use hedging devices that go from the inclusive use of *we* to avoid a strong authoritative voice, to performatives with *may*.

4. Historical evolution of directives

The historical evolution of my results does not render a clear curve but rather peaks: Jacson (1804) and Gregory (1895) show higher scores in directives with *we*, but Withering (1776) would have surpassed them, in whole number of directives during the previous century. Conversely, judging from the verbs used in the imperatives – *see*, combinations with *let* for textual acts– and the scarce number of directives for physical acts, it cannot be said that a strong authorial voice is expressed in Botany through the use of directives. All the authors avoid giving "orders" to the readers, their approach being quite soft even if their texts are intended to be didactic and instruct the reader on how to recognise, collect, dissect and classify plants. One of the reasons for this could be the type of addressee authors had in mind, occasionally children or women. In this sense, if we pay attention to the titles of the works, these can be quite revealing. While male authors reinforce their authority in the field with lengthy titles with Latin terms, that even in the case of Withering mention his knowledge of other scholars like the *celebrated Linnaeus*, or the number of plants and plates included, female works are just: *curious*, *by a Lady*, *familiar*, about *flowers* or *wild flowers* (see Table I). These titles only show a change in Emily L. Gregory: *Elements of Plant Anatomy*. The use of directives, even if not very high in all the authors, concords with these titles: men use more imperatives and other types of directives with deontic modals like *must*, whereas women use fewer imperatives and prefer the modal *may*. Again, an indication of change can be perceived in Gregory who uses both modal forms: "we *may* now *pass* on to the consideration/we *must* therefore *pass* to the next class" (1895: 107/117).

A secondary objective of this study which was marked in the introduction was to compare my results, with all the necessary cautions, with those obtained by Kohnen (2008: 307) in the Early Modern English sections of the Helsinki Corpus. His findings show that, among all the genders studied, the highest number of *let us/ let's* directives appear in Handbooks (4.8), followed by Comedy (4.3), Sermon (4.3), Philosophy (3.5), Bible (2.5), Education (2.1), Trial (1.4), Fiction (1.4), Private Letters (0.8) and Biography (0.6).

For my purpose the most adequate genres to contrast the data would be Handbooks and Education. Transformed into Kohnen parameters (in 10,000 words) my results render a total of 2.3 for *let us/ let's* directives, slightly lower than Handbooks and higher than Education. Considering that Kohnen (2008: 307) indicates the highest scores correspond to the most interactive genres and particularly that: "Handbooks were interactive because the setting was usually a dialogue", our results would not be so different if limited to Education. Although letters and dialogues were usual formats in Botany (Shteir 1990: 312), it is not the case of any of the works selected in my subcorpus. Further comparison for performatives is not possible since Kohnen (2008: 308) uses sermons, letters and prayers that differ from my corpus; had they been analogous, the general apparent decrease he suggests after the seventeenth century in their use, would also have differed from my results. There is complete coincidence with his conclusions when he manifests the limits of the "inferrability" (2008: 309) of indirect directives, since there is no clear pattern to be elicited from a corpus and by hand searches would be necessary.

5. Conclusion

Even though it might be difficult to elicit all types of directives as engagement markers from a corpus, authority can be measured through their use.

Male writers in the eighteenth century seem to use more directive forms than their female colleague, Blackwell (1737). During the nineteenth century, female writers appear to engage with their readers through the use of the first-person plural but preferring forms that mitigate any type of requests with modals like *may*. Withering (1776) would still surpass them all in his use of directives and *must* constructions. In general, all authors seem to be quite "delicate" when addressing the readers to obtain a response on their part.

The use of directives, even for textual or cognitive acts, is not so strong or in a considerable amount so as to say female writers are using them to defend their authorial voice, quite on the contrary, perhaps because of the type of descriptive science Botany was, women had more possibilities to excel in it without exposing themselves. Notwithstanding, the tendency, as engagement markers show, could have commenced to change since an upturn can be perceived in Gregory. Such evidence may encompass the idea that female scientists started flourishing thanks to Botany.

Works cited

Collins Dictionary Online. 2020. Botany: Terms Used in Botany. Word List: Terms Used in Botany. Retrieved January 20, 2020, from https://www.collinsdictionary.com/word-lists/botany-terms-used-in-botany

Biber, Douglas; Johansson, Stig; Leech, Geoffrey; Conrad, Susan, and Finegan, Edward. 1999. *Longman Grammar of Spoken and Written English*. London: Longman.

Easterby-Smith, Sarah. 2017. Botanical Collecting in 18th-century London. *Curtis's Botanical Magazine*, 34/4: 279–297. https://doi.org/10.1111/curt.12205

Easterby-Smith, Sarah. 2018. *Cultivating Commerce: Cultures of Botany in Britain and France, 1760–1815*. Cambridge: Cambridge University Press.

George, Sam. 2005. Linnaeus in Letters and the Cultivation of the Female Mind: 'Botany in an English Dress'. *Journal for Eighteenth-Century Studies*, 28/1: 1–18. First published Online 2008. https://doi.org/10.1111/j.1754-0208.2005.tb00313.x

George, Sam. 2014. Carl Linnaeus, Erasmus Darwin and Anna Seward: Botanical Poetry and Female Education. *Science & Education*, 23: 673–694. https://doi.org/10.1007/s11191-014-9677-y

Hyland, Ken and Jiang, Feng (Kevin). 2016. 'We Must Conclude that…': A Diachronic Study of Academic Engagement. *Journal of English for Academic Purposes*, 24: 29–42. https://doi.org/10.1016/j.jeap.2019.100809

Hyland, Ken. 2002. Authority and Invisibility: Authorial Identity in Academic Writing. *Journal of Pragmatics*, 34: 1091–1112. https://doi.org/10.1016/S0378-2166(02)00035-8

Hyland, Ken. 2005. Stance and Engagement: A Model of Interaction in Academic Discourse. *Discourse Studies*, 7/2: 173–192. https://doi.org/10.1177/1461445605050365

Jacson, Maria Elizabeth. 1804. *Botanical Lectures. By a Lady. Altered from* "Botanical Dialogues for the use of schools", *and adapted to the use of persons of all ages, by the same author*. London: printed for J. Johnson by T. Bensley.

Kohnen, Thomas. 2008. "Tracing Directives through Text and Time: towards a methodology of corpus-based diachronic speech-act analysis". In Jucker, Andreas and Taavitsainen, Irma (eds.), *Speech Acts in the History of English*. Amsterdam: John Benjamins. 295–310. https://doi.org/10.1075/pbns.176.16koh

Lareo, Inés; Monaco, Leida Maria; Esteve-Ramos, María José and Moskowich, Isabel (comps.). 2020. *The Corpus of English Life Sciences Texts (CELiST)*. A Coruña: Universidade da Coruña. https://doi.org/10.17979/spudc.9788497497848

Lincoln, Almira Hart Phelps. 1832. *Familiar Lectures on Botany, including practical and elementary Botany, with generic and specific Descriptions of the most common native and foreign Plants and a Vocabulary of botanical Terms. For the use of Higher schools and academies.* Hartford: J. F. Huntington.

Mele-Marrero, Margarita. 2019. "Neither *I* nor *we*: Unexplicit Authorial Voice in 18thC Academic Texts" In Moskowich, Isabel; Crespo, Begoña; Puente-Castelo, Luis and Monaco, Leida Maria. *Writing History in Late Modern English: Explorations of the Coruña Corpus*, 216–234.

Moskowich, Isabel; Camiña, Gonzalo; Lareo, Inés and Crespo, Begoña (comps.) 2016. *Corpus of English Philosophy Texts (CEPhiT)*. Amsterdam: John Benjamins. https://doi.org/10.17979/spudc.9788497497077

Moskowich, Isabel; Lareo, Inés; Camiña, Gonzalo and Crespo, Begoña (comps.). 2012. *Corpus of English Texts on Astronomy (CETA)*. Amsterdam: John Benjamins. https://doi.org/10.17979/spudc.9788497497084

Moskowich, Isabel; Lareo, Inés; Lojo Sandino, Paula and Sánchez-Barreiro, Estefanía (comps.) 2019. *Corpus of History English Texts (CHET)*. A Coruña: Universidade da Coruña. https://doi.org/10.17979/spudc.9788497497091

Pérez Hernández, Lorena and Ruiz de Mendoza, Francisco J. 2002. Grounding, Semantic Motivation, and Conceptual Interaction in Indirect Directive Speech Acts. *Journal of Pragmatics*, 34: 259–284. https://doi.org/10.1016/S0378-2166(02)80002-9

Searle, John R. 1976. A Classification of Illocutionary Acts. *Language in Society*, 5/1: 1–23. https://doi.org/10.1017/S0047404500006837

Shteir, Ann B. 1990. Botanical Dialogues: Maria Jacson and Women's Popular Science Writing in England. *Journal for Eighteenth-Century Studies*, 23: 301–317. https://doi.org/10.2307/2738798

Stearn, William T. 1966. *Botanical Latin. History, Grammar, Syntax, Terminology and Vocabulary.* London: David & Charles Newton Abbot.

University of Saskatchewan, Canada. *Glossary of Botanical Terms*. Retrieved February 25, 2020 from https://www.usask.ca/biology/rareplants_sk/root/htm/en/researcher/5_gloss.php

Van Olmen, Daniël and Heinold, Simone (eds.). 2017. *Imperatives and Directive Strategies from a Functional-Typological Perspective*. Amsterdam: John Benjamins. https://doi.org/10.1075/slcs.184

Zou, Hang J. and Hyland, Ken. 2019. 'Think about How Fascinating this is': Engagement in Academic Blogs Across Disciplines. *Journal of English for Academic Purposes*, 43: 1–12. https://doi.org/10.1016/j.jeap.2019.100809

CHAPTER 8

Linguistic indicators of persuasion in female authors in the Corpus of English Life Sciences Texts

Begoña Crespo
University of A Coruña

1. Introduction

This study describes the way in which women authors used linguistic devices of persuasion in Late Modern scientific writing, looking at samples from the Coruña Corpus of English Scientific Writing, and also, more specifically, at introductory material to these works such as prefaces and dedications. Whilst scientific writing is considered to be presented in an objective and rigorous tone (what in 1993 Tannen referred to as "considerate style"), prefaces can be said to evince a somewhat more intimate and persuasive tone ("high-involvement style"), in that the structural rhetoric of such texts in principle allows for a direct relationship between author and target readership. The features denoting persuasion, argumentation and interaction to be analysed here include first and second person pronouns, predictive and necessity modals, conditional subordination, suasive verbs, and *to*-infinitives, all commonly cited in the literature as serving persuasive functions (Biber 1988; Biber and Conrad 2001; Prelli 1989; Mischke 2005; Arakelyan and Muradyan 2016, among others).

This analysis will allow me to discover how frequently and under what discursive circumstances (communicative format used, lexico-grammatical context) female writers were present in their own works as a means of capturing their readers' attention, as well as to see the kind of linguistic devices writers preferred and how this can be interpreted in terms of writing style.

Classical authors were keenly aware of the power of persuasion. Aristotle, for example, described it as encompassing three main principles: *ethos, pathos* and *logos*. The first referred to the speaker or communicator, the second related to the emotional state of the audience, and the third referred to how the message was conveyed, how it was constructed, and how arguments intertwined to produce a particular effect on the audience (Perloff 2003).

https://doi.org/10.1075/z.237.08cre

In this paper I will focus on the author/writer as persuader (*ethos*), but also on the linguistic manifestation of persuasion (*logos*). I will try to show how some of the linguistic elements that are thought to influence the target readership are used by female authors of science. I will also consider context to see whether further elements enhance or mitigate persuasive strategies. This will require a detailed analysis of some excerpts, both those from main texts and those from the introductory material of works.

2. Women scientists, prefaces and persuasion

With the emergence of the new science in the late seventeenth century, a change in the style of writing science occurred, one in accordance with the paradigms of clarity of expression, brevity, objectivity, and the avoidance of ornamentation. These were some of the broad stylistic tenets that came to dominate late seventeenth- and eighteenth-century scientific writing. While this was in keeping with the puritanical spirit of the period, writers still had to argue in defence of their findings and opinions, to show reliability, trustworthiness and the veracity of facts, and for this reason the author relied on the use of appropriate linguistic mechanisms and rhetorical structures. For any writing to influence its readership, authors have to build up well-organised texts focusing on the audience and using these mechanisms and structures in a more or less explicit way. In terms of Late Modern scientific writing, audience-oriented communication might be expected from those whose ultimate goal was to transmit the observation of nature in all its manifestations, as well as to convince the reader of the truthfulness of the claims therein. In fact, the effectiveness of authors' endeavours here could be strengthened by incorporating the negotiation of views, attitudes and opinions with the addressee. All this, then, became part of the dissemination of scientific knowledge.

I will address here two different forms of discourse. On the one hand, prefaces, front matter, advertisements or any other sort of introductory material; on the other, the main body of the texts themselves. Prefaces are considered to be more emotional than main texts, given that they may contain direct appeals to the audience, sometimes even mentioned explicitly as the identified readers of the work (Pratt 1840). On top of this, we can also find rhetorical formulae such as acknowledgements, dedications, expressions of gratitude or indebtedness to mentors or sponsors ("It remains to say, that my acknowledgments are due to Dr. Shaw and Mr. Barbut, as the principal sources from which I have collected my materials" (Wakefield 1816: v). These are normally well-known members of the learned community or those from the higher echelons of society. It is curious to note that in the current data, manifestations of gratitude in prefaces written by women are addressed exclusively to men. Prefatory material can also function as a guide for the reader to understanding the content of the work, its organisation, or the purpose of the book (Gregory 1895). The writer may also decide to open herself up to her readership and explain the motivations underlying the publication of the work. All these intentions form part of the communicative goal of introductory material (Bhatia 1997). As Smellie (1790), one of the authors included in CEPhiT, Corpus of English Philosophy Texts (Moskowich *et al.* 2018), claimed in his introduction to *Philosophy of Natural History*, and which Lincoln would repeat in 1832, "A PREFACE, should contain an account of the circumstances and motives which induced an author to write upon that particular subject." Prefaces are considered to be more emotional than main texts, given that they may contain direct appeals to the audience, sometimes even mentioned explicitly as the identified readers of the work (Pratt 1840). On top of this, we can also find rhetorical formulae such as acknowledgements, dedications, expressions of gratitude or indebtedness to mentors or sponsors ("It remains to say, that my acknowledgments are due to Dr. Shaw and Mr. Barbut, as the principal sources from which I have collected my materials" (Wakefield 1816: v). These are normally well-known members of the learned community or those from the higher echelons of society. It is curious to note that in the current data, manifestations of gratitude in prefaces written by women are addressed exclusively to men. Prefatory material can also function as a guide for the reader to understanding the content of the work, its organisation, or the purpose of the book (Gregory 1895). The writer may also decide to open herself up to her readership and explain the motivations underlying the publication of the work. All these intentions form part of the communicative goal of introductory material (Bhatia 1997). As Smellie (1790), one of the authors included in CEPhiT, Corpus of English Philosophy Texts (Moskowich *et al.* 2018), claimed in his introduction to *Philosophy of Natural History*, and which Lincoln would repeat in 1832, "A PREFACE, should contain an account of the circumstances and motives which induced an author to write upon that particular subject."

Taking all this into consideration, it seems reasonable to assume that prefaces will contain more persuasive features from the various available lexico-grammatical resources than the main texts. In the latter, persuasion of this kind, if it exists at all, is likely to be subtle, less clearly in evidence, in that writers need to focus primarily on facts. On the contrary, when seeking approval and acceptance by the epistemic community, and whoever is the type of assumed reader (layman or scholar), the use of specific linguistic features is very much expected, where and when these can fit into the framework of desired objectivity. Hence, the style in prefaces exhibits traces of involvement whilst texts are more informational.

Two further factors can help distinguish these two communicative formats: first, the level of formality in the writing, with prefatory material typically being less formal than main texts; second, the proximity to orality. Introductions and dedications often contain cues of orality, embodying a conversational tone and marking the inclusion of others, yet such cues disappear once the scientific content itself begins to be explored (Argamon *et al.* 2003). Given that women are claimed to be more cooperative and participatory in their conversational tone (Tannen 1993), some traces of this may be detectable in prefatory material. Moreover, as Cameron has claimed in her study of women's language, "there is a characteristic register or 'women's language' consisting of certain linguistic features connoting tentativeness, deference, and lack of authority" (Cameron 1992: 3).

For Lakoff (1975) there is a notable relationship between femininity and unassertive speech, which she defines as 'women's speech.' According to her, in a male-dominated society, women are pressured into showing the feminine qualities of weakness and subordinance towards men. Thus, "it is entirely predictable, and given the pressure towards social conformity, rational, that women should demonstrate these qualities in their speech as well as in other aspects of their behavior" (Cameron, McAlinden and O'Leary 1989: 76).

In previous studies on female writers of science several tendencies have been detected. In Crespo (2016) I found that women science writers in general used a higher proportion of persuasive strategies than men, but that there were some distinctions when the strategies were considered individually. Thus, necessity modals, conditional subordination and suasive verbs were employed more frequently by women, whereas predictive modals seemed to be more characteristic of male writing. Bearing in mind that suasive verbs were the least frequent feature by women, the use of these can perhaps be seen as the corollary of the social pressure exerted by eighteenth-century society on women, compelling them to seek a position in which they stood out from the male-oriented norm and thus to have any chance of their voices being heard. So, whereas they were obliged to comply with the principles of objectivity, clarity and conciseness, they also had to be subtle. For this reason, we should talk about veiled and not overt persuasion (Crespo 2016).

Female authors seem to follow the general writing trends imposed on the community of practice, shifting towards the depersonalised discourse of modern science in line with what has been claimed by authors such as Halliday (1988: 166) and Taavitsainen (1994: 339), who date this form of discourse to the nineteenth century. Yet women could not yet play an active role in scientific life (Knight 1986) and thus they probably adhered to the ideological constructs inherent in the expression of objectivity as a means of achieving affiliation and recognition (Crespo 2019). Indeed, they did so with astonishing regularity in comparison to men, which can perhaps be explained on the grounds that women followed male patterns in an exaggerated way as a means of attaining self-assuredness (Moskowich and Crespo 2014).

In Section 3 I will describe the material used for analysis here and the methodology.

3. Material and methodology

My data will consist of samples taken from a subcorpus of the Coruña Corpus of English Scientific Writing, CELiST (Corpus of English Life Sciences Texts) (Lareo *et al.* 2020). This subcorpus compiles a total of 40 10,000-word samples, two per decade from the eighteenth and nineteenth centuries. The whole body of texts, then, amounts to 400,305 words. However, I will limit my analysis to texts written by women, which total 80,187 words, only 20% of the subcorpus. Detailed information of these texts is set out in Table 1:

Table 1. Material: Texts and prefaces

Author	Year	Title	Communicative format	Sample words	Preface words
Blackwell, Elizabeth	1737	*A Curious Herbal, containing five hundred cuts, of the moſt uſeful plants, which are now uſed in the practice of Physick. Engraved on folio copper plates after drawings, taken from the LIFE. To which is added a short description of ye plants and their common uses in PHYSICK. In Two Volumes. Vol. I.*	Catalogue	10,019	67
Jacson, Maria Elizabeth	1804	*Botanical Lectures. By a Lady. Altered from "Botanical Dialogues for the use of schools", and adapted to the use of persons of all ages, by the same author.*	Lecture	10,051	133
Wakefield, Priscilla	1816	*An introduction to the natural history and classification of insects, in a series of familiar letters*	Letter	10,008	309
Lincoln, Almira Hart Phelps	1832	*Familiar lectures on botany, including practical and elementary botany*	Textbook	10,013	792
Pratt, Anne	1840	*Flowers and their associations*	Guide	10,021	317
Agassiz, Elizabeth	1859	*A First Lesson in Natural History,*	Letter	10,035	51
Lankester, Phebe	1879	*Wild flowers worth notice … for their beauty uses and associations*	Guide	10,020	2,837
Gregory, Emily	1895	*Elements of Plant Anatomy*	Textbook	10,020	621
			Total	**80,187**	**5,127**

As we can see here, the prefatory material included in these works amounts only to 5,127 words of the data, distributed thus:

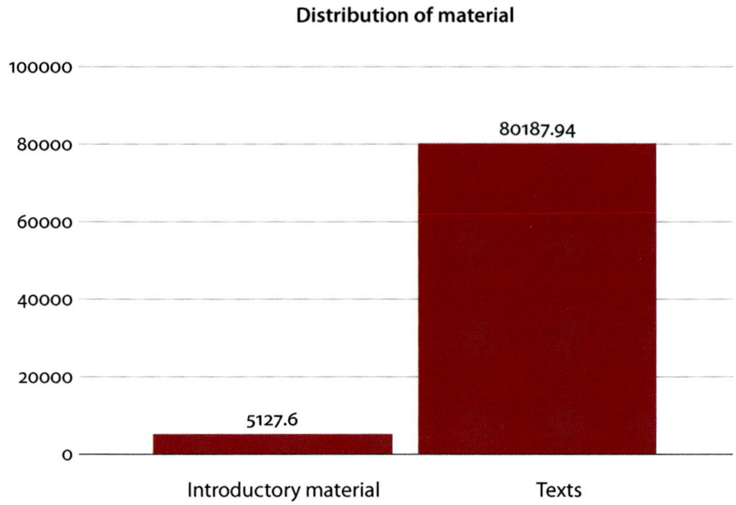

Figure 1. Words per kind of material

Given that the representation of both types of material (94% texts and 6% prefaces) is clearly unbalanced, I will normalise figures to 1,000 to provide easier comparisons. The introductory material of these texts ranges from advertisements and introductions to prefaces; samples from the main texts are not taken from the same parts of the book, and may have been extracted from the beginning, middle, or final sections, this to avoid linguistic idiosyncrasies in relation to the structure of the work (see Moskowich 2021). Women's works are represented by a variety of formats: a catalogue, a lecture, two letters, two guides and two textbooks.

The features under analysis have been taken as indexical of persuasion by a number of authors (Biber 1988; Biber and Conrad 2001; Prelli 1989; Mischke 2005 and Arakelyan and Muradyan 2016), and are given in Table 2.

The data were extracted using the Coruña Corpus Tool (CCT), a retrieval tool specifically designed to obtain information from the Coruña Corpus of English Scientific Writing (Barsaglini-Castro and Valcarce 2020). For the preface material I used the free software tool, Antconc (Anthony 2018) which allows for the creation of concordances, wordlists, frequency lists, keywords and n-grams once original files are converted into txt files.

First and second person pronouns included in the searches were the singular and plural forms in the nominative and accusative cases, plus reflexive pronouns. In the case of predictive modals, contracted forms were also searched for. Manual disambiguation was necessary when analysing suasive verbs and *to*-infinitive

Table 2. Linguistic features marking persuasion

Predictive modals	Necessity modals	Conditional subordinators	Suasive verbs	1st pp	2nd pp	*To*-inf
would	*must*	*if*	*agree, allow, arrange, ask, beg, beg, command, decide, decree, demand, desire, determine, enjoin, entreat, grant, insist, instruct, intend, move, ordain, order, pledge, pray, prefer, pronounce, propose, recommend, request, require, resolve, rule, stipulate, suggest, urge, vote*	*I*	*You*	
shall	*ought*	*unless*		*Me*	*yours*	
will	*should*			*mine*	*yourself*	
contracted forms				*Myself*	*yourselves*	
				We		
				Us		
				Ours		
				ourselves		

constructions. The suasive verbs considered were those proposed by Biber *et al.* (1999) and Quirk *et al.* (1985), including their corresponding morphological forms, and on revising each individual concordance it was necessary to exclude hits which, due to the conversion process, could have corresponded to grammatical categories other than the verb. Thus, I excluded nouns in *order, move, command, request* or *-ing* forms when used with the term *demand*, and *-ed* forms when used with *prefer, intend* and *determine*. For the sake of rigour, individual concordances of *to*-infinitives have been also revised to eliminate prepositional phrases (*to the thickness*, in Blackwell 1737).

4. Data analysis and discussion

Table 3 shows raw figures on the general occurrence of linguistic features of persuasion in the data:

Table 3. Raw numbers of linguistic features

Features	Forms
1st person pronouns	524
2nd person pronouns	256
Predictive modals	257
Necessity modals	114
Conditional subordinators	112
Suasive verbs	141
To-infinitives	942
Total	**2,346**

There are a total of 2,346 instances of the linguistic features under scrutiny in the data, representing 2.75% of the total number of words.

In what follows I will describe the distribution of these features according to their occurrence in both prefaces and main texts. The relative frequencies of the features in both types of material, looking specifically at each century, will be examined later in the paper.

4.1 Prefaces and bodies: General data

A preliminary overview of the data suggests that persuasion is more frequently manifested in prefaces than in the works themselves, as hypothesised. As noted above, figures have been normalised to 1,000 words to better express their frequency of occurrence, as shown in Figure 2:

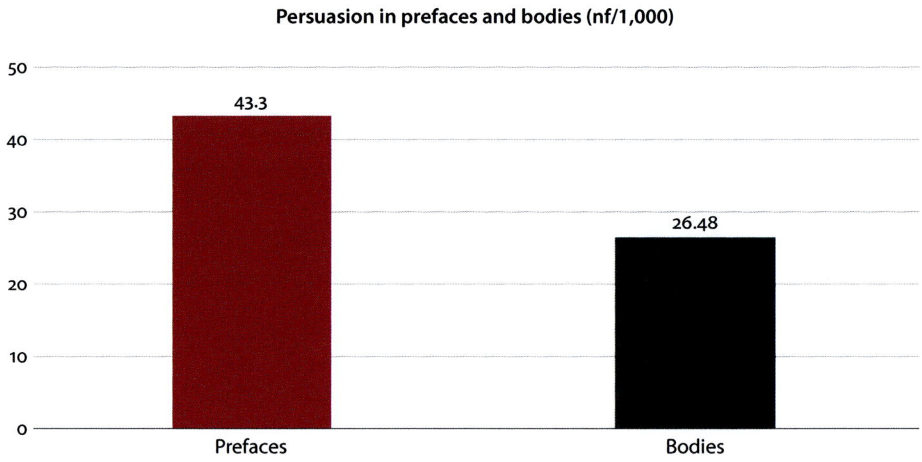

Figure 2. Persuasion markers in prefaces and bodies

Richard Perloff (2003: 8) defines persuasion as "a symbolic process in which communicators try to convince other people to change their attitudes or behavior regarding an issue through the transmission of a message, in an atmosphere of free choice". Yet it may be the case that the purpose is simply to raise awareness of an issue without requiring any kind of explicit response. When comparing prefaces and main texts we find that the frequency of occurrence of persuasion markers searched for is far higher in the former (43.3) than the latter (26.48), as might have been expected. This means that prefaces tend to be more interactive, aiming to engage with the reader's emotions and to attract their interest.

Examples (1) to (4) below illustrate authorial presence and the expression of the writer's thoughts in seeking the reader's engagement:

(1) I found the want of a suitable book for beginners, and prepared for the use of my pupils a sketch, of which the following pages are but the filling up.

(2) With respect to botanical facts, I have no claim to any discoveries, neither have I ventured to make any innovations upon the science itself.

(Lincoln 1832, preface)

(3) In answer to this objection, I would remark, that from experience in teaching others, and from observation of the operations of my own mind, I am led to believe that books most remarkable for a concise style, are not the most favourable for the development of the mind. (Lincoln 1832, preface)

(4) But much as I have desired to aid the youthful mind in acquiring the elements of knowledge, still more have I desired to lead that mind to the fountain of all knowledge, to teach it to behold the providence of God as ever active, and watchful over all, even the least of his works. (Lincoln 1832, preface)

In most of these examples the use of the first singular pronoun is accompanied by verbs of thought (*venture, believe*) and also suasive verbs such as *desire* and *to*-infinitive constructions (*to aid, to lead, to remember, to interest, to become, to surmount*). The target audience is addressed by means of certain nominal referents: *beginners, pupils, the student, the youthful mind, readers*. Although, in principle, this might be seen to be a distant relationship, they work like vocatives for the target readership in that they capture and hold readers' attention.

These examples, then, try to arouse the emotional interest of the reader. The overt presence of the author manifesting sincerity and honesty has the goal of being credible, and this in turn to show trustworthiness as well as expertise (Dillard 2014). This is one of the prototypical characteristics noted by O'Keefe (1990) that all persuasive texts should have; another is the explicitness of the author's viewpoint, which again is transmitted by the linguistic constructions mentioned above.

(5) To those readers who have just commenced the study of botany, or who may intend to pursue it at some future time, the author hopes these pages may not be unacceptable. (Pratt 1840, preface)

(6) It is hoped that the associations connected with many flowers, which are in this volume presented to the readers, will enable them to remember the names of several plants, and serve to interest them in the study of nature.

(Pratt 1840, preface)

This clustering of linguistic features is a clear manifestation of a writer's intention to maximise their influence through making a direct approach to the target readership. The reasons behind the selection of the topic, the format chosen, or the personal motivation for writing the work itself are some of the explanations that justify the more or less conversational tone employed in an attempt to gain the addressee's acceptance. The external reasons for writing the work are thus negotiated in this prefatory material (Swales 1990).

(7) I have, therefore, endeavoured to compose a complete elementary system, which may enable the student, of whatever age, to surmount those difficulties, which hitherto have too frequently impeded the perfect acquirement of this interesting science; (Jacson 1804, preface)

(8) I therefore say to all, observe, collect, and preserve the wild flowers you find; arrange and name them scientifically, if possible; but if that be not in your power, still they will always be interesting to you as a pleasing record of "times and places, and old familiar faces," which one day you will value. When you have become well acquainted with the wild plants of your own neighbourhood, there are always *rare ones* to look for, and great is the interest and excitement attending the discovery of a rare species in an unexpected locality; but before you can expect to become a discoverer, you must be a tolerable botanist. (Lankester 1879, preface)

In Example (8) the tone between author and audience is one of proximity, and instead of vocatives we find the use of the second person pronouns, another feature indicative of persuasive intent. According to Perloff (2003), pronouns are extraordinarily significant in persuading readers. This is especially so in the case of the pronoun *you*, which he claims is used mainly to shorten the distance in a communicative act between persuader and receiver.

Persuasion markers are also present in main texts, although to a lesser extent. In these cases persuasion is more in line with an informative style that transmits facts through conditions that guarantee the truthfulness of these facts (*if, as the pupil proceeds…*) and through providing evidence of what is going to be done (*you will not find it difficult; it is easy to watch; desiring to possess; difficult to determine; to recommend it for this purpose*) or must be done (*must be clearly described*). Sometimes, other word categories that have not initially been considered, such as adjectives, can add a further air of persuasion (*easy, desiring, difficult*). This may also occur with verbs: *watch, described, interested.* Taken together, such uses represent a bundle of strategies that focuses the audience's attention and guides the readership in their interpretations (Hyland 2005, 2015). This can be seen in Examples (9) to (15):

(9) entertainment in botany however instruction and amusement may be united <if> as the pupil proceeds he examines and compares the different (Jacson 1804: 48)

(10) development of these side axes is often very difficult to <determine> in the pteridophytes there is much uncertainty respecting the manner (Gregory 1985: 101)

(11) the very trouble entailed in collecting the material seeming to <recommend> it for this purpose in a church where bodily sacrifices (Lankester 1879: 112)

(12) this is one of our most poisonous native plants and <must> be clearly described in order to be avoided it belongs (Lankester 1879: 104)

(13) close if the account of them has interested you you <will> not find it difficult to keep many of these animals (Agassiz 1859: 81)

(14) they float excites them into unusual brilliancy it is easy <to> watch the action of this singular quality in the jelly-fishes (Agassiz 1859: 60)

(15) love done was law to the lord the lady desiring <to> possess some of the bright blue blossoms caused her faithful (Lankester 1879: 100)

These examples also serve to illustrate that any analysis of the linguistic features involved must take into account the context in which they occur: the surrounding elements and expressions can contribute to grounding the line of argumentation and persuasive nature of a text.

In addition to this, a persuasive tone can be manifested through overt mention of the author or through reference to the reader. In (16) the inclusive *we* embodies the addressor and the addressee, both participants in the communicative process being invoked in a single pronoun. Thus, the supposition here is that they are engaged in the process of sharing knowledge and are seen to be on the same level, with no hierarchies, hence underlining values of proximity and solidarity. Proximity, in turn, can be interpreted as an indicator of persuasion.

(16) and jaundice the germans put it into their wine as <we> do borrage & burnet as a great cordial and comforter (Blackwell 1737: 15)

In Example (17), from the work by Lincoln, the first-person plural collocates with a private verb, *trust*, and a second person referent:

(17) of the mosses are now in blossom these we trust <you> have learned to consider as presenting much that is interesting (Lincoln 1832: 281)

This *we* refers to the whole epistemic community in which the reader, directly addressed in the second person pronoun, is included. Besides, *trust*, as a verb of feeling entails emotional overtones. Indeed, persuasion here pervades the entire sentence.

a. *Prefaces and bodies: Specific data*

Not all authors exhibit the same kind of persuasive style when writing their introductory material. Figure 3 indicates an overall decline over time in the normalised frequency of occurrence of the features analysed in the prefaces.

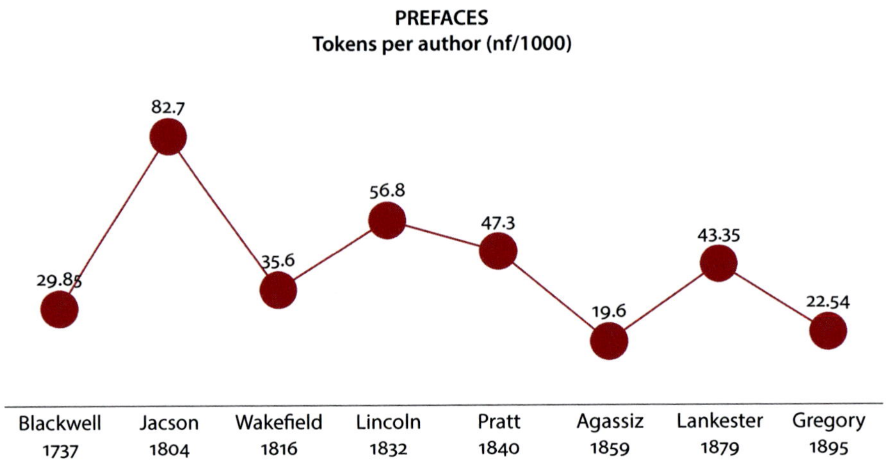

Figure 3. Persuasion tokens in prefaces

However, there are several outliers with notably high frequencies of persuasive elements. Jacson's introductory material, for instance, contains by far the highest frequency of forms (82.7 per 1,000 words). It is worth noting that this takes the form of an advertisement which very much reinforces the goal of the text and the explicitness of the author (see Example (9) above as well):

(18) and I flatter myself that the following work, in Botanical Lectures, will be found an easy introduction to the use of the translated system of vegetables, the only English work from which the pupil can become a Linnean, or universal botanist. (Jacson 1804, preface)

Two first-person pronouns, a predictive modal (*will*) and a nominal referent (*pupil*) replacing *you*, as well as the adjective *easy*, contribute to the persuasive tone of this fragment, and serve to elicit an abstract response from the reader.

In turn, Agassiz's preface contains the lowest normalised figures of these features:

(19) THIS little book, which it is hoped may be interesting for children, and perhaps of some use to parents whose children share the general juvenile delight in Aquariums, has been prepared... (Agassiz 1859, preface)

However, the example does illustrate one of the moves mentioned by Bhatia (1997) as typical in this kind of genre: the author's promotional intent to the target readership is marked through the use of subjective forms such as the adjective *interesting* and the noun *delight*.

In the main texts we perceive a slightly increasing tendency in the use of the persuasive markers under study. Such a tendency is not uniform, and at least two outliers are worth commenting on here.

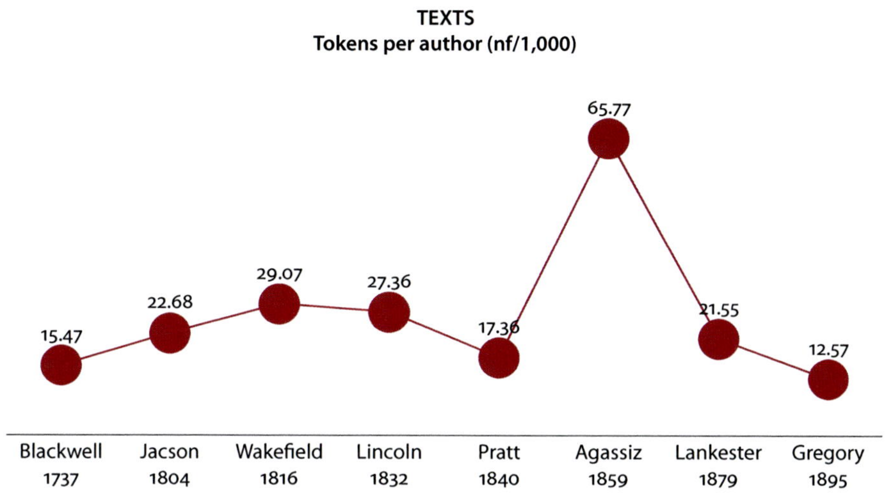

Figure 4. Persuasion tokens in texts

The more striking of these is found in Agassiz's sample (65.77 nf/1,000). It contains the highest number of the second-person pronouns found (181 out of 256 in the prefatory data) and a considerable number of first-person pronouns (181 out of 524). Predictive modals and *to*-infinitives are also abundant (92 out of 235, and 111 out of 852, respectively).

The notable frequency of *you* forms points to the dialogic nature of the text, which is conversational in tone, and hence reader-oriented. The author's aims are reinforced by self-mentions through instances of the first person. The pronominal use merges with other argumentative elements such as the predictive modals indicating logical possibility (*will, shall*) and even hypothetical modality (*would*). The argumentative style is paired by well-structured sentences in which *to*-infinitive constructions contribute to the management of discourse integration. In many of the examples the infinitive clauses complement adjectives or other verbs to mark the author's stance with regard to the proposition encoded in the clause.

(20) I have said that the coral animals grow in clusters, but thus far I have only described the single animal that begins the coral stock. Now I will show you how he multiplies himself, till, instead of one animal there are countless multitudes living together in one community. The adjoining figure shows you a part of such a community. When the first coral animal has undergone the changes I have described, and assumed its permanent character, it begins to spread and grow taller... (Agassiz 1859)

At the other end of the scale, the sample of the text by Gregory shows the least number of persuasive markers: An example can be seen in (13):

(21) THE transition from thallus to stem is said to take place in the Hepaticae of the second series, or the one formerly referred to as culminating in the Jungermanniaceae. Exactly what is meant by this expression, "transition from thallus to stem," is seldom clearly explained. It may be safe to state here that some morphologists claim that the present living forms show how such a transition may actually have taken place. The process may be described something as follows. (Gregory 1895)

The lack of personal references, yet with the mention of authorities, plus the use of a great many aseptic, passive constructions, leads to coldness and distance, but also to the objectivity characteristic of the later twentieth-century scientific writing.

In an attempt to confirm the two tendencies noted above, I have applied the statistical measure R^2, which in this case indicates a decrease in persuasive features in prefaces, this largely over the course of the nineteenth century (see Figure 5).

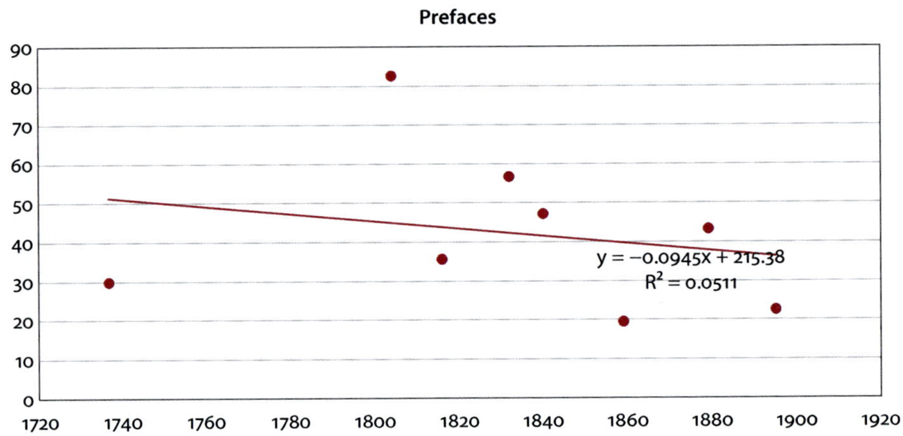

Figure 5. Time variable in prefaces

This would contrast with the tendency in main texts (see Figure 6). In both cases, the R^2 readings are far from 1, which indicates that the results are not significant. In fact, the time factor can explain only 5.11% of variability in prefaces and 2.9% in texts. These percentages point to a minimal variability which can also be interpreted as indicating stability over the period under scrutiny.

Atkinson (1999: 123–125) detected a tendency towards a less overt expression of persuasion in the articles of the *Philosophical Transactions* of the Royal Society between 1675 and 1975, with persuasive devices becoming less numerous the closer we came to contemporary English. This might be seen as symptomatic of the shift from an author-centred to an object-centred kind of prose, as well as to the increasingly more abstract level of texts over time. If we compare Examples (11) and (12), above, we can see how Atkinson shows us the right path to follow.

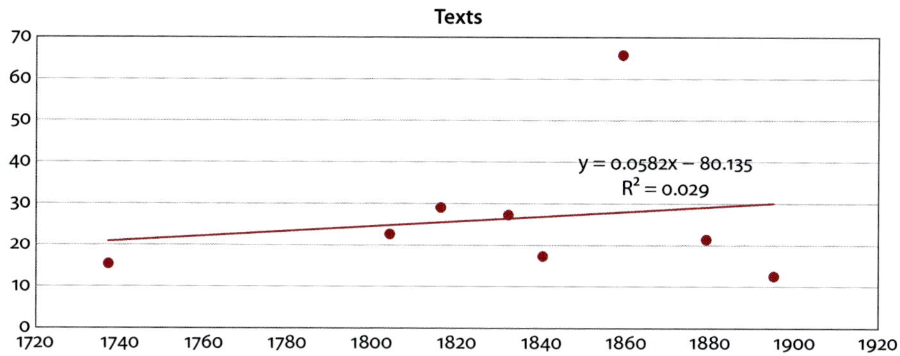

Figure 6. Time variable in main texts

Having considered the frequency of occurrence of features in both prefaces and main texts, I will now take a closer look at linguistic forms indexical of persuasion.

b. *Linguistic features*

As can be seen in Figure 7 below, which includes all the markers of persuasion under survey here, the one with greatest use is the *to*-infinitive construction, with 40% of all occurrences.

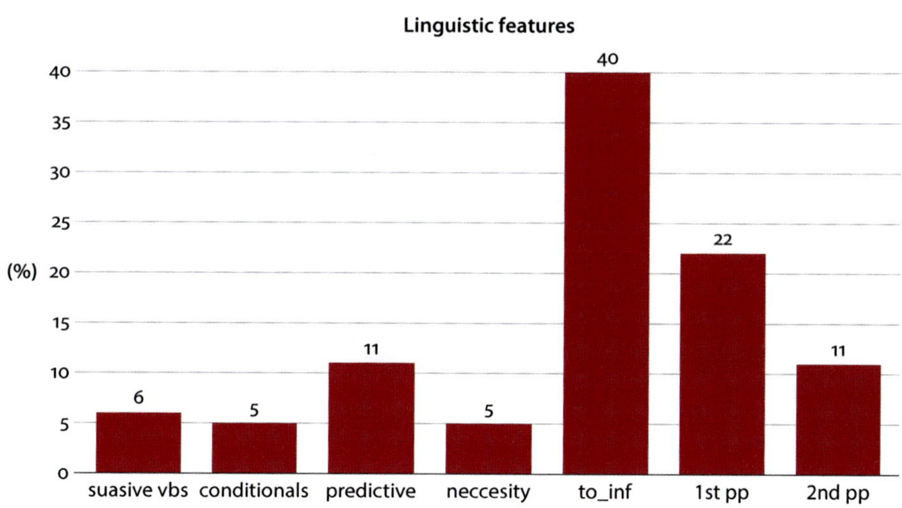

Figure 7. Overall use of linguistic features

This overall analysis well illustrates women writers' preferences in terms of their persuasive strategies. As shown in Figure 7, *to*-infinitive constructions predominate notably (40%), which might also indicate an argumentative style which relies on logical consequences intended to persuade the addressee (see Examples (10) and

(14), above). The second most common feature is first-person pronouns (22%) (Examples 1 to 4; Example (16)). These pronouns can perform several functions, one of which is organising the structure of the text. Second person pronouns and predictive modals both represent 11% of the persuasive elements under study (Examples 8, 13, 17, 20); the former indicates proximity to the reader, putting the emphasis on the receivers of the discourse, whereas the latter expresses certainty and logical argumentative discourse. Suasive verbs follow, at 6% (Examples (10) and (11)). This feature most clearly manifests an overt expression of persuasion, in that the lexical content of suasive verbs is persuasive *per se*. In fact, as Biber (1988: 111) has claimed, they "imply intentions to bring about certain events in the future". Conditionals and necessity modals are the least frequent features, at 5% (Examples (9) and (12)). Conditional sentences indicate the necessary conditions for actions to become true, and necessity modals express different degrees of obligation for certain things to occur. Some of these features are not always clearly persuasive, yet they can be included in the kind of pragmatic argumentation employed by authors.

There are similarities and differences in the use of the various linguistic features that have been traced in both formats. It should be noted that although the frequency of occurrence of most features is generally higher in prefaces than in main texts, there are a couple of exceptions.

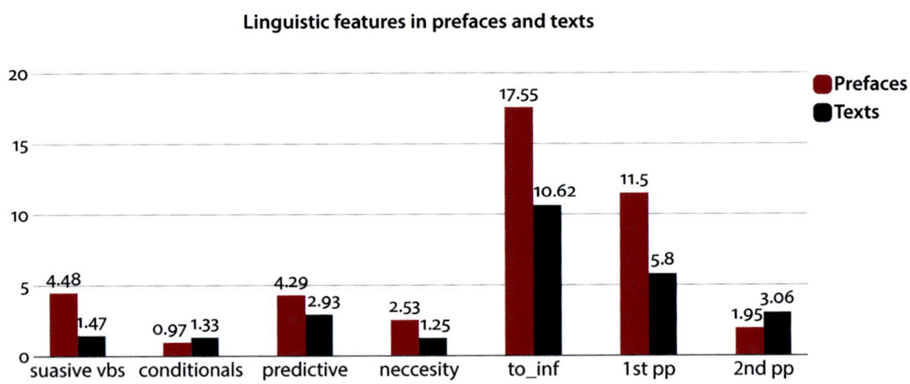

Figure 8. Linguistic features in prefaces and texts

Overall, *to*-infinitive constructions (17.55 vs 10.62 in prefaces and main texts, respectively) is the most abundant feature in both formats, followed by first-person pronominal forms (11.5 vs 5.8), suasive verbs (4.48 vs 1.47) and predictive modals (4.29 vs 2.93).

Despite not being very frequent features in the overall data, conditionals (1.33 vs 0.97) and second-person pronouns (3.06 vs 1.95) predominate in the main text samples. Such an abundance of *you* forms in main texts is a notable finding here, providing these works with an unusual sense of intimacy.

(22) likewise from a vast number of scales of that colour *if* you search the fields and gardens attentively it is not (Wakefield 1816: 40)

(23) also to be found at this season the dandelion too *you* well know is found among the earliest flowers of spring (Lincoln 1832: 281)

Example (22) contains both features which characterise the argumentative side of persuasive discourse. In (23) the speaker's attitude is felt through the personal pronoun together with the adverb *well* which is modulating the discourse, plus another private verb, a personal one, *know*, with the aim of raising awareness of the proposition of the sentence.

It is indeed in the use of personal pronouns where a contrast can be found. Although the prominence of both would be expected in prefaces, since these address the readership to explain what the work is about and seek to convince readers of its usefulness and appropriateness, second person pronouns are more often attested in the main texts of works. This predominance can be seen most clearly in the samples by Wakefield, Lincoln and Agassiz, which can be explained on the grounds of the communicative format used: the format Letter in the case of Wakefield and Agassiz and that of Textbook in the case of Lincoln.

(24) I had fondly promised myself the pleasure of participating with you in all my pursuits and had planned many agreeable schemes (Wakefield 1816: 1)

(25) have flowers and are considered as of any importance but you must recollect that they are the workmanship of his hand (Lincoln 1832: 282)

(26) to see swimming about in your aquarium last summer have you ever heard of a sea-anemone don't fancy from its name (Agassiz 1859: 8)

The use of *if*-clauses is similar in most samples, with two exceptions: those of Agassiz and Wakefield (see Examples (27) and (28)):

(27) a cavity hollowed out of the substance of the body *if* we look at the slender stalk with a microscope we (Agassiz 1859: 45)

(28) where these insects are numerous it may be heard especially *if* the weather be warm almost every hour of the day (Wakefield 1816: 33)

A different communicative format, Jacson's Lecture, also shows a good number of conditional cases:

(29) applied to every kind of hairyness which exists on plants *if* the young parts of plants be examined by a microscope (Jacson 1804: 39)

Constructions that create argumentative structures are to be expected in scientific writings. It is perhaps more striking to detect such direct addresses to the reader through the second-person pronoun. Nevertheless, the communicative format selected in these cases substantiates this use. All the features under survey enrich the persuasive nature of any work to a greater or lesser extent (Connor and Upton 2003).

5. Concluding remarks

In this study I have explored the ways in which women writers employed a particular writing style permeated by certain persuasive features. By focusing on persuasion, I have been able to show how female authors interact with their readership when they introduce their own works. They do so by appealing to solidarity and the emotional side of language or any other feature that, in collocating with persuasive elements, might suggest closeness to the reader, structural guidance, or argumentation.

Through an analysis of the ways in which female authors of science attempt to persuade the unseen but imagined audience of the relevance and usefulness of their writing, I have noted that the formality of written texts is not in itself an obstacle for a writer's interaction with their readership. This presumed distance is not reflected in introductory material such as prefaces, forewords or even direct introductions to a work. In fact, the role of these is generally to facilitate communication among participants in the discourse, as well as the negotiation of meaning.

In light of the results obtained, we can state that more features indexical of persuasion have been traced in prefaces than in the samples themselves. However, we cannot claim an absence of subjectivity and authorial presence transmitted through persuasive devices in the main texts here. Whereas prefaces are direct and emotional, and are pervaded by an interpersonal tone which is in most cases reinforced by the language context, persuasion in main texts might better be classified as indirect, hidden in the textual argumentation and within the lines of informative style.

Analysing the use of features by various authors, there was a slightly decreasing frequency of use of these in prefaces over time, and, on the contrary, a gradual increase in texts. The low values in R^2 in both prefaces and texts indicate stability across time.

A general overview of the frequency of occurrence of the set of features showed that all of these are more frequent in prefaces except the second-person pronoun *you* and *if*-clauses, which are more abundant in main texts. The informal nature or the target reading public of the communicative format in which these instances have been found (letters, lectures and textbooks) might serve to explain this prevalence.

These findings confirm that both linguistic elements and external/contextual factors are inextricably related, and if we can approach them with this in mind, scientific writings from the past can tell us a different story.

Works cited

Anthony, Laurence. 2018. AntConc (Version 3.5.7) [Computer Software]. Tokyo, Japan: Waseda University. Retrieved January 20, 2020, from http://www.laurenceanthony.net/software

Arakelyan, Rouzanna and Muradyan, Gevorg. 2016. Language as an Influential Tool for Persuasion. *Armenian Folia Anglistika*, 1/15: 39–45. https://doi.org/10.46991/AFA/2016.12.1.039

Argamon, Shlomo, Moshe Koppel; Fine, Jonathan and Shimoni, Anat Rachel. 2003. Gender, Genre, and Writing Style in Formal Written Texts. *Text*, 23/3: 321–346.

Atkinson, Dwight. 1999. *Scientific Discourse in Sociohistorical Context: The Philosophical Transactions of the Royal Society of London, 1675–1975*. Mahwah, NJ: Lawrence Erlbaum.

Barsaglini-Castro, Anabella, Valcarce, Daniel. 2020. The Coruña Corpus Tool: Ten Years On. *Procesamiento del Lenguaje Natural*, 64: 13–19.

Bhatia, Vijay K. 1997. Genre-mixing in academic introductions. *English for Specific Purposes*, 16/3: 181–195. https://doi.org/10.1016/S0889-4906(96)00039-7

Biber, Douglas and Conrad, Susan. 2001. "Register variation: A corpus approach". In Schiffrin, Deborah; Tannen, Deborah and Hamilton, Heidi (eds.), *The handbook of discourse analysis*. Oxford: Blackwell. 175–96.

Biber, Douglas; Johansson, Stig; Leech, Geoffrey; Conrad, Susan and Finegan, Edward. 1999. *Longman Grammar of Spoken and Written English*. Essex: Longman.

Biber, Douglas. 1988. *Variation across Speech and Writing*. Cambridge, UK: Cambridge University Press. https://doi.org/10.1017/CBO9780511621024

Cameron, Deborah. 1992. *Feminism and Linguistic Theory*. Palgrave Macmillan. https://doi.org/10.1007/978-1-349-22334-3

Cameron, Deborah, McAlinden, Fiona and O'Leary, Kathy. 1989. "Lakoff in context: the social and linguistic functions of tag questions". In Cameron, Deborah. and Coates, Jennifer (eds.), *Women in Their Speech Communities: new perspectives on language and sex*. London; New York: Longman. 74–93.

Connor, Ulla. and Upton, Thomas. 2003. "Linguistic Dimensions of Direct Mail Letters". In Leystina, Pepi and Meyer, Charles F. (eds.), *Corpus Analysis. Language Structure and Language Use*. Amsterdam: Rodopi. 71–86.

Crespo, Begoña. 2016. On writing Science in the Age of Reason. *Revista Canaria de Estudios Ingleses (RCEI)*, 72: 53–78.

Crespo, Begoña. 2019. "How intimate was the tone of female history writing in the Modern period? Evidence from the Corpus of History English Texts". In Moskowich, Isabel; Crespo, Begoña; Puente-Castelo, Luis and Monaco, Leida Maria (eds.), *Writing history in Late Modern English: Explorations of the Coruña Corpus*. Amsterdam/Philadelphia: John Benjamins. 186–213. https://doi.org/10.1075/z.225

Dillard, James Pryce. 2014. "Language style and persuasion". In Holtgraves, Thomas (ed.), *The Oxford Handbook of Language and Social Psychology*. Oxford: Oxford University Press. 177–187.

Gregory, Emily Lovira. 1895. *Elements of Plant Anatomy*. Boston, London: Published by Ginn & company.
Halliday, Michael Alexander Kirkwood. 1988. "On the Language of Physical Science". In Ghadessy, Mohsen (ed.), *Registers of Written English: Situational Factors and Linguistic Features*. (OLS). London: Pinter. 162–178.
Hyland, Ken. 2005. Stance and engagement: a model of interaction in academic discourse. *Discourse Studies*, 7/2: 173–192. https://doi.org/10.1177/1461445605050365
Hyland, K. 2015. Genre, Discipline and identity. *Journal of English for Academic Purposes*, 19: 32–43. https://doi.org/10.1016/j.jeap.2015.02.005
Knight, Dan. (ed.). 1986. *The Age of Science. The Scientific World-View in the Nineteenth Century*. Oxford: Basil Blackwell.
Lakoff, Robin. 1975. *Language and Women's Place*. New York: Harper and Row.
Lareo, Inés; Monaco, Leida Maria; Esteve-Ramos, María José and Moskowich, Isabel (comps.). 2020. *The Corpus of English Life Sciences Texts (CELiST)*. https://doi.org/10.17979/spudc.9788497497848
Mischke, G. Elizabeth. 2005. *Analysing involvement in distance-education study-guides: an appraisal-based approach*. UNISA. Retrieved May 20, 2020, from http://uir.unisa.ac.za/bitstream/handle/10500/1864/00front.pdf?sequence=3
Moskowich, Isabel. and Crespo, Begoña. 2014. Stance is present in scientific writing, indeed. Evidence from the Coruña Corpus of English Scientific Writing. *Token. A Journal of English Linguistics*, 3: 91–114.
Moskowich, Isabel. 2021. "The making of CELiST, a bunch of disciplines". In Moskowich, Isabel; Lareo, Inés and Camiña, Gonzalo (eds.), *"All families and genera": Exploring the Corpus of English Life Sciences Texts*. Amsterdam/Philadelphia: John Benjamins. 1–19. https://doi.org/10.1075/z.237
Moskowich, Isabel; Camiña-Riobóo, Gonzalo; Lareo, Inés and Crespo, Begoña (comps.) 2018. *Corpus of English Philosophy Texts (CEPhiT)*. A Coruña: Universidade da Coruña. https://doi.org/10.17979/spudc.9788497497077
O'Keefe, Daniel J. 1990. *Current communication: An advanced text series, Vol. 2. Persuasion: Theory and research*. Newbury Park, CA: Sage Publications, Inc.
Perloff, Richard M. 2003. *The Dynamics of Persuasion: Communication and Attitudes in the Twenty-First Century*. UK: Lawrence Erlbaum Publishing.
Pratt, Anne. 1840. *Flowers and their Associations*. London: Charles Knight and Co.
Prelli, Lawrence J. 1989. The rhetorical construction of scientific ethos. In Simon, Herbert W. (ed.), *Rhetoric in the human science*. London: Sage. 87–104.
Quirk, Randolph; Greenbaum, Sidney; Leech, Geoffrey and Svartvik, Jan. 1985. *A Comprehensive Grammar of the English Language*. London: Longman.
Smellie, William. 1790. *The philosophy of natural history*. Vol. I. Dublin: printed by William Porter. https://doi.org/10.1037/14228-000
Swales, John. 1990. *Genre Analysis English in Academic and Research Settings*. Cambridge, UK: Cambridge University Press.
Taavitsainen, Irma. 1994. "On the Evolution of Scientific Writings between 1375 and 1675: Repertoire of Emotive Features". In Fernández, Francisco et al. (eds.), *Papers from the 7th International Conference on English Historical Linguistics*. Valencia, Sept. 1992 Amsterdam/Philadelphia: John Benjamins. 329–342. https://doi.org/10.1075/cilt.113.32taa

Tannen, Deborah. 1993. "The Relativity of Linguistic Strategies: Rethinking Power and Solidarity in Gender and Dominance". In Tannen, Deborah (ed.), *Gender and Conversational Interaction*. Oxford: Oxford University Press. 165–188.

Wakefield, Priscilla. 1816. *An introduction to the Natural History and Classification of Insects, in a series of familiar Letters. With Illustrative Engravings*. London: printed for Darton, Harvey and Darton. https://doi.org/10.5962/bhl.title.157681

CHAPTER 9

Persuasion in English scientific writing

Exploring suasive verbs in the Corpus of English Life Sciences Texts and Posthumanism English Texts

Anabella Barsaglini-Castro
University of A Coruña

1. Introduction

Scientific writing has often been regarded as an impersonal form of discourse because of the degree of objectivity that it demands of its authors. However, contrary to the findings of earlier studies, in which impersonality was usually considered the most direct persuasive strategy, or even an unmediated way of demonstrating the veracity of statements (Hyland 2002), recent work has shown that linguistic features such as modal verbs (see Alonso-Almeida and Álvarez-Gil 2019) and pronouns (see Moskowich and Crespo forthcoming) reflect the author's presence in the text. Such objectivity, understood in the sense of detachment from statements, can thus be questioned or considered to be a relative value depending on the writer-reader relationship or the way that authors communicate with their target audience. From here, we can infer that depersonalised prose is not in itself feasible if we assume that every text entails an identity, however minimal, and also that it is precisely the author's stance, viewed as a possible expression of involvement, that we assume produces or contributes to the persuasion of the reader.

This chapter describes a pilot study on persuasion strategies in Late Modern and Contemporary English which compares the Corpus of English Life Sciences Texts (CELiST) (Lareo *et al.* 2020), a subcorpus of the Coruña Corpus of English Scientific Writing (CC), and Posthumanism English Texts (PET), a corpus of my own.[1] Although these corpora contain non-fiction samples from the eighteenth and nineteenth centuries (CELiST), and fiction and non-fiction from the twentieth and twenty-first (PET), I have selected samples from the nineteenth century in the case of the former, and non-fiction from the twentieth and twenty-first the case of the latter, this in order to facilitate comparison. As detailed below, the two

1. This corpus is still under compilation at the moment of writing this chapter.

https://doi.org/10.1075/z.237.09bar

corpora share the specific characteristics that will be used in the analysis, including information about the sex of the author, the period, and the subject matter dealt with by the texts. I will focus on linguistic features generally considered to indicate persuasion and argumentation, particularly suasive verbs (Quirk *et al.* 1985; Biber 1988, 1995; Biber and Conrad 2009), looking at their frequency of occurrence in the texts under scrutiny. Also, given that not all samples have the same word count, both quantitative and qualitative methods will be applied in order to allow for richer comparisons of the results.

In what follows, Section 2 provides an outline of the main theoretical issues in relation to the current study. Section 3 describes the material selected for analysis and the methodology used. Section 4 presents the analysis and findings, looking mainly at the variables of the author's sex and the period in which the different texts were published. Concluding remarks will be provided in Section 5.

2. Persuasion, stance and suasive verbs in scientific writing

Scientific texts can be said to be distinguishable by their clarity, precision, verifiability, universality and objectivity. Although these characteristics contrast notably with the concept of stance, they all entail a shade of persuasion. As early as the seventeenth century authors such as Bacon and Boyle defended the notion of impersonality as the most direct persuasive strategy (Collier and Toomey 1997), yet contemporary authors such as Hyland (2005); Toolan (2010) and Locher (2017) have claimed that stance makes a significant contribution to the persuasion of the reader.

Whereas the earlier writers here focused on the most abstract and argumentative aspect of texts, in recent years there has been a greater focus on more subtle and deeper level of writing. For this reason, we might usefully provide here a brief discussion and exploration of the main concepts used in this study: persuasion, stance and suasive verbs.

According to the *Oxford English Dictionary* (OED), the concept of persuasion involves the addressing of arguments or appeals to a person in order to induce cooperation, submission, or agreement. In other words, it is a strategy, used either in speech or writing and is intended to induce a belief or an action in the listener or reader through argumentation and/or emotion. The combination of these features is already present in Aristotle's sources of persuasion, *ethos, pathos* and *logos* (see Aristotle 1984), due to their influence in terms of justified argumentation. In fact, we can briefly summarise Aristotle's theory by saying that *ethos, pathos* and *logos* create the author's profile or image, the reader's state of mind, and provide evidence from data, respectively. This is complemented by the other two components of argumentation: language and the organisation of speech, which provide specific wording and structure depending on the content, the topic or field, the issue of intentionality, and the target audience. Thus, depending on the expected audience and the genre, writers may convey their ideas, content or knowledge in many different ways. These can be more or less direct (as well as conscious) and can influence readers and hence constitute an act of persuasion. Toolan (2011: 16) also suggests that the act of persuading consists of providing previously known information and optionally adding extra information in order to manipulate or change the addressee's point of view. However, if we consider the dominant canon established by Boyle in 1669, according to which scientific knowledge had to be transmitted in a plain, concise, unembellished and self-censored way (Boyle 1669), it seems that texts seeking to encourage a certain attitude in the reader tended to be more audience-oriented and were also well structured (see Crespo 2015). In this sense, by using rhetorical mechanisms of persuasion and argumentation in a more or less explicit way, authors managed to argue in defence of their opinion and findings in order to convince readers of the veracity of facts and events, and to prove and show their truthfulness.

Regardless of the differing terminology used in the literature, the concept of stance, which I believe to be closely related to persuasive strategies, refers to the way in which writers communicate with readers through their texts (Moskowich 2017). Stance also manifests itself as the open expression of an author's attitude or commitment to the message, and hence, its analysis covers the linguistic choices that include those ways in which authors mention themselves through giving personal opinions. Such authorial presence has become a crucial aspect in the study of language due to the diverse kinds of assessments or judgements that it may convey, and also because of how these are transmitted to the addressees.

Following Locher's (2017: 489) theory, according to which stance influences the perception of stories,[2] and also taking into account that readers are somehow 'guided' by the precise wording of the text (Hyland 2005; Toolan 2010), it seems logical to think that authorial presence is a significant aspect of persuasion. After all, it is the writer who decides which discourse markers to use, and when and how[3] these might best be employed to influence the reader (see also Crespo 2021).

2. Although the original theory refers to fictional stories, in this study it will be applied to non-fiction (scientific) texts.

3. Either directly, by addressing the reader through personal pronouns or imperative verbs; or indirectly, using adjectives, adverbs or expressions that show the author's attitude or opinion on the topic.

We should also note here that despite the difference that may exist between engagement and stance, authors' commitment to their ideas, the confidence they show, and the identity they project all serve to reinforce their credibility (Hyland 2002: 1091, 2005: 173). Last but not least, suasive verbs seem to be a suitable means of expressing argumentation and persuasion due to their formulation as commands, suggestions or directives (dealt with in Mele-Marrero 2021).

The current study analyses the use of suasive verbs as persuasion markers in scientific texts and their relationship to author stance. This will be done through asking whether the "neutrality" – seen as the degree of abstractness and the high level of objectivity – in the scientific texts under scrutiny can be considered as a determining factor in persuasive strategies. The results will be considered within the framework of Biber's multidimensional analyses (1988, 1995), specifically dimension 4, "Overt expression of persuasion" (Biber and Conrad 2009).

3. Corpus and methodology

As noted above, my examination of persuasion strategies will draw on Biber's Dimension 4 'Overt expression of persuasion'. Although this dimension includes a number of features, such as conditional subordinators, predictive and necessity modals and suasive verbs, only the latter are considered here. Thus, all the forms of the following suasive verbs were searched for in the corpus:

> *agree, arrange, ask, beg, command, decide, demand, grant, insist, instruct, ordain, pledge, pronounce, propose, recommend, request, stipulate, suggest, urge*

The data used have been drawn from the Corpus of English Life Sciences Texts (CELiST), a subcorpus of the Coruña Corpus of English Scientific Writing (CC), and Posthumanism English Texts (PET), a corpus I am compiling at present.

CELiST has been compiled following the same principles as the other corpora in the CC in order to ensure optimal representativeness (Crespo and Moskowich 2010). It contains samples from texts dealing with Biology, Zoology, Botany and other disciplines published between 1700 and 1900, and comprises a total of ca. 400,000 words. For the present study, however, I have used only the nineteenth-century section, which contains twenty texts amounting to a total of 200,085 words.

The PET corpus consists of two subcorpora: Corpus of Fiction Posthumanism English Texts (CoFiPET) and Corpus of Non-Fiction Posthumanism English Texts (CoNFiPET). Taken from a carefully planned selection of fifty texts published from 1950 to 2017, PET includes eighteen complete science fiction novels and thirty-two non-fiction texts related to transhumanism, posthumanism, transcendence, technology and artificial intelligence. For the present study, the non-fiction subcorpus (CoNFiPET) has been used. The samples therein vary in length, but amount to a total of 252,967 words, which is broadly comparable to CELiST.

Although the samples in CELiST and CoNFiPET differ slightly in size due to their respective compilation criteria, both corpora meet the requirements of representativeness and balance. Of all the material to be analysed, 200,085 words (44%) pertain to the field of Life Sciences, and 252,967 (56%) to that of Posthumanism and related areas. The total number of words under examination, then, is 453,052, with a distribution shown in Figure 1:

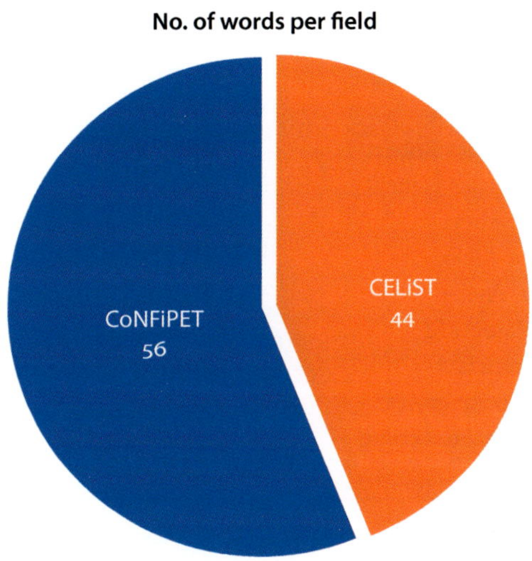

Figure 1. Distribution of words per field

One of the main differences between these subsets of texts is related to their comparative length internally. Whereas CELiST includes texts with an almost balanced word count, that is, ca. 10,000 words per text, the samples in CoNFiPET vary between 1,500 and 15,900 words. Because of this disparity in word count, and also in order to guarantee the reliability of comparisons, findings have been normalised to 10,000 words. The exact number of words in each sample, as well as the information relative to authors' names and dates of publication, are indicated in Tables 1 and 2 below:

CELiST

Table 1. Distribution of words in CELiST

Year	Author	Words
1804	Jacson, Maria Elizabeth	10,051
1808	Wilson, Alexander	10,008
1816	Wakefield, Priscilla	10,008
1819	Lawrence, William	10,000
1824	Jenner, Edward	9,771
1828	Godman, John Davidson	10,011
1832	Lincoln, Almira Hart Phelps	10,013
1835	Jardine, Sir William	10,007
1840	Pratt, Anne	10,021
1848	Dalyell, Sir John Graham	10,010
1859	Agassiz, Elizabeth	10,035
1859	Darwin, Charles Robert	10,014
1863	Huxley, Thomas Henry	10,048
1867	Spencer, Herbert	10,001
1876	Macalister, Alexander	10,023
1879	Lankester, Phebe	10,020
1880	Balfour, Francis M	10,004
1889	Galton, Francis	10,019
1895	Gregory, Emily Lovira	10,020
1898	Packard, Alpheus Spring	10,001
Total		**200,085**

CoNFiPET

Table 2. Distribution of words in CoNFiPET

Year	Author	Words
1973	Boden, Margaret A.	6,250
1984	Glover, Jonathan	4,523
1990	Pollock, John	12,683
1991	Haraway, Donna	14,538
1992	Harris, Vicky	4,249
1996	Sharkey, Amanda J. C.	7,333
1997	Ansell-Pearson, Keith	13,161
1999	Hayles, N. Katherine	15,602
1999	Kirby, Vicki	5,925
2000	Clough, Patricia Ticineto	9,549
2002	Fukuyama, Francis	5,512
2005	Turkle, Sherry	12,453
2007	Colleen Stevenson, Melissa	10,002
2007	Doucet, Hubert	3,437

Table 2. (*continued*)

Year	Author	Words
2009	Doede, Bob	11,957
2009	Hauskeller, Michael	6,635
2010	Jotterand, Fabrice	1,630
2012	Gagnon, Philippe	6,357
2012	Herzfeld, Noreen	5,174
2013	Bostrom, Nick	15,817
2013	Braidotti, Rosi	14,570
2013	More, Max	7,358
2013	Neill, Daniel B.	2,192
2014	Ferrando, Francesca	4,961
2014	Rothblatt, Martine	7,914
2014	Vita-More, Natasha	2,554
2014	Wellington, Naomi	5,936
2015	Klichowski, Michal	2,361
2016	Habibi, Don	13,956
2016	Schneider, Susan	6,515
2016	Skågeby, Jörgen	7,952
2017	Holm, Søren	3,911
Total		**252,967**

Since the two main variables to be considered in the analysis are the sex of the author and the period of the sample – including the nineteenth, twentieth and twenty-first centuries – Figures 2 and 3 set out the number of words corresponding to each group.

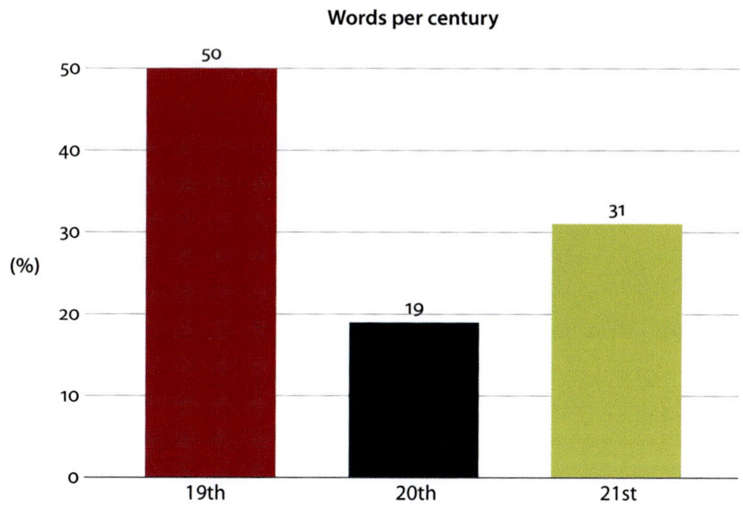

Figure 2. Distribution of words per century

As seen in Figure 2, the nineteenth century alone represents half of the total data, whereas the material published in the twentieth and twenty-first centuries constitutes the other half.

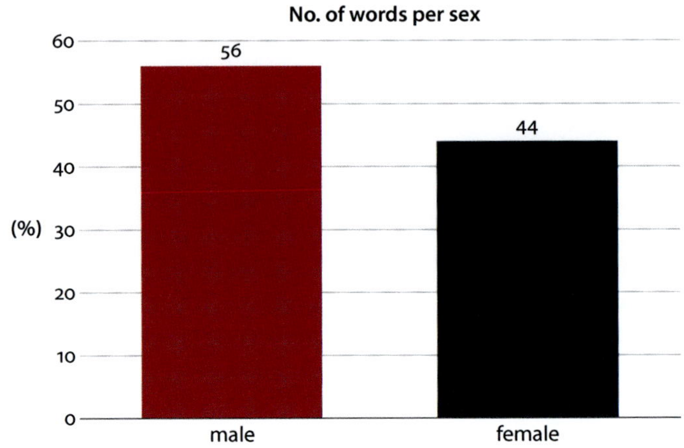

Figure 3. Distribution of words per sex

With a difference of only 12% between the texts written by male and female authors, my data appear to be quite balanced in terms of sex. Raw word counts here are 11,215 (56%) and 8,785 (44%) for men and women, respectively. Figure 4 below, in turn, illustrates the distribution of words per sex and corpus.

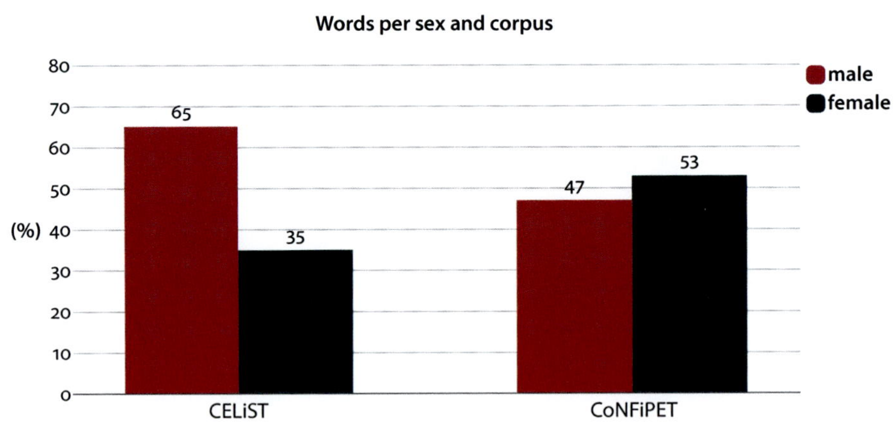

Figure 4. Distribution of words per sex and corpus

Given that the second variable to be considered in this study is that of the sex of the author, it should be noted that the volume of texts written by women is in general not very large. Indeed, the number of words written by women in my CELiST material (70,168), which represents 35%, is almost half that of men with 129,917 words (65%), as might be expected for the Late Modern English period, whereas in CoNFiPET the number of samples selected has deliberately been the same for both sexes, and hence there is only a slight difference (6%) in terms of word count.

The linguistic features listed above as typical of the overt expression of persuasion were retrieved by using two different tools. The texts from CELiST were searched using the 2019 version of the Coruña Corpus Tool (CCT) (Barsaglini-Castro and Valcarce 2020), whereas those from CoNFiPET were examined with AntConc (Anthony 2019). The queries included base and third person singular forms (*-s*), as well as the past simple (*-ed*) and the present and past participle (*-ing*) forms. However, not all such forms corresponded to the lexical category of interest to us here. This is the case with *grant* and *decide*, for instance, and manual disambiguation was therefore carried out where necessary to distinguish the verbal form (Examples (1) and (3)) from a proper noun (2), or even from an adjectival form (4), among others.

(1) In an increasingly technological, computerized world, information is a prime commodity, and when it is used in biological theorizing it is <granted> a kind of atomistic autonomy as it moves from place to place, is gathered, stored, imprinted and translated. (Kirby 1999)

(2) It is in this sense that it has been truthfully said: technology is the ontology of the age, (<Grant> 1986 p. 32). (Doede 2009)

(3) Owning a body gives the right to manage one's body and get rid of it in the name of existing in a robot or a system. Everyone can thus <decide> on their own if, and what, transhumanist treatment they will undergo. (Klichowski 2015)

(4) When we compare the body of a wasp or bee with that of a worm, we see that there is a <decided> transfer of parts headward; this process of cephalization so marked in the Crustacea likewise obtains in insects. (Packard 1898)

In a second step, aspects such as the syntactic functions and the semantic level of the forms were also considered, and manual disambiguation was also required here. The following examples illustrate some of the forms that have been discarded:

(5) But let me here remark – how ill would the singing of birds <agree> with the general appearance of winter – the leafless tree, – the snowy mead, – the frozen rivulet! Yet it must be noticed here, that these rigors, in the midst of this dreary season, are sometimes suddenly softened, and a temperate state of the air succeeds. (Jenner 1824)

(6) A Claſs is the firſt and higheſt diviſion of every ſyſtem. It may be compared to a dictionary, in which all the words having the ſame initial letter are <arranged> together, every word may be compared to a genus; the claſſic character is conſtituted from a ſingle circumſtance, as the words are arranged by a ſingle letter; this one circumſtance muſt be poſſeſſed alike by every plant admitted into the Claſs, how different ſoever they may be in other reſpects. (Jacson 1804)

(7) Those who are old enough remember the time when things were different as a kind of golden age, an age when a programmer was a skilled artisan who was given a problem and <asked> to conceive of and craft a solution. (Turkle 2005)

As can be seen, Examples (5), (6) and (7) cannot be considered as suasive verbs, in that their meanings are "coincide", "organised" or "put in order" and "requested", respectively.

All the findings obtained from these data will be presented and analysed in more detail in the following section.

4. Analysis of data

The texts under study amount a total of 453,052 words, of which 200,085 are from nineteenth-century Life Sciences samples and 252,967 words from twentieth- and twenty-first-century Posthumanism texts. Although the dataset is thus not especially large,[4] it can be considered to be of an adequate size for a study of the use of suasive verbs as persuasion markers, given the very specific nature of the form under scrutiny and the texts. Thus, it is expected that these data can indeed provide a picture of how writers influence readers in scientific writing.

Searches of CELiST and CoNFiPET revealed that not all the types in my initial list were found. Moreover, as a result of the disambiguation process, only 425 tokens with argumentative and persuasive shades of meaning remained from the initial 687 that had been retrieved automatically. Table 3 lists the raw number of occurrences corresponding to the different types, plus their percentages in all the material.

4. Specialised corpora are usually smaller than general corpora (see Nurmi 2002).

Table 3. Presence of suasive verbs in CELiST and CoNFiPET

Suasive verbs	No. of tokens	Percentage
agree	24	6%
arrange	3	1%
Ask	42	10%
Beg	4	1%
command	1	0%
decide	34	8%
demand	18	4%
grant	18	4%
insist	35	8%
instruct	7	2%
ordain	1	0%
pledge	0	0%
pronounce	2	0%
propose	50	12%
recommend	20	5%
request	1	0%
stipulate	4	1%
suggest	147	35%
urge	14	3%
Total	**425**	**100%**

Of the 19 suasive verbs included in the search, *suggest* (147), *propose* (50) and *ask* (42) were the most frequently used, and *arrange* (3), *beg* (4) and *stipulate* (4) the least used; meanwhile *pledge* occurred in neither corpus Figure 5 below shows the distribution of tokens in the two corpora. As can be seen, some types had very low frequencies in both.

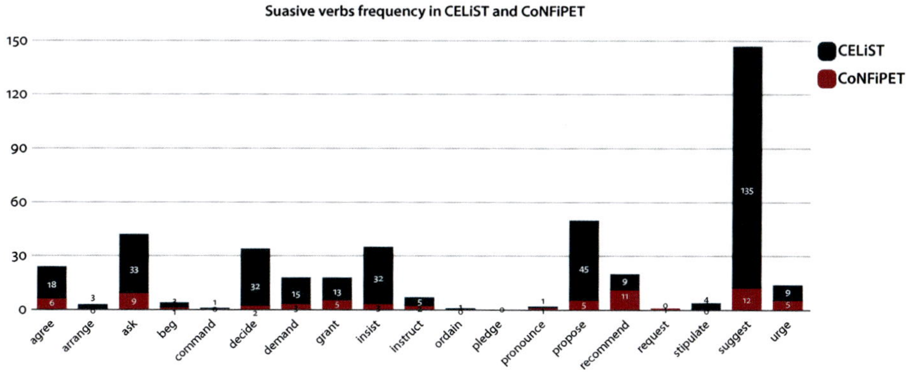

Figure 5. Suasive verbs frequency in CELiST and CoNFiPET

CELiST yielded no tokens for *arrange, command, ordain, pledge* or *stipulate*, and the same was the case for *pledge* and *request* in CoNFiPET. This might reflect different rhetorical fashions over time, thus explaining the different absences here. As for the remaining verbs, they seem to have a relatively moderate use in both corpora, although clearly higher in CoNFiPET texts. For instance, *agree, decide* and *insist* are attested just 6, 2 and 3 times, respectively, in CELiST, but 18, 32, and 32 times in CoNFiPET. On the contrary, *recommend* is the only item with a higher number of occurrences in CELiST than in CoNFiPET, although the counts differ by only 2 tokens.

In the material scrutinised I have found all the structures Quirk *et al.* mentioned as typical of suasive verbs, including *suggest, propose, ask, insist* and *demand.* These authors claimed (1985: 1182–1183) that such verbs are usually followed by a *to*-infinitive, as in Examples (8) and (9), by a *that*-clause, as in Examples (10) and (11), or even by a *that*-clause with a "putative" *should*, as in Example (12), among others.

(8) I envisage the task of implementing human reasoning in OSCAR as progressing through a number of stages, each stage associated with a certain group of problems, and it seems a useful exercise to list these problems in the tentative order in which I <propose> *to* undertake them. (Pollock 1990)

(9) Here I <propose> merely *to* enumerate a few of our native species of simple Ascidiæ, and that, as before said, as a prelude to further comment on some singular points in the history of others, and especially of those consisting of a multitude of individuals united indissolubly in compound masses. (Dalyell 1848)

(10) This 'factor X' "is the human essence, the most basic meaning of what it is to be human" (2002, 150). Fukuyama <insists>, nevertheless, *that* this human essence cannot be found in any particular property like language, conceptual thinking, self-consciousness, or the ability to act in accordance with moral principles. (Hauskeller 2009)

(11) Many socialist regimes abolished private property, weakened the family, and <demanded> *that* people be altruistic to mankind in general rather than to a narrower circle of friends and family. But evolution did not shape human beings in this fashion. (Fukuyama 2002)

(12) We do not know for sure, but this kind of anomaly <suggests> that we *should* not take people's stated "preferences" about how long they would wish to live too seriously, and that revealed preferences might be a more reliable index of their guiding values. (Bostrom 2013)

Similarly, and based on the results obtained in this study, it should be noted that there is a clear relationship between the lexical meaning of these verbs and their effect as overt markers of persuasion. Thus, meaning plays an important role for writers in terms of engaging readers or even eliciting their agreement with the stance of a text. For instance, if we consider the meanings of *suggest* and *propose* – which are very close – we observe that in addition to maintaining a certain pattern (i.e. *suggest* [sth], *suggest* that [sb] do [sth], *propose* that, or *propose* [sth]), they both introduce an idea or plan for the addressees to think about and decide on.

In addition, and although not the main concern of this study, we might note the presence of pronouns and other features denoting involvement. Thus, the use of pronouns ((13) and (14)) as well as proper nouns ((15) and (16)) preceding suasive verbs is certainly a relevant feature in conveying authorial identity, values and engagement. These are also a direct way of addressing the readership in seeking acceptance, and in showing commitment to the message, which underlines the idea that stance contributes to the persuasion of the reader.

(13) *I* would <suggest> that in the formation of the polar cells part of the constituents of the germinal vesicle, which are requisite for its functions as a complete and independent nucleus, is removed, to make room for the supply of the necessary parts to it again by the spermatic nucleus. (Balfour 1880)

(14) My mother says, the best remedy for the loss of your company, is, to establish a regular correspondence on some particular subject, in which both feel an equal interest; and, as we derived so much amusement and advantage, last summer, from the study of Botany, *she* <recommends> our continuing that, combined with one still more interesting, because it treats of animated beings; I mean Entomology, or the knowledge of Insects, a pursuit of which she is extremely fond, and has kindly offered to give me lessons, that I am to communicate to you. (Wakefield 1816)

(15) In working through the 'real problem' of man, *Nietzsche* <insists> on making a distinction between the 'actual instruments' of culture and the 'virtual bearers' of culture. (Ansell-Pearson 1997)

(16) *Transhumanists* <agree> with Russell's claim that humanity 'is the product of causes that had no prevision of the end they were achieving' – this is precisely the reason why it is incumbent upon humans to devise technologies that will enable us to have a say in our future development. (Doede 2009)

Therefore, when authors use suasive verbs in their writing, they not only reinforce their argumentative discourse but also, as Biber (1988: 242) has put it, "imply intentions to bring about some change in the future" (e.g. *command* and *stipulate*). Such intentionality on the author's part can be formulated through suggestions, commands or directives that are ultimately employed as an explicit mechanism of argumentation and persuasion in order to emphasise their opinions or validate their claims.

When grouped by subject matter (Figure 6), it is clear that the use of suasive verbs is considerably higher in CoNFiPET (14.19 nf) than in CELiST (3.30 nf), which is not surprising if we consider that the corpora themselves, and hence the data, reflect two separate periods (Figure 2). Consequently, we can infer that the more frequent use of suasive verbs in CoNFiPET is mainly due to the fact that the texts belong to the twentieth and twenty-first centuries, and that authors are increasingly aware of their usefulness to make scientific discourse more attractive and convincing. On the contrary, the lower value in CELiST corresponds to the texts written in the nineteenth century.

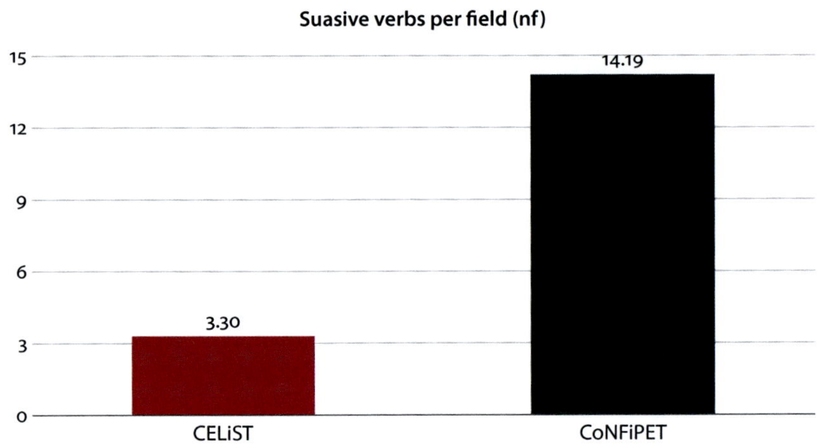

Figure 6. Suasive verbs per field

Accordingly, and on the same lines as work by Crespo and Moskowich (2010), among others, my data confirms the scant use of suasive verbs in nineteenth-century scientific texts, regardless of their degree of objectivity/subjectivity or subject matter. This, in turn, means that authors can use either direct or indirect directives, or even other features such as modal verbs or pronouns, as a means of empowering their arguments. Such use implies an authorial presence that, contrary to Bacon and Boyle's canon, is subjective and therefore more convincing and likely to influence the audience.

As previously mentioned, in comparing the values obtained across the two periods (Figure 7) I have normalised frequencies to 10,000, since the three sections in my material contain a different number of words. Thus, we find that the twentieth century, with the lowest raw number of words (see Figure 2 above) is, curiously, the one showing the highest frequency of suasive verbs (7.79 nf), at a rate more than twice that of the nineteenth century and also somewhat higher than the twenty-first century. On the contrary, the nineteenth century, with the highest raw number of words (i.e. half of the material) sees the lowest frequency of suasive verbs (3.30 nf). The twenty-first century turns out to be fairly balanced, in that it has a greater raw word count than the twentieth century and a smaller one than the nineteenth century, and also that its frequency of suasive verbs is lower than that of the twentieth century but almost twice as high as that of the nineteenth century. As noted above, the samples for both the twentieth and the twenty-first centuries are drawn from to the Posthumanism texts in CoNFiPET, whereas those for the nineteenth century are from the Life Sciences texts in CELiST. In other words, the increase in the use of suasive verbs, as a resource authors use to persuade their readers, may not be completely dependent on time but rather on subject-matter.

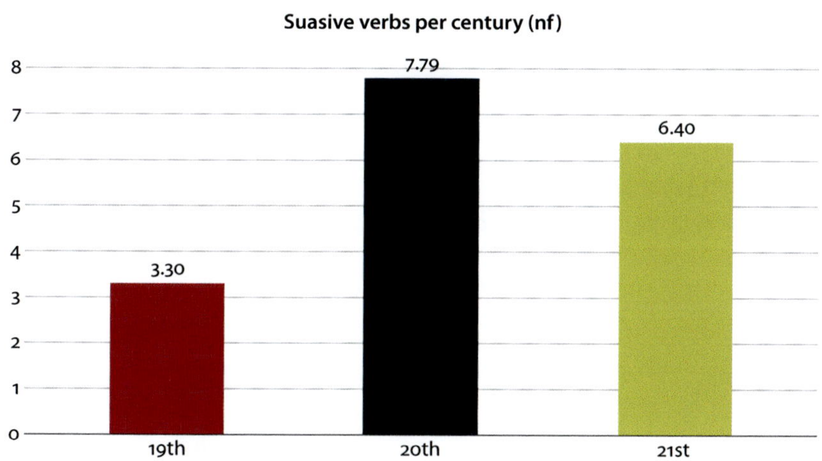

Figure 7. Distribution of suasive verbs per century

So, after a consideration of the variable of time period, we observe that the results seem to indicate that the authors in the nineteenth century did not make as much use of the suasive verbs as those of the twentieth and twenty-first centuries. Since suasive verbs denote indirect directives, they are used to project the authors' 'open' opinion and reveal their presence in the texts. However, authors also use other mechanisms, such as modal verbs or pronouns (see Moskowich and Crespo forthcoming), through which the tone of the text becomes more subjective, and hence persuasive.

The second variable considered here is the sex of the authors, in that differences in the linguistic behaviour of male and female authors may be observed when these try to persuade their readership. As previously shown in Figure 4, the number of words written by men and women is quite different, more so in the case of CELiST, due to the fact that in the nineteenth century women did not often participate openly in science and CELiST tries to capture this situation. Given the difference in the raw word counts corresponding to these two groups of authors, frequencies have again been normalised. As shown in Figure 8 below, the overall count for suasive verbs used by women is 9.94 (nf) and 7.55 (nf) for men.

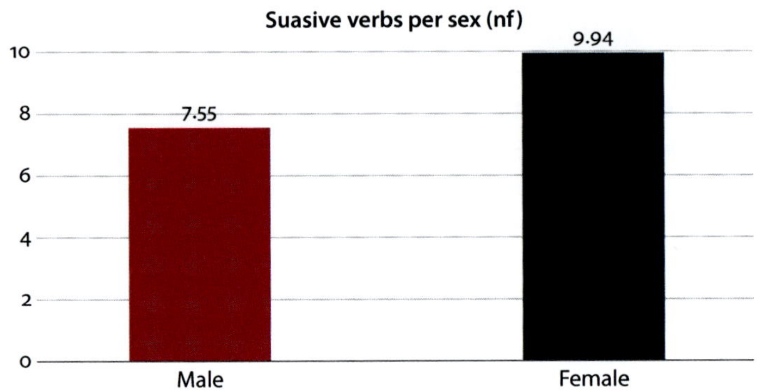

Figure 8. Distribution of suasive verbs per sex

On analysing these results in more detail, focusing in each subcorpus, we find that the distribution of the forms under scrutiny varies, as set out in Figure 9.

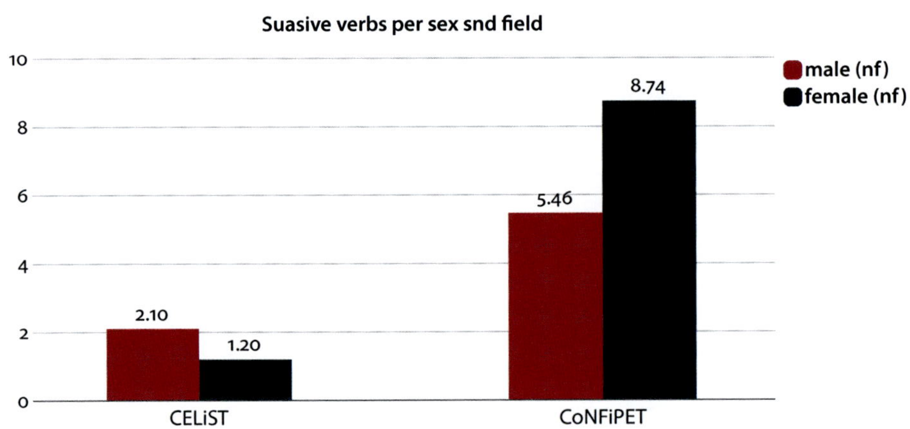

Figure 9. Distribution of suasive verbs per sex and field

The use of suasive verbs by male writers (2.10 nf) in CELiST almost doubles that of female (1.20 nf). The values for CoNFiPET show that the women in this corpus use suasive verbs much more frequently (8.74 nf) than men (5.46 nf), contrary to what is seen in CELiST. In fact, when comparing both sexes in both corpora, we can see that the use of suasive verbs by male authors in CoNFiPET almost triples that in CELiST. Something similar happens with female authors, whose use of suasive verbs in CELiST is nearly eight times less than in CoNFiPET. Although this may be affected to a greater or lesser extent by variables such as the period or the sex of the author, it could be argued that the increase in the use of suasive verbs by female authors in Contemporary texts is due to the notable presence of women in scientific discourse and their audacity to *suggest, propose* or *insist*, as well as to the subject matter itself. After all, technological development and science are now increasingly being studied, discussed and questioned, and this perhaps leads to such a notable finding.

In sum, although the corpora used here are not very large, the slight increase in the frequency of suasive verbs in general, and over the three time periods, in this preliminary study seems to indicate that scientific writing does not continue to follow the same patterns, and that certain tendencies can be seen to be changing. Authors tend to break with the old canon – one of impersonal style – and begin to establish a much more direct kind of communication with their target audience in order to achieve their aim, which implies a perceptible increase in the use of these persuasion strategies, as well as of authorial presence.

5. Concluding remarks

The research presented in this chapter has focused on the analysis of the use of suasive verbs in Late Modern and Contemporary English scientific texts by comparing the Corpus of English Life Sciences Texts (CELiST), a subcorpus of the CC, and the Corpus of Non-Fiction Posthumanism English Texts (CoNFiPET), a part of PET. The main aim was to describe the use of suasive verbs (in apparently objective texts) in light of the following variables: the sex of the author, the period in which the sample was written, and its subject matter.

Despite the fact that only one of the linguistic features included in Biber's Dimension 4 "Overt expression of persuasion" has been searched for in the corpora, some preliminary conclusions can be drawn from the results. Of the 19 suasive verbs searched, one of the most remarkable findings was that of the dissimilarity of frequencies across the subject-matter or field, with texts on Posthumanism showing higher rates of occurrence than those on Life Sciences. This not only suggests that the use of suasive verbs is increasing in Contemporary texts, but also that they play an important role in the persuasion of the reader.

The analysis of the behaviour of suasive verbs over time showed that those texts published in the twentieth century had the highest rates of occurrence, probably because of their relation to the specific topic or field concerned, or even due to the authors' awareness of the need to present their findings in the most compelling way possible as a means of obtaining the acceptance of their target audience. In addition, findings revealed that contemporary female authors tend to make more use of suasive verbs than men, which may be explained both by the transparent and object-oriented style previously established by Bacon and Boyle, and the rise of the role of women in society and the impact that this might have had on their linguistic habits. It is also noticeable here that the analysis using the variable of sex not only reinforces my initial expectations regarding female writing but also sheds some light on the different linguistic uses that male and female writers make of suasive verbs in the nineteenth, twentieth and twenty-first centuries. Although other linguistic features relevant to persuasion, such as personal pronouns or modal verbs, have not been quantified, I have observed that they may be said to behave in a way which is comparable to a greater or lesser extent with suasive verbs, and it is probably their combination with the latter that reinforces the communicative results achieved.

Thus, the subjectivity implicit in the use of suasive verbs counterbalances the supposed objectivity of scientific writing, and also the idea that stance – understood as an audience-engagement mechanism – is essential to the construction of persuasive argumentation. In other words, the rhetoric of science has become a very practical resource to render scientific discourse more attractive.

Works cited

Alonso-Almeida, Francisco and Álvarez-Gil, Francisco J. 2019. "Modal verb categories in CHET". In Moskowich, Isabel; Crespo, Begoña; Puente-Castelo, Luis and Monaco, Leida Maria (eds.), *Writing history in Late Modern English: Explorations of the Coruña Corpus*, 151–167. https://doi.org/10.1075/z.225

Ansell-Pearson, Keith. 1997. *Viroid Life: Perspectives on Nietzsche and the Transhuman Condition*. London/New York: Routledge. https://doi.org/10.4324/9780203047033

Anthony, Laurence. 2019. AntConc (Version 3.5.8) [Computer Software]. Tokyo, Japan: Waseda University. Retrieved December 12, 2019, from https://www.laurenceanthony.net/software

Aristotle. 1984. "Rhetoric". In Barnes, John (ed.), *The Complete Works of Aristotle. The Revised Oxford Translation*. Vol. 2. Princeton University Press: Princeton-New Jersey. 2152–2269.

Barsaglini-Castro, Anabella and Valcarce, Daniel. 2020. The Coruña Corpus Tool: Ten Years On. *Revista del Procesamiento de Lenguaje Natural (sePLN)*, 64: 13–19.

Biber, Douglas. 1995. *Dimensions of Register Variation: A Cross-linguistic Comparison*. Cambridge, UK: Cambridge University Press. https://doi.org/10.1017/CBO9780511519871

Biber, Douglas. 1988. *Variation across Speech and Writing*. Cambridge, UK: Cambridge University Press. https://doi.org/10.1017/CBO9780511621024

Biber, Douglas and Susan Conrad. 2009. *Register, Genre and Style*. Cambridge: Cambridge University Press. https://doi.org/10.1017/CBO9780511814358
Bostrom, Nick. 2013. "Why I Want to be a Posthuman When I Grow Up". In More, M. and Vita-More, N. (eds.), *The Transhumanist Reader*. Wiley Blackwell. 28–53. https://doi.org/10.1002/9781118555927.ch3
Boyle, Robert. 1669. *Certain physiological essays and other tracts written at distant times, and on several occasions by the honourable Robert Boyle; wherein some of the tracts are enlarged by experiments and the work is increased by the addition of a discourse about the absolute rest in bodies*. London: Henry Herringman.
Collier, James H. and Toomey, David M. 1997. *Scientific and technical communication: theory, practice, and policy*. Thousand Oaks, Calif: Sage.
Crespo, Begoña. 2021. "Linguistic indicators of persuasion in female authors in the Corpus of English Life Sciences Texts". In Moskowich, Isabel; Lareo, Inés and Camiña, Gonzalo (eds.), *"All families and genera": Exploring the Corpus of English Life Sciences Texts*. Amsterdam/Philadelphia: John Benjamins. 147–167. https://doi.org/10.1075/z.237.08cre
Crespo, Begoña and Moskowich, Isabel. 2010. "CETA in the Context of the Coruña Corpus". *Literary and Linguistic Computing*, 25/2: 153–164. https://doi.org/10.1093/llc/fqp038
Crespo, Begoña. 2015. Women writing science in the eighteenth century: some hints about their language use. *Anglica. An International Journal of English Studies*, 24/2: 103–128.
Doede, Robert. 2009. Transhumanism, technology, and the future: Posthumanity emerging or sub-humanity descending? *The Appraisal Journal*. Vol. 7. 39–54.
Fukuyama, Francis. 2002. *Our Posthuman Future: Consequences of the Biotechnology Revolution*. New York: Farrar, Straus and Giroux.
Hauskeller, Michael. 2009. Prometheus unbound: Transhumanist arguments from (human) nature. *Ethical Perspectives*. 16. 3–20. https://doi.org/10.2143/EP.16.1.0000000
Hyland, Ken. 2002. Authority and invisibility: Authorial identity in academic writing. *Journal of Pragmatics*, 34: 1091–1112. https://doi.org/10.1016/S0378-2166(02)00035-8
Hyland, Ken. 2005. *Metadiscourse: Exploring interaction in writing*. London/New York: Continuum.
Kirby, Vicki. 1999. Human Nature. *Australian Feminist Studies*, 14:29, 19-29. https://doi.org/10.1080/08164649993308
Klichowski, Michał. 2015. "Transhumanism and the idea of education in the world of cyborgs". In Krauze-Sikorska, H. and Klichowski, M. (eds.), *The Educational and Social World of a Child. Discourses of Communication, Subjectivity and Cyborgization*. Poznan: Adam Mickiewicz University Press. 431–438.
Lareo, Inés; Monaco, Leida Maria; Esteve-Ramos, María José and Moskowich, Isabel (comps.). 2020. *The Corpus of English Life Sciences Texts (CELiST)*. A Coruña: Universidade da Coruña. https://doi.org/10.17979/spudc.97884974997848
Locher, Miriam A. and Jucker, Andreas H. 2017. *Pragmatics of Fiction*. Berlin: De Gruyter Mouton. Newcastle: Cambridge Scholars. https://doi.org/10.1515/9783110431094
Mele-Marrero, Margarita. 2021. "Engagement in the botanists of the Corpus of English Life Sciences Texts: Flourishing female scientific writing". In Moskowich, Isabel; Lareo, Inés and Camiña, Gonzalo (eds.), *"All families and genera": Exploring the Corpus of English Life Sciences Texts*. Amsterdam/Philadelphia: John Benjamins. 133–146. https://doi.org/10.1075/z.237.07mel

Moskowich, Isabel and Crespo, Begoña. Forthcoming. "… but be fure you let it fettle": Late Modern authors' presence in scientific texts.

Moskowich, Isabel. 2017. "Pronouns as stance markers in the Coruña Corpus: An analysis of the CETA, CEPhiT and CHET". In Alonso-Almeida, Francisco (ed.), *Stancetaking in Late Modern English Scientific Writing. Evidence from the Coruña Corpus*. Colección Scientia [Applied Linguistics]. Valencia: Servicio de Publicaciones de la Universidad Politécnica de Valencia. 73–91.

Nurmi, Arja. 2002. Does size matter? The Corpus of Early English Correspondence and its sampler. *Variation Past and Present: VARIENG Studies on English for Terttu Nevalainen. Mémoires de la Société Néophilologique de Helsinki*, 61: 173–84.

OED. 1989. *Oxford English Dictionary*. Retrieved January 20, 2020, from http://www.oed.com/

Pollock, John. 1990. Philosophy and Artificial Intelligence. *Philosophical Perspectives*, 4, 461–498. https://doi.org/10.2307/2214201

Quirk, Randolph; Greenbaum, Sydney; Leech, Geoffrey N. and Svartvik, Jan. 1985. *A Comprehensive Grammar of the English Language*. London: Longman.

Toolan, Michael J. 2010. *Narrative Progression in the Short Story: A corpus stylistic approach*. Amsterdam: John Benjamins. https://doi.org/10.1075/lal.6

Toolan, Michael J. 2011. "How does Literary Language Move Us? On Being 'Immersed' and Emotionally Engaged by Literary Narratives." In Penas Ibañez, Beatriz; Muñoz, Micaela and Conejero, Marta (eds.), *Con/Texts of Persuasion*. Kassel: Reichenberger. 15–25.

Turkle, Sherry. 2005. "Personal Computers with Personal Meanings". In Turkle, S. and Ebrary, Inc. (eds.), *The second self: Computers and the human spirit*. Cambridge, Mass: MIT Press. 155–181.

CHAPTER 10

"If you will take the trouble to inquire into it rather closely, I think you will find that it is not worth very much"

Authorial presence through conditionals and citation sequences in late modern English life sciences texts

Luis Puente-Castelo
University of A Coruña

1. Introduction

If, as Hunston (1993: 57) claims, "[t]he production of a written text is a social process, both in the sense that it represents the interaction between a writer and a reader (Sinclair 1981) and in the sense that the text plays a role in a particular social system (Halliday and Hasan 1985)", the study of a text may reveal or help to confirm the nature of the relations among the different actors and the social system in which the text itself is inscribed (*cf.* Biber and Conrad 2009).

One of the main elements to consider in such an approach is the presence of the author in the text, and the role they choose to adopt. This has been the object of extensive scholarly interest, looking particularly at contemporary academic writing; it is also the focus of the present paper.

The aim here is to consider how two different linguistic strategies (the use of certain conditional structures and the sequences employed to introduce citations of other authors' words) help authors to express their attitude towards their content, during a period – the eighteenth and nineteenth centuries – in which some of the rhetorical strategies which influence these uses were still developing. To do so, the study draws on data from CELiST (Lareo *et al.* 2020), the subsection on life sciences of the Coruña Corpus of English Scientific Writing.

In what follows, Section 2 will analyse the evolution of the role of authors in scientific texts during this period, whilst Section 3 will briefly re-examine some studies on authorial presence and a number of problems therein. Sections 4 and 5 will look into the uses of conditionals and citation sequences with this function, respectively, and Section 6 will present the corpus and methodology used. Section 7 will present and discuss the results, followed by some concluding remarks in Section 8.

https://doi.org/10.1075/z.237.10pue

2. The author in the text in late modern English scientific writing

The eighteenth and nineteenth centuries together constitute the period in which science finally transformed into what we nowadays understand as such. These two centuries witnessed the definitive substitution of the medieval scholastic paradigm with a new model of scientific practice, one which was influenced by empiricist authors and characterised as having the real world, rather than the written word, at its core. This led to an intense development of scientific knowledge, improving its accessibility and making it more useful to society.

Intertwined with these changes, both influencing them and being influenced by them, was the development of a new way of writing science, a definite scientific register. Its emergence can be attributed to two parallel movements.

First, the focus of scientific texts shifted during this period. The logocentric model characteristic of Scholasticism (Taavitsainen and Pahta 1998: 167), in which the content of texts written by classical and medieval "authorities" was considered undisputable, was abandoned, to be replaced by the so-called "rhetoric of immediate experience", characterised by profuse descriptions and an extensive authorial presence in the text (Atkinson 1996: 335). This would later evolve into an object-centred approach (Atkinson 1996: 341), in which the presence of the author diminished drastically, and the importance of rigorous methods and evidence increased. This led to the emergence of a series of linguistic features, such as the use of nominalisations, objectifications, specific terminology and the use of the passive voice, which are now considered to be stereotypical characteristics of scientific writing.

Second, the new scientific societies founded during this period actively promoted scientific debates and experimentation among their members, and thus they began to consider science through the lens of social conventions, reconfiguring "science" as a social phenomenon (Hyland 1996, 2000). This in turn influenced the language used in the scientific register, since from that time authors increasingly sought to achieve the best possible reception of their claims by their peers (Bazerman 1988; Myers 1989; Swales 1990). This led to the emergence of a series of rhetorical strategies such as avoiding confrontation, recognising different points of view and contributions from other researchers, striving for consensus, displaying humility and politeness, and avoiding categorical statements, which correspond at the linguistic level to the use of strategies such as modal verbs, *verba dicendi*, distancing verbs, or probability adverbs (Allen, Qin and Lancaster 1994; Hyland 1996, 1998).

However, the development of these two vectors at the same time is also problematic in terms of the presence of the author in the text. This is the object of the next section.

3. An overview on studies of authorial presence

It is well known that some of the characteristics of scientific texts, such as the use of the passive voice or nominalisations, contribute to authors' self-effacement. However, from a rhetorical point of view, authors are required to be present in the text in order to perform several crucial actions, including the carving out of a "niche" for their research, justifying the relevance of their studies, and projecting themselves as members of the scientific communities (Swales 1990; Hyland 1999).

It is also frequently necessary for authors to express their opinions about the content of their texts, be it their own findings, or those of others. This sometimes obliges authors to tread a difficult path between opposing needs: showing humility and refraining from using categorical statements, but at the same time avoiding the dangers of undermining one's own case, and, in the case of expressing doubts about someone else's material, steering clear of confrontation.

In recent years, there has been extensive scholarly interest in the study of the expression of authors' own opinions in texts and the solutions used to overcome the difficulties in reconciling different rhetorical needs (Salager-Meyer 1994; Hyland 1996, 1999; Hunston and Thompson 1999; Charles 2003, 2006, among many others). However, several frameworks, sometimes competing ones, have been used here. Chafe and Nichols (1986) studied the attitude of authors towards their content as part of *evidentiality*, an expression of "the kinds of evidence a person has for making factual claims" (Anderson 1986: 273); meanwhile, Hunston and Thompson include these phenomena as part of *evaluation*, "the broad cover term for the expression of the speaker or writer's attitude or stance towards, viewpoint on, or feelings about the entities or propositions that he or she is talking about" (1999: 5). Biber and Finegan (1988) and Conrad and Biber (1999) distinguished several types of *stance*, such as *epistemic stance*, in terms of the degree of certainty or uncertainty an author feels. Some related concepts, such as *involvement, hedging* (Myers 1989; Crompton 1997; Hyland 1998), and *metadiscourse* (Hyland 2005) have also been used.

Clearly, then, this is an area of the study of language characterised by its complexity and conflicting conceptualisations, with a whole array of different terms for concepts which to some extent overlap, and with some of these terms being used synonymously or contrastingly, a problem identified by several authors (Hunston and Thompson 1999: 2–5; Alonso-Almeida 2012: 202).

A further problem is that the treatment in the literature of authorial presence has been commonly reduced to the study of the degree of certainty or commitment of the author towards an element of content (with concepts such as *epistemic modality* (Stubbs 1996; Hyland 1998; Alonso-Almeida 2012), *doubt stance markers* (Biber and Finegan 1988: 98) or Halliday's (1994) *modalization*), whereas in fact authorial presence goes beyond this axis, expressing other nuances such as the formal correctness of a term or the noteworthiness of a content, as will be described below.

Finally, regarding the linguistic elements analysed, studies on authorial presence have frequently focused on the use of the first person (Tang and John 1999; Moskowich 2016; Mele-Marrero 2019, see also Crespo 2021) and of modality (Biber and Finegan 1988; Hunston and Thompson 1999; Alonso-Almeida 2012), sometimes in relation to phenomena such as hedging, the display of humility and politeness, and various others.

However, less attention has been paid to some other linguistic resources, such as the use of certain types of conditional structures (Warchal 2010), or particular structures used when writers introduce quotations from other authors (Hunston 1993; John 2013), by means of which writers can convey their opinions whilst also expressing other attitudes. The following sections deal with these uses.

4. Conditionals and authorial presence

Conditional structures are characterised by their formal and functional versatility, in that they can fulfil a significant number of different functions (Dancygier 1998: 2). Indeed, they are a linguistic feature through which authors can express their attitude towards not only the content of a claim but also its form or relevance, thus contributing to the expression of politeness and the avoidance of criticism.

The present study will focus on three types of conditionals featuring authorial presence: Non-committal conditionals, Metalinguistic conditionals, and Relevance conditionals (Puente-Castelo 2017).

Non-committal conditionals are used by authors to express their uncertainty about the veracity of a claim. Such claims might be the writer's own, as in (1) below, or those of other authors (2). By using these conditionals, authors are able to present claims which they nevertheless avoid endorsing, thus distancing themselves and protecting their face, as well as being polite.

(1) If we are correct in suggesting that there is an isomorphism between the Helmholtze-Gibson debate and the debate about linguistic knowledge, then a computational approach to language seems promising. (Warchal 2010: 147)

(2) Chomsky's views cannot be reconciled with Piaget's, if I understand both correctly. (Quirk *et al.* 1985: 1096)

Non-committal conditionals comprise both Warchal's (2010: 145) *reservation conditionals* (akin to (1), in that they refer to the doubts of an author towards their own claims), and Quirk *et al*'s *uncertainty conditionals* (1985: 1096), which, as in (2), convey uncertainty about the statements of other authors.

Metalinguistic conditionals are used to express the uncertainty that an author has about the form or the choice of words used in their own discourse. Thus, in (3) below, the author is expressing doubts as to whether "florid" is indeed the right term to describe "his style". It is interesting to note, however, that these conditionals can also sometimes be used with ironic overtones, thus leading to possible alternative interpretations.

(3) His style is florid, if that's the right word. (Quirk *et al.* 1985: 1096)

Metalinguistic conditionals appear with the same name in both Quirk *et al.* (1985: 1096), and Warchal's classifications (2010: 145), whilst Gabrielatos (2010: 239) calls them *comment conditionals.*

Finally, *Relevance conditionals* are commonly used by authors to express the circumstances under which a statement is to be considered relevant. For instance, in (4) below, readers should only take the comment into account if they consider it to be important, thus enabling them to disregard it if they do not.

(4) Finally (if this is important), the S1 meaning can be converted into an S meaning to recover a more intuitive object to represent the meaning of the original sentence. (Warchal 2010: 148)

These conditionals express the opinion of the author as to the noteworthiness of a comment, in that they feel the need to justify its insertion by noting that it might or might not be considered relevant, thus avoiding the criticism of those who might deem it irrelevant. However, these conditionals are often used as a purely rhetorical device to allow authors to introduce information which they themselves do not consider essential, similar to a side note. Relevance conditionals appear in both Gabrielatos' (2010: 239) and Warchal's (2010: 144–145) classifications under the same name, and are also the prototypical examples of Sweetser's (1990: 121) *Speech Act conditionals.*

5. Expressing opinions by means of citation sequences

Reporting the words of other authors through the use of citations is a well-known context in which writers[1] can "lay claim to their own academic credibility" (Hunston and Thompson 1999: 8), because citations not only contribute to the recognition of peers' work but also "influence writer identity due to the inherent evaluative quality

1. It has become common practice in works on citations to refer to the person whose words are being quoted as "author" and the person quoting those words as "writer" (Thompson and Ye 1991: 366; John 2013: 189). In the current study, author and writer are in general used interchangeably, unless the distinction is relevant, as in this section, where the convention will be followed.

of some types of citations and how these are used in the writer's argument". (John 2013: 187). Thus, citations are a prime means by which writers can express their opinion about a content.

It has traditionally been assumed that a writer can agree, disagree or remain neutral about a statement (John 2013: 189), but in citations it has also been considered necessary to take into account the original attitude of the quoted author towards their own words (Myers 1989: 4). Thus, Hunston (1993) distinguishes three possible options: (a) both the original author and the writer citing their work agree with the information, (b) both remain uncommitted, or (c) the author agrees with it, whilst the citing writer remains neutral. These frameworks, however, focus exclusively on the certainty-uncertainty axis, ignoring all the range of other opinions, such as noteworthiness or appropriateness, that may also be expressed by means of citation sequences.

Moreover, the attribution of particular linguistic features to each of the abovementioned points of view is not always univocal. Verbs such as *point out, suggest,* and *claim*, have been identified, respectively, as central examples for each of Hunston's points of view identified above. However, as illustrated by John (2013: 196), the ascription of a particular meaning to each citation sequence is influenced by several linguistic factors. Thus, *suggest*, as in (5), normally indicates neutrality on the part of the writer, but if *suggest* is used as part of an *as* structure (6), it tends to indicate agreement with the cited author. Similarly, a modifying adverb, such as *aptly* (7), may reinforce or mitigate the position of the writer towards the quoted content.

(5) [AUTHOR][2] suggests that when an author is writing a text, he or she does not write it for real readers out there, but for an "ideal reader" who is constructed in his or her mind. [SMY:F:17] (John 2013: 194)

(6) As [AUTHORS] suggest, claquing can also cultivate a sense of exoticism. In the case of the translation of The True story of Ah Q, the claquing of the literary references and quotation in the ST do give the TT a sense of cultural foreignness … [SMY:F:52–53] (John 2013: 196)

(7) However, as [AUTHOR] aptly points out the difficulty of this definition is that "in a language …" (1996: 57). [SMY:F:5] (John 2013: 196)

All this shows how the attitude of a writer towards a quoted content is (a) multifaceted, with several possible interpretations for a single act of quotation, making it difficult to directly classify them; and (b) expressed through a combination of different linguistic resources. This makes it necessary to analyse each case individually, considering its linguistic context as a whole in order to identify the opinion the writer is trying to express. The next section will explore this issue further.

2. Authors named in John's corpus have been anonymised to ease comprehension.

6. Corpus and methodology

This study uses the full CELiST corpus, totalling 400,305 words. In order to extract and analyse the data, different approaches were followed for each of the linguistic features.

To identify instances of conditionals, selected conditional particles (including not only *if* and *unless*, but also less common conditional particles, such as *providing*, and inversion operators with conditional interpretation, see Puente-Castelo 2017: 53) were searched for in CELiST using the Coruña Corpus Tool (CCT) (Parapar and Moskowich 2007; Barsaglini-Castro and Valcarce 2020). These instances were then recorded with sufficient context and manually disambiguated in order to discard those cases in which conditional particles were used in other contexts, thus eliminating interrogative uses of *if*, uses of *providing* and *supposing* as full verbs, and all cases in which inversion operators did not introduce a conditional, among others.

Each example was then classified according to its function in discourse, following the typology presented in Puente-Castelo (2017: 107), so that cases of the three conditionals identified above (non-committal, metalinguistic, and politeness conditionals) could be detected. These were then further classified according to a series of parameters, namely the formal type of conditional, the order of their constituents, and the verb form combinations.

Regarding citation sequences, a query for the tags identifying quotations ("\[quotation\]",[3] "\[quotlat\]", "\[quotgreek\]", "\[quotfrench\]" and "\[quotwelsh\]") was introduced in the CCT. Results were then recorded with an extensive context in order to be able to identify the general attitude of the writer towards the content. Cases were then analysed and classified according to the type of citation sequence used and the attitude of the writer, distinguishing, by means of close reading, between sequences in which the writer expresses their agreement, disagreement, or neutrality, plus isolated uses.

7. Analysis of the results

7.1 Conditionals

777 cases of conditionals were found in the corpus. Of these, there are 31 cases (or 76.73 per million) of relevance conditionals, 21 (51.98 per million) of non-committal conditionals, and 4 of metalinguistic conditionals, as shown in Figure 1 below. Thus,

3. These searches feature backward slashes, because square brackets are metacharacters in the CCT and, in order to search for them, it is necessary to use the designated escape metacharacter in the environment, the backward slash.

the total number of cases is just 56, meaning that an analysis of the data according to socio-historical parameters is unfeasible, in that there are insufficient cases from which to identify trends.

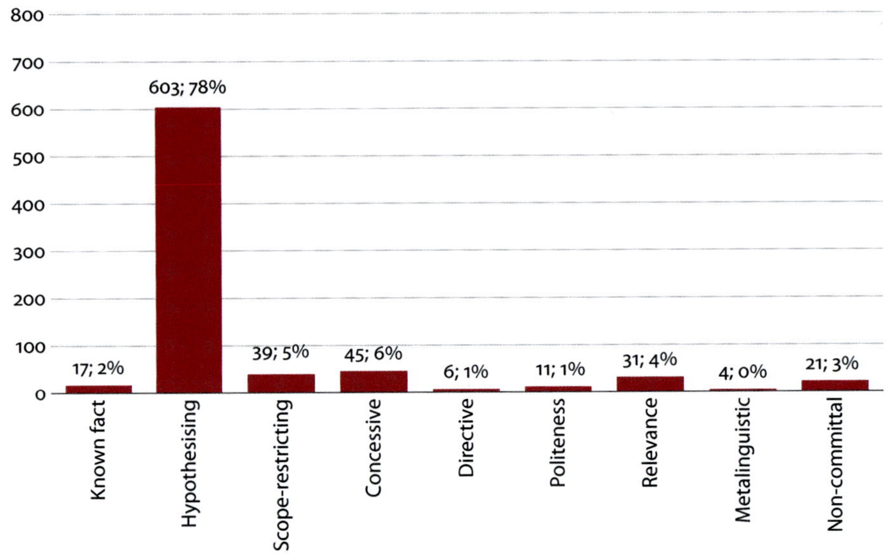

Figure 1. Cases of conditional per function in CELiST

a. *Relevance conditionals*

As noted above, relevance conditionals are used by authors to express the conditions under which the related argument is to be considered relevant by readers. This use can be clearly seen in (8) below, in which the author expresses the conditions in the protasis, under which the information of the apodosis is deemed relevant.

(8) If you like this little lesson in Natural History, I hope, at some future time, to write another one for you about animals of another kind, which are constructed on an entirely different plan. (Agassiz 1859 [72 (12984)])

As with most types of conditionals, relevance conditionals are mostly expressed by means of the conditional particle *if*, which, as shown in Figure 2 below, accounts for almost three quarters of all the uses.

However, it is also noticeable that other particles account for a good proportion of the uses. For instance, there are two uses with *lest*, as in (9) below, and three with *should* as an inversion operator.

(9) It seems scarcely necessary to describe it botanically; but, lest it should be confounded with other species of the same genus, it may be well to say that it belongs to the family Campanulaceæ. (Lankester, 1879 [77 (4084)])

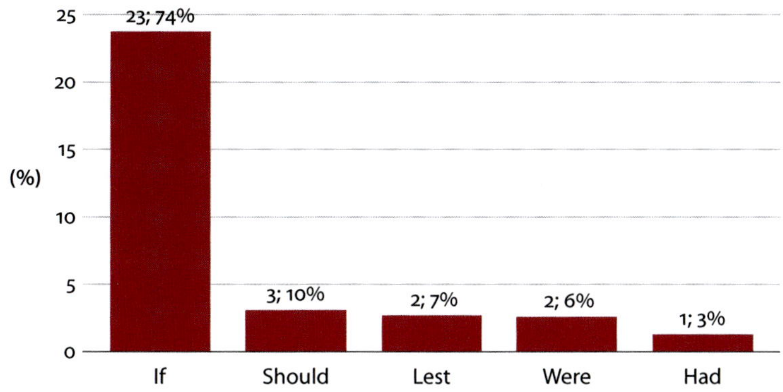

Figure 2. Relevance conditionals per type of conditional

Regarding the order of constituents, most cases of relevance conditionals (27 out of 31) present a protasis-first order, whilst there are three with a final protasis and one with a protasis in the middle of the apodosis.

Finally, verb-form combinations in relevance conditionals are shown in Table 1 below. The most common combination is present simple-present simple, but the high proportion of use of *should* in the protasis and imperatives in the apodosis are both noticeable.

Table 1. Verb-form combinations in relevance conditionals

Verb form combinations	Number of cases
Present simple, present simple	7
Past simple, would	4
Would, imperative	2
Should, may	2
Present simple, imperative	2
Present simple, will	2
Other 12 combinations	12 total cases

Examples with these uses are provided in (10) and (11) below, respectively.

(10) Should it be said that this is but another aspect of the antagonism already set forth, since these undeveloped forms are also the smaller forms; the reply is that though in part true, this is not wholly true. (Spencer, 1867 [75 (6975)])

(11) If you have a mind for an excellent Herring-Pye, proceed as follows: (Dodd 1752 [52 (8372)])

b. *Non-committal conditionals*

As noted above, non-committal conditionals are used by authors to express their uncertainty about a claim, be it their own or that of another author. In this sense, they are the type of conditional which best captures the expression of the prototypical certainty-uncertainty axis in the study of authorial presence. A good example of these uses is seen in (12) below, in which the author states clearly "if we may trust the beautiful researches of Bütschli", thus expressing their uncertainty about them.

(12) the original nucleus which, if we may trust the beautiful researches of Bütschli, takes place during conjugation in Infusoria as a preliminary to the formation of a fresh nucleus. (Balfour 1880 [78 (7896)])

Conditionals of this type are mostly expressed by means of *if*, with just a single case in which a different conditional particle is found, the inversion operator *should*, shown in (13) below. It is also interesting to note that non-committal conditionals show a higher than average use of middle and final protasis orders, with 8 (38%) and 4 (19%) cases, respectively.

(13) But from the description and peculiarities here given, others can easily restore it to its proper place, should I be mistaken. (Dalyell 1848 [71 (5679)])

Regarding the verb-form combinations, most cases of non-committal conditionals use *present simple-present simple* combinations, as shown in Table 2. However, it is interesting to note some unusual combinations, such as present simple-past simple and present subjunctive-*will*.

Table 2. Verb-form combinations in non-committal conditionals

Verb form combinations	Number of cases
Present simple, present simple	3
Present simple, past simple	2
Present subjunctive, will	2
Other 14 combinations	14 total cases

Finally, of note is that all 21 uses of non-committal conditionals appear in texts by male rather than female authors, despite the latter accounting for more than 18% of the words in the corpus. This disparity may have to do with the more confrontational nature of non-committal conditionals, as previous studies have found that women prefer subtler mitigation strategies, such as modals (Crespo 2011).

c. *Metalinguistic conditionals*

Metalinguistic conditionals are used to express an author's doubt about the form of their discourse, its choice of words, style, etc. Only four of these conditionals have been found in CELiST, one example being that seen in (14) below. All of these use *if*, with no other conditional particle or inversion operator found.

(14) Perhaps the Name of Merry-wing, applied to another Fly, from the Noiſe it makes with its Wings, would be more properly applied to this, if the Acceptation of the Word had not rendered it almoſt univerſal, at leaſt among Engliſh Writers. (Hughes 1750 [51 (9879)])

Further interesting results relate to the order of the constituents and the verb-form combinations. Two uses each of protasis in final and middle position were found, whereas not a single case of the most common order of constituents, protasis-apodosis, was attested. The preferred verb-form combination is *may*-present simple, exemplified in (15) below, accounting for three of the four cases.

(15) and ſeem to be the Spawn juſt vivified (if I may uſe that Expreſſion). (Dodd 1752 [52 (1016)])

7.2 Citation sequences

275 instances of quotations were found in the corpus (680.75 per million words). These cases were not distributed equally; as shown in Figure 3, there is notable diachronic variability in the number of cases, a variability which also seems to appear at the sample level, with some samples presenting a high number of uses and many others presenting none at all. Although part of this variability could perhaps be explained by the section of the text from which particular samples are extracted,[4] the results hint at a scenario in which quotations are much less widespread than at present.

4. The section of the text from which each of the samples is extracted is taken into account during corpus compilation to ensure representativeness (see also Moskowich 2021). However, it has not been implemented as a parameter of selection of samples in the CCT, for samples occasionally include more than one section.

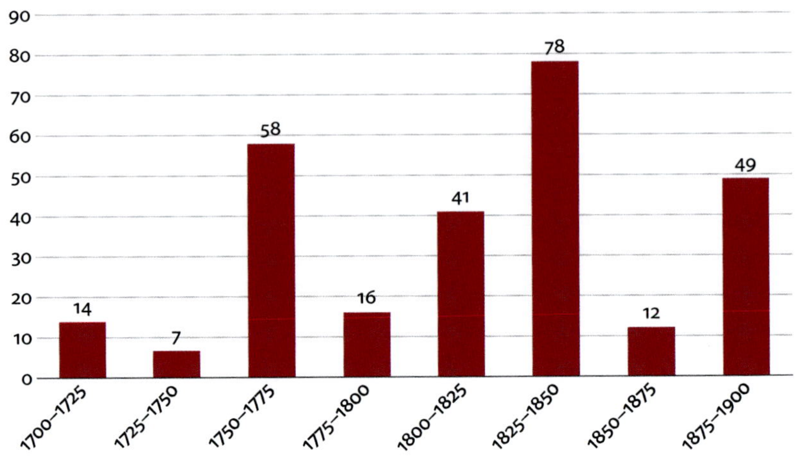

Figure 3. Cases of quotations over time

Regarding the general attitude conveyed by the writer towards the quotation, as shown in Figure 4 in most cases (66.91%) the writer appears to remain neutral towards the content of the quotation, although this neutrality in fact embraces several attitudes, as will be explained below. The writer conveys agreement towards the quotation in 16.73% of the cases, which contrasts with only 1.82% (5 cases) in which the author expresses their open disagreement.

The remaining cases (14.55%) are what have been called "isolated" cases. That is, cases in which the quotation appears isolated, with no citation sequence and consequently no linguistic context from which to infer the attitude of the writer. These will also be dealt with in more depth below.

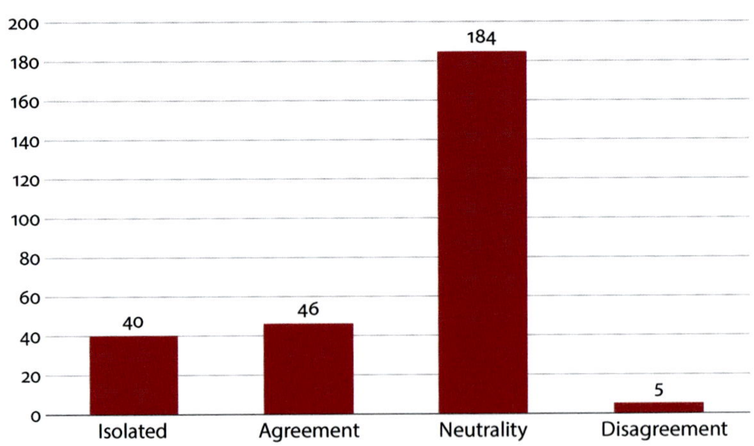

Figure 4. Number of quotations per attitude of the writer towards the content

These attitudes are not distributed evenly, but, as can be seen in Figure 5 below, there are clear and important differences across the period: At the start of the timeframe, most of the few quotations used form part of the "isolated" group. Quotations of this kind then rapidly disappear, except for a peak during the last quarter of the eighteenth century, this explained by their extensive use in one text, by Edward Donovan (1794). The second quarter shows only quotations with which the writer agrees, but from 1750 onwards most uses show neutrality. The first citation sequences expressing disagreement appear in the 19th century.

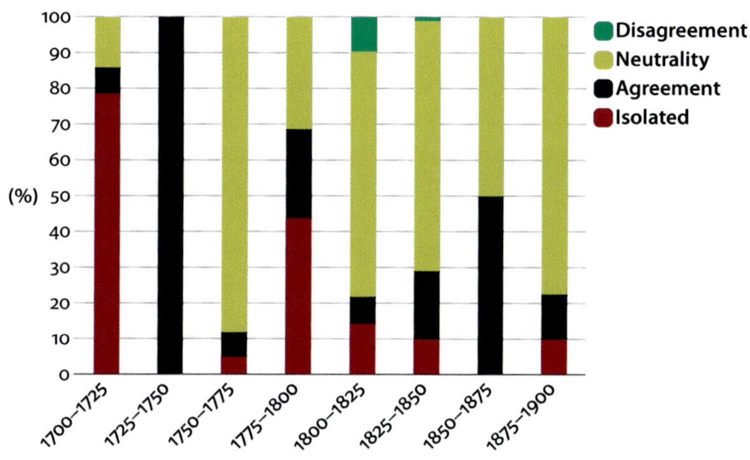

Figure 5. Proportion of use of the different types of quotations over time

These data conform to a model which corresponds neatly with the evolution of science and the authorial presence of writers in the text during the period. At the beginning of the eighteenth century, quotations are still few and far between, and they are mostly used in isolation. A rapid change then occurs, first to an author-centred model, in which quotations are mostly used as a tool to reinforce one's own accounts, but are still used infrequently; later on, there is a shift to the object-centred model. It is then, with its onset from the middle of the eighteenth century, that citation becomes widespread, as it becomes essential to recognise the contributions and points of view of others. And just as the scientific register becomes more and more objective, most quotations are presented with a neutral evaluation on the part of the writer.

In what follows, each of the different types is analysed in more depth.

a. *Isolated quotations*

Isolated quotations account for 14.55% of all the uses of quotations in the corpus, and, as explained above, they are particularly prevalent at the start of the period. They are frequently used as a prompt, helping the author introduce a new topic to be commented on, or, as in (16) below, as part of a text commentary.

(16) [quotation]
This is evident: Becaufe in the ftrongeft cohefion of Particles touching one another, we find that the Weight of fome Bodies will pull the Particles afunder.
(Keill 1717 [044 (1879)])

However, it is also common that quotations of this kind introduce miscellaneous elements, such as extracts of other works (17) or even poems. They also include quotations used as introductions with little or no relation with the content, as happens today with dedications.

(17) We cannot better conclude this difcourfe, than by fubjoining the following extract from the preface of that work. [quotation]
(Donovan 1794 [060 (7652)])

b. *Citation sequences expressing agreement*

Quotations towards which writers express their agreement account for 16.73% of the cases. They are also frequent at the start of the period, when they are used to cite authorities, as in (18) below.

(18) So that the Words of Virgil may be juftly applied to thefe: [quotlat]
(Hughes 1750 [051 (9462)])

Table 3. Introductory elements in agreement quotations

Introductory elements	Number of cases
as [author] says…	4
[author] observes…	4
No intro	4
[author] says…	4
Other 25 introductory elements	30 total cases

Regarding the introductory elements to these quotations, shown in Table 3 above, there is not a single introductory element linked to the expression of agreement, as no string shows more than four cases, with *as x says, observe* and *say* as the most frequent introductory sequences, together with cases with no introductory element.[5]

5. Cases have been classified as having no introductory element when quotations appear directly in the text with no previous reference in the context.

However, there are some elements which make the expression of agreement more likely. These include *as*, present in eight of the 46 cases, and modifying adverbs, such as *justly* (as in Example (19) below) which appear in seven cases, something which was also identified by John (2013: 196); other elements include *for, in the words of* (20), and *thus* + an introductory verb (21 below).

(19) Pliny juftly obferves, [quotlat] (Hughes 1750 [051 (5488)])

(20) And all, in the Words of the great Milton, [quotation] (Hughes 1750 [051 (9462)])

(21) And in some whole groups of animals and in certain members of other groups, the embryo does not at any period differ widely from the adult: thus Owen has remarked in regard to cuttle-fish, [quotation] and again in spiders, [quotation] (Darwin 1859 [073 (6226)])

c. *Citation sequences expressing neutrality*

Neutrality quotations are the most common, representing 66.91% of the cases. As such, they show a higher number of instances and also more notable variability. As shown in Table 4 below, quotations with no clear introductory element are the most common, followed by uses with *say, add,* and *regard,* uses with the expression *the [quot] of [author],* and uses in footnotes.

Table 4. Introductory elements in neutrality quotations

Introductory elements	Number of cases
No intro	36
[author] says	34
[quot], says [author]	8
Footnote	8
[quot] of [author]	7
[author] adds	5
[author] regards sth as	5
Other 65 introductory elements	81 total cases

These are also the kind of quotations showing the most versatility in their uses, as the expression of a writer's neutrality towards a content may indicate all types of attitudes. Neutrality is sometimes an expression of concealed scepticism, but can also be a purely rhetorical device deployed in the immediate environment of a quotation, with the real opinion appearing at some distance from it. This can be seen in (22) below, in which the real opinion, "not worth very much", appears only following several hedges.

(22) People say, [quotation] This is very commonly assumed to be a fact, and it is an argument that is commonly brought forward as conclusive; but if you will take the trouble to inquire into it rather closely, I think you will find that it is not worth very much. (Huxley 1863 [074 (8489)])

However, neutrality quotations go beyond these uses, including also those cases which go beyond the certainty-uncertainty axis, such as citation sequences commenting on the noteworthiness of an account (23), akin to relevance conditionals.

(23) The only illustrative facts I have met with which seem noteworthy, are some named by [Mr]. Gould in his work on The Birds of Australia. He says: [quotation] (Spencer 1867 [075 (8723)])

There are also some metalinguistic uses, with citation sequences being used to introduce translations or definitions, as in (24), or terminology, particularly with the expressions *[author] calls x [quot]* and *the [quot] of [author]*, as in (25). This environment can also be used to comment on the style of an author, as in (26).

(24) Linnæus has divided the animal kingdom into six classes, and comprised insects in the fifth, with this single definition: [quotation] which are the horns affixed to the head. (Wakefield 1816 [064 (1410)])

(25) This is the [quotlat] of Artedi, [No]. 23, page 86. (Borlase 1758 [053 (2861)]

(26) Linnæus, in his bold and graphic language, said, [quotation] (Lincoln 1832 [068 (6200)])

Finally, neutrality conditionals are also prominent in one of the main features of the new scientific writing that appeared in this period: the literature review. The first instances of this new use in the data were found in Dodd's 1752 text.

d. *Citation sequences expressing disagreement*

There are only five examples of quotations towards which the writer expresses disagreement, all in nineteenth-century data. This suggests that the expression of open disagreement continued throughout the period to be considered too categorical and to be avoided. One of these uses is shown in (27) below:

(27) The assertion of LINNEUS, [quotlat] is not only unsupported by any authentic testimony concerning animals of the monkey kind, but directly contradicted by all the well-ascertained facts relating to those which most nearly resemble us in stature. (Lawrence 1819 [065 (3177)])

8. Concluding remarks

The above analysis has illustrated how conditionals and citation sequences act as vehicles for the presence of the author in the text, but at the same time it also shows how the opinion they express towards the content in question goes beyond the simple certainty-uncertainty axis.

These other opinions, such as the style or appropriateness of the wording, the noteworthiness, relevance or trustworthiness of statements, or the writer's confidence in the content, can be expressed by means of both conditionals and citation sequences, since both strategies can express these nuances.

The results here also show that, at least with citation sequences, differences in the use of these strategies over time can be interpreted as evidence of the broader changes in the socio-historical environment of the text, that is, the evolution of scientific practice and the scientific register.

Finally, we must note that conditionals and citation sequences do sometimes appear together. This is the case with the example in the title, already presented above as (22):

(22) People say, [quotation]. This is very commonly assumed to be a fact, and it is an argument that is commonly brought forward as conclusive; but if you will take the trouble to inquire into it rather closely, I think you will find that it is not worth very much. (Huxley 1863 [074 (8489)])

These uses show an even more forceful authorial presence, and thus allow us to discern how authorial presence might be best explained as a system in which several devices act at the same time, contributing, sometimes in opposite directions, to a common goal. Further research, including other linguistic features and more corpora, will be needed to prove this.

Acknowledgements

I would like to thank Dr. Suganthi John (University of Birmingham) for extensive and illuminating discussions on the original idea and content of this chapter.

Funding

The research reported here has been funded by the *Programa de axudas á etapa posdoutoral da Xunta de Galicia (Consellería de Cultura, Educación e Ordenación Universitaria).* This grant is hereby gratefully acknowledged.

Works cited

Allen, Bryce; Qin, Jian and Lancaster, Frederik Wilfrid. 1994. Persuasive Communities: A Longitudinal Analysis of References in the Philosophical Transactions of the Royal Society, 1665–1990. *Social Studies of Science*, 24/2: 279–310.

Alonso-Almeida, Francisco. 2012. "An analysis of hedging in eighteenth century English astronomy texts". In Moskowich, Isabel and Crespo, Begoña (eds.), *Astronomy "playne and simple". The Writing of Science between 1700 and 1900*. Amsterdam: John Benjamins. 199–220. https://doi.org/10.1075/z.173.10alo

Anderson, Lloyd B. 1986. "Evidentials, paths of change and mental maps: typologically regular assymetries". In Chafe, Wallace L. and Nichols, Johanna (eds.), *Evidentiality and the linguistic coding of epistemology*. Norwood, NJ: Ablex. 273–312.

Atkinson, Dwight. 1996. The Philosophical Transactions of the Royal Society of London, 1675–1975: A sociohistorical discourse analysis. *Language in Society*, 25: 333–371. https://doi.org/10.1017/S0047404500019205

Barsaglini-Castro, Anabella and Valcarce, Daniel. 2020. The Coruña Corpus Tool: Ten Years On. *Procesamiento del Lenguaje Natural*, 64: 13–19.

Bazerman, Charles. 1988. *Shaping Written Knowledge: The Genre and Activity of the Experimental Article in Science*. Madison: University of Wisconsin Press.

Biber, Douglas and Finegan, Edward. 1988. Adverbial stance types in English. *Discourse Processes*, 11: 1–34. https://doi.org/10.1080/01638538809544689

Biber, Douglas and Conrad, Susan. 2009. *Register, genre, and style*. Cambridge: Cambridge University Press. https://doi.org/10.1017/CBO9780511814358

Chafe, Wallace L. and Nichols, Johanna. 1986. *Evidentiality and the linguistic coding of epistemology*. Norwood, NJ: Ablex.

Charles, Maggie. 2003. 'This mystery …': A corpus-based study of the use of nouns to construct stance in theses from two contrasting disciplines. *Journal of English for Academic Purposes*, 2/4: 313–326. https://doi.org/10.1016/S1475-1585(03)00048-1

Charles, Maggie. 2006. The construction of stance in reporting clauses: A cross-disciplinary study of theses. *Applied Linguistics*, 27/3: 495–518. https://doi.org/10.1093/applin/aml021

Conrad, Susan and Biber, Douglas. 1999. "Adverbial marking of Stance in speech and writing". In Hunston, Susan and Thompson, Geoff (eds.), *Evaluation in Text: Authorial Stance and the Construction of Discourse*. Oxford: Oxford University Press. 56–73.

Crespo, Begoña. 2011. Persuasion markers and ideology in eighteenth century Corpus of English Philosophy Texts (CEPhiT). *Revista de Lenguas para Fines Específicos*, 17: 199–228.

Crespo, Begoña. 2021. "Linguistic indicators of persuasion in female authors in the Corpus of English Life Sciences Texts". In Moskowich, Isabel; Lareo, Inés and Camiña, Gonzalo (eds.), *"All families and genera": Exploring the Corpus of English Life Sciences Texts*. Amsterdam/Philadelphia: John Benjamins. 147–167. https://doi.org/10.1075/z.237.08cre

Crompton, Peter. 1997. Hedging in academic writing: some theoretical aspects. *English for Specific Purposes*, 16: 271–289. https://doi.org/10.1016/S0889-4906(97)00007-0

Dancygier, Barbara. 1998. *Conditionals and Prediction: Time, knowledge and causation in conditional constructions*. Cambridge: Cambridge University Press.

Gabrielatos, Costas. 2010. A corpus-based examination of English if-conditionals through the lens of modality: Nature and types. Unpublished PhD dissertation. Lancaster: Lancaster University.

Halliday, M. A. K. 1994. *An Introduction to Functional Grammar (2nd edn.)* London: Edward Arnold.

Halliday, M. A. K. and Hasan, Ruqaiya. 1985. *Language, Context and Text. Aspects of language in a social-semiotic perspective*. Victoria: Deakin University Press.

Hunston, Susan. 1993. "Evaluation and ideology in scientific writing". In Ghadessy, Mohsen (ed.), *Register analysis*. London: Pinter. 57–73.

Hunston, Susan and Thompson, Geoff. 1999. *Evaluation in Text: Authorial Stance and the Construction of Discourse*. Oxford: Oxford University Press.

Hyland, Ken. 1996. Writing without conviction? Hedging in science research articles. *Applied Linguistics*, 17/4: 433–454. https://doi.org/10.1093/applin/17.4.433

Hyland, Ken. 1998. *Hedging in Scientific Research Articles*. Amsterdam: John Benjamins. https://doi.org/10.1075/pbns.54

Hyland, Ken. 1999. Academic attribution: Citation and the construction of disciplinary knowledge. *Applied Linguistics*, 20/3: 341–367. https://doi.org/10.1093/applin/20.3.341

Hyland, Ken. 2000. *Disciplinary Discourses: Social interactions in academic writing*. Harlow, Essex: Pearson Education.

Hyland, Ken. 2005. *Metadiscourse: exploring interaction in writing*. London: Continuum.

John, Suganthi. 2013. "Identity Without the 'I': A Study of Citation Sequences and Writer Identity in Literature Review Sections of Dissertations". In Tang, Ramona (ed.), *Academic Writing in a Second or Foreign Language: Issues and Challenges Facing ESL/EFL academic writers*. London: Continuum. 186–203.

Lareo, Inés; Monaco, Leida Maria; Esteve-Ramos, María José and Moskowich, Isabel (comps.). 2020. *The Corpus of English Life Sciences Texts (CELiST)*. A Coruña: Universidade da Coruña. https://doi.org/10.17979/spudc.9788497497848

Mele-Marrero, Margarita. 2019. "Neither I nor we. Inexplicit authorial voice in eighteenth century academic texts". In Moskowich, Isabel; Crespo, Begoña; Puente-Castelo, Luis and Monaco, Leida Maria (eds.), *Writing History in Late Modern English: Explorations of the Coruña Corpus*. Amsterdam: John Benjamins. 215–234. https://doi.org/10.1075/z.225

Moskowich, Isabel. 2016. "Pronouns as stance markers in the Coruña Corpus: an Analysis of the CETA, CEPhiT and CHET". In Alonso-Almeida, Francisco (ed.), *Stancetaking in Late Modern English Scientific Writing. Evidence from the Coruña Corpus*. València: Universitat Politécnica de València. 73–91.

Moskowich, Isabel. 2021. "The making of CELiST, a bunch of disciplines". In Moskowich, Isabel; Lareo, Inés and Camiña, Gonzalo (eds.), *"All families and genera": Exploring the Corpus of English Life Sciences Texts*. Amsterdam/Philadelphia: John Benjamins. 1–19. https://doi.org/10.1075/z.237

Myers, Greg. 1989. The pragmatics of politeness in scientific articles. *Applied Linguistics*, 10/1: 1–35. https://doi.org/10.1093/applin/10.1.1

Parapar, Javier and Moskowich, Isabel. 2007. The Coruña Corpus Tool. *Revista del Procesamiento de Lenguaje Natural*, 39: 289–290.

Puente-Castelo, Luis. 2017. On conditionality: A corpus-based study of conditional structures in Late Modern English scientific texts. Unpublished PhD Dissertation. A Coruña: Universidade da Coruña.

Quirk, Randolph; Greenbaum, Sidney; Leech, Geoffrey and Svartvik, Jan. 1985. *A comprehensive grammar of the English language*. London: Longman.

Salager-Meyer, Françoise. 1994. Hedges and textual communicative function in medical English written discourse. *English for Specific Purposes*, 11/2: 93–113. https://doi.org/10.1016/0889-4906(94)90013-2

Sinclair, John. 1981. "Planes of Discourse". In Rizvi, S. N. A. (ed.), *The Two-Fold Voice: Essays in honour of Ramesh Mohan*. Saltzburg: Universität Saltzburg. 70–89.

Stubbs, Michael. 1996. *Text and Corpus Analysis: Computer-Assisted Studies of Language and Culture*. Oxford: Blackwell.

Swales, John. 1990. *Genre analysis: English in academic and research settings*. Cambridge: Cambridge University Press.

Sweetser, Eve. 1990. *From Etymology to Pragmatics*. Cambridge: Cambridge University Press. https://doi.org/10.1017/CBO9780511620904

Taavitsainen, Irma and Pahta, Päivi. 1998. Vernacularisation of medical writing in English: A corpus-based study of Scholasticism. *Early Science and Medicine*, 3/2: 157–185.

Tang, Ramona and John, Suganthi. 1999. The 'I' in identity: Exploring writer identity in student academic writing through the first person pronoun. *English for Specific Purposes*, 18: S23–S39. https://doi.org/10.1016/S0889-4906(99)00009-5

Thompson, Geoff and Yiyun, Ye. 1991. Evaluation in the reporting verbs used in academic papers. *Applied Linguistics*, 12/4: 365–382. https://doi.org/10.1093/applin/12.4.365

Warchal, Krystyna. 2010. Moulding interpersonal relations through conditional clauses: Consensus-building strategies in written academic discourse. *Journal of English for Academic Purposes*, 9: 140–150. https://doi.org/10.1016/j.jeap.2010.02.002

CHAPTER 11

"This ingenious hypotheſis hath a great appearance of truth"

The expression of true facts in the Corpus of English Life Sciences Texts

María José Esteve-Ramos
Universitat Jaume I

1. Introduction and objectives

This chapter addresses the way scientists of the Enlightenment made their assumptions and new theories sound more truthful and reliable. Many resources were at hand for this purpose, but the use of adverbs – and more specifically adverbs of certainty – was one of them, and this study aims at showing that the writers of these important works used them to support their attempts to sound truthful, in a society which was the receptor of new and unprecedented knowledge, as part of newly established rules of a more clear and plain English.

Experimental disciplines gradually developed in the period known as Late Modern English, coinciding with the ideological currents and newly promoted methods of the so-called Age of Reason. Specialised discourse begins to acquire its most distinctive features, and the expression of assertion appears as a reinforcement of the individual contribution to science. Life Sciences, however, present a line of discourse that moves between a more humanistic approach in describing nature and a more analytical one when it comes down to explaining the more experimental part. In both cases, e.g. when referring to the description of new species or when presenting new results, the writer will necessarily address the reader in order to pursue the belief that what is hitherto being exposed is a true fact–persuasively, in line with classical rhetorical goals. Following the same approach as in previous research (Esteve-Ramos and Lareo 2019), this chapter looks into how adverbs of certainty are used in scientific texts, and compares the results with the ones already obtained in the Corpus of English History Texts (CHET).

https://doi.org/10.1075/z.237.11est

There have been many studies describing the language of science of this period, and some of the claims come straight from contemporary authors, as Crespo (2011: 201) writes:

> Following Boyle's claim in the second half of the seventeenth century that scientific knowledge had to be transmitted in a concise and unornamented way, with experiments to be reported as objectively as possible, the language used to convey observations on nature by the new men of science gradually acquired the characteristics of clearness and simplicity of expository prose.

However, the scientific and/or academic prose of the period presented some differences from equivalent present-day practices. In this chapter, as already stated above, we are interested in showing how adverbs play an important part in signalling the intention of the writer as to the ultimate purpose of convincing the reader of the true elements of his/her discourse. One of the reasons may be found in what Crespo (2011: 202) explains: "(…) the scholar had to address both two groups of the epistemic community: other authors and learners. He had to be convincing about his theories and findings for both elements in his target audience." But as she continues: (2011: 202) "The reliability, trustworthiness and indeed veracity of facts depended on other mechanisms used by the author, such as claiming witnesses or the use of reference to other works to support statements, rather than expecting them to be taken at face value". This chapter shows some of the distinctive rhetoric contemporary science-writers deployed to persuade their readers of their intellectual *bona fides*.

In order to tackle with the data – which will be dealt with in the next two sections – we need to contextualise the emergence of these texts in the appropriate historical and disciplinary context. The discipline of Biology did not exist as such, as it is explained in chapter 1 in this book. Moskowich (this volume) points out that it is not until the end of the eighteenth century that we can find a definition of Biology which entails "The branch of science that deals with living organisms as objects of study (…) (OED)". It has been the methodological practice in the conception and work of the Coruña Corpus that disciplines have been limited and named using the classification of sciences established by the UNESCO as a starting point. However, it is clear that a twenty-first-century delimitation of sciences does not correspond to eighteenth-century ones, and the case of Life sciences is very much a good example of this. William Hunter's collection, kept at the University of Glasgow, is a fine example. The decision was finally taken considering that the disciplines (…) are certainly varied, although related in some ways, at least in that they all deal with the study of life and organisms. You could then make a link to the emergence of separate disciplines that is such a feature of nineteenth-century science (Moskowich 2021). This subcorpus, CELiST (Lareo *et al.* 2020), encompasses many subjects related to the natural world, which would find their own independent way as the period moved along, becoming new specialised areas of knowledge. In a way, this will be a reflection of the new scenario where the fundamentals of contemporary science would emerge.

In order to demonstrate the role of adverbs in eighteenth-century scientific rhetoric, this chapter is organised as follows: Section 2 presents the theoretical background and lists a selection of examples attesting this hypothesis for each of the selected adverbs. Section 3 presents a table with full information of the data obtained, and organises the material so that other factors as age, adverb type or sex can be further interpreted in Section 4, where a brief analysis of the most outstanding evidence will be carried out. Finally, a preliminary analysis is offered, followed by a conclusion and some further research lines to be pursued.

2. On adverbs and examples: Expressing true facts in Life Sciences

Adverbs and their evolution have been studied from different perspectives and in different periods (Conrad and Biber 2000; Downing 2002 or Simon-Vanderbergen and Aijmer 2007), where their presence was an important marker of discourse. Simon-Vanderbergen (2007: 69) has been already a relevant source for the beginning of this enquiry, where it helped frame the previous study based on the history sub-corpus (Esteve-Ramos and Lareo 2019). This study gave us an interesting insight, although their research was not based on historical corpora. It relates to the semantic field of modal certainty and their base for selecting these is explained like this: "that the adverbs should express the speaker's strong commitment to the truth of the proposition" (Simon-Vanderbergen and Aijmer 2007: 69). The appearance in the recent decades of more research related to pragmatics – and more specifically, historical pragmatics – cannot be diminished as it is fundamentally linked to the interpretation of language within a specific context. The appearance of pivotal work reflected in many publications (Jucker and Taavitsainen 2010 or Taavitsainen, Jucker and Tuominen 2014, among others) launched a new impulse. More recently, seminal works like the latest by Jeremy J. Smith (2020) puts forward new "reimagined" (as he himself labels it) frameworks of understanding this wider conception of language study. On this we will rely on Section 4 for the interpretation of the results.

As in Esteve-Ramos and Lareo (2019), the theoretical framework for the current study is Quirk′s *et al.* (1985: 615, 627). We decided again to focus on type A, "degree of truth" adverbs, group (i) for the sake of comparison with CHET, the Corpus of History English Texts (Moskowich *et al.* 2019), as this constitutes the main objective of this chapter. However, group (iii) of Quirk's classification also relates to the initial question, and should be also included in further studies. This additional survey would give a more complete description of this linguistic and pragmatic feature in the corpora studied, albeit we do have to restrict this study to type A due to the limitation in the chapter extension:

Group i. Express conviction, either as a direct claim (e.g. *undeniably*) or as an appeal to general perception (e.g. *evidently*). The adverbs included in this group are:[1] "*admittedly* (0), *assuredly* (4), *avowedly* (0), *certainly* (41), *decidedly* (4), *definitely* (6), *incontestably* (1), *incontrovertibly* (0), *indeed* (84), *indisputably* (0), *indubitably* (0), *surely* (3), *unarguably* (0), *undeniably* (0), *undoubtedly* (0), *unquestionably* (1), *clearly* (30), *evidently* (23), *manifestly* (5), *obviously* (7), *patently* (0), *plainly* (27), *undeniably* (0).

Not all the adverb types searched show occurrences, but for those that do, we have selected a set of illustrative examples, thirteen in total:

(1) Why here should be the two modes – why this re-invigoration should be required on the part of the female element we do not know; but it is most *assuredly* the fact, and it is presumable, that, however long the process of asexual multiplication could be continued, – I say there is good reason to believe that it would come to an end if a new commencement were not obtained by a conjunction of the two sexual elements (Huxley 1863: 84)

(2) As to a hole being left in the side for the young to be fed and void their excrements through, as Pennant and others relate, it is *certainly* an error: I have never met with any thing of the kind in the nest of the Baltimore. (Wilson 1808: 25)

(3) That all mammalia, besides the human subject, posses this bone, is not so *decidedly* ascertained, as that man has it not. (Lawrence 1819: 174)

(4) A strong support for this hypothesis would be afforded were it to be *definitely* established that a polar body is not formed in the Arthropoda and Rotifera. (Balfour 1880: 63)

(5) The unanimous reports of all travellers, as well as the specimens of such animals exhibited in Europe, prove *incontestably*, that the manlike simiæ, whether the orang-utang of Borneo, or chimpansé of Angola, as well as the long-armed monkey or gibbon, are widely different from the human subject in this respect." (Lawrence 1819: 185)

(6) This ingenious Hypothefis hath a great Appearance of Truth, and is *indeed* the beft that can be given in a Northern Climate, where it is evident that their Food can only be found in Summer Months: (…) (Hughes 1750: 74)

1. This sequence includes all the adverbs listed in each group and type by Quirk *et al.* (1985). The numbers in brackets represent the number of tokens (instances) found in CELiST.

(7) With reſpect to the ſynonyms, which made [Mr]. Davies uncertain about it, they are *unqueſtionably* wrong, as the ſynonyms of Wulfen (otherwiſe an excellent botaniſt) too often are. (Smith 1795: 246)

(8) But I will make it appear that they are ſo far from proving any thing againſt this Doctrine of Secretions, that they are the greateſt Arguments that could poſſibly be urged for the truth of it. Nothing does more *evidently* demonſtrate the Intentions of Nature in her Operations than the various Methods ſhe is ſometimes forced to take to bring the ſame thing about. (Keill 1717: 132)

(9) If they are both like the same thing, they ought *manifestly* to be like each other! The best authorities, however, tell you that it is quite different (Huxley 1863: 109)

(10) The anterior and posterior limbs in each member of the vertebrate and articulate classes are *plainly* homologous. We see the same law in comparing the wonderfully complex jaws and legs in crustaceans (Darwin 1859: 436)

(11) I know you will approve this proposal; for, though some inattentive observers have despised the science as trifling and useless, they have exposed their ignorance by such an opinion. *Surely*, nothing can be more suited to a creature endowed with reason, than to investigate the works of God. (Wakefield 1816: 2)

(12) On my view of characters being of real importance for classification, only in so far as they reveal descent, we can *clearly* understand why analogical or adaptive character, although of the utmost importance to the welfare of the being, are almost valueless to the systematist (Darwin 1859: 427)

(13) Its importance is increased by the discovery by Strasburger of the existence of an analogous process amongst plants. Two questions about it *obviously* present themselves for solution: What are the conditions of its occurrence with reference to impregnation? (Balfour 1880: 63)

It becomes clear that these examples show the preoccupation of the writer for attaining a persuasive, convincing role on the reader. Section 4 below will offer an analysis.

3. Evidence in context: data from CELiST

Data are drawn from the Corpus of English Life Sciences Texts (CELiST), the subcorpus of the Coruña Corpus to which this book is devoted. This subcorpus has 400,305 words. The word count for eighteenth-century material is 200,220 and the number of words for the nineteenth century is 200,085. For each century twenty text samples by different authors are provided, and when necessary, numbers have been normalised to 10,000 words for comparison. Table 1 presents a summary of the findings:

Table 1. Main findings

Date/words	Sex	Author	Age	Genre	Type/ tokens	Adverbs/ tokens
1707/10005	M	Douglas, James	32	Guide	plainly/1	1
1707/10002	M	Sloane, Hans	47	Treatise	plainly/3 certainly/1	4
1717/10014	M	Keill, James	44	Essay	indeed/3 evidently/5 certainly/2	10
1720/10000	M	Gibson, William	40	Guide	plainly/2 indeed/1 clearly/1	4
1723/10018	M	Blair, Patrick	43	Treatise	plainly/2 indeed/3 manifestly/1	6
1743/10012	M	Edwards, George	49	Treatise	plainly/1 indeed/1 certainly/1	3
1750/10005	M	Hughes, Griffith	43	Treatise	plainly/1 indeed/1	2
1752/10006	M	Dodd, James Solas	31	Essay	plainly/3 indeed/4 aſſuredly/1 manifestly/1	9
1758/9998	M	Borlase, William	62	Treatise	plainly/2 indeed/6 certainly/1	9
1766/10004	M	Pennant, Thomas	40	Treatise	indeed/1 certainly/2	3
1769/10009	M	Bancroft, Edward	25	Letter	indeed/3	3
1774/10035	M	Goldsmith, Oliver	44	Treatise	plainly/1 indeed/4	5
1776/10003	M	Withering, William	35	Guide	evidently/1 clearly/3	4
1786/10017	M	Speechly, William	63	Treatise	indeed/1 certainly/1	2
1794/10019	M	Donovan, Edward	26	Guide	clearly/1 certainly/1	2
1795/9999	M	Smith, Sir James Edward	36	Catalogue	unqueſtionably/1 surely/1 indeed/3 evidently/2	7
				Total		**74**

Table 1. (*continued*)

Date/words	Sex	Author	Age	Genre	Type/ tokens	Adverbs/ tokens
1804/10051	F	Jacson, Maria Elizabeth	49	Lecture	indeed/1 evidently/2 certainly/1	4
1808/10008	M	Wilson, Alexander	42	Treatise	indeed/5 evidently/2 certainly/3	10
1816/10008	F	Wakefield, Priscilla	66	Letter	surely/1 plainly/2 clearly/3 certainly/1	7
1819/10000	M	Lawrence, William	36	Treatise	indeed/4 incontestably/1 decidedly/1 clearly/2 certainly/1 manifestly/1	10
1824/9771	M	Jenner, Edward	74	Article	surely/1 plainly/1 obviously/1 indeed/10 clearly/3 certainly/2	18
1828/10011	M	Godman, John Davidson	34	Treatise	obviously/1 indeed/1 evidently/1 certainly/2	5
1832/10013	F	Lincoln, Almira Hart Phelps	39	Textbook	indeed/1	1
1835/10007	M	Jardine, Sir William	35	Treatise	plainly/1 indeed/2 certainly/1	4
1840/10021	F	Pratt, Anne	34	Guide	indeed/2 certainly/1	3
1848/10010	M	Dalyell, Sir John Graham	73	Treatise	obviously/1 indeed/3 evidently/3 definitely/1 certainly/4	12
1859/10035	F	Agassiz, Elizabeth	37	Letter	indeed/2	2
1859/10014	M	Darwin, Charles Robert	50	Treatise	plainly/4 indeed/2 clearly/5 assuredly/2 certainly/3 manifestly/1	17

(*continued*)

Table 1. (*continued*)

Date/words	Sex	Author	Age	Genre	Type/ tokens	Adverbs/ tokens
1863/10048	M	Huxley, Thomas Henry	38	Lecture	indeed/8 assuredly/1 certainly/6 manifestly/1	16
1867/10001	M	Spencer, Herbert	47	Treatise	obviously/1 indeed/5 definitely/1 clearly/2	9
1876/10023	M	Macalister, Alexander	32	Textbook	obviously/1 indeed/1 certainly/1	3
1879/10020	F	Lankester, Phebe	54	Guide	indeed/3 evidently/1 decidedly/1 clearly/1	6
1880/10004	M	Balfour, Francis M	29	Treatise	obviously/2 definitely/3 clearly/2 certainly/4	11
1889/10019	M	Galton, Francis	67	Treatise	indeed/3 decidedly/1 clearly/3	7
1895/10020	F	Gregory, Emily Lovira	54	Textbook	plainly/2 clearly/3	5
1898/10001	M	Packard, Alpheus Spring	59	Textbook	plainly/1 evidently/6 definitely/1 decidedly/1 clearly/1 certainly/1 certainly/1	12
					Total: 236	**162**

This table shows date of publication, number of words of the sample, sex of the authors, their name and results with the specification of which adverbs and with what frequency they appear. All these different aspects deserve particular attention. After the searches we have found a total of 236 certainty adverbs in the 400,305 words under analysis. When looking at the data in some more detail, we can see that these forms are distributed differently according to century and sex of the author as set out in Figure 1, where tokens have been normalised per 10,000 in order to offer a valid source for comparison.

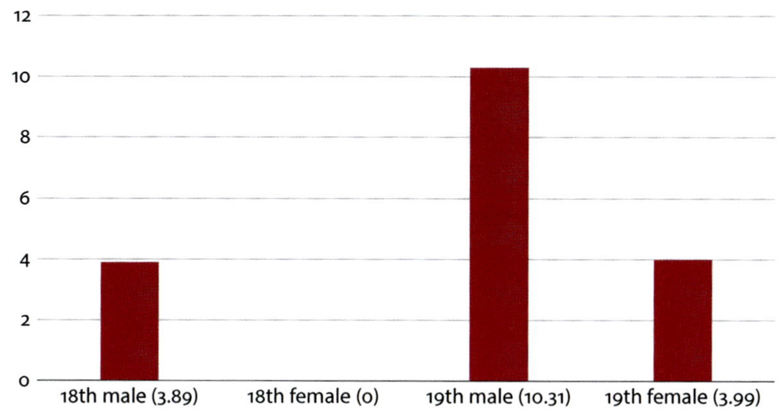

Figure 1. Distribution of certainty adverbs per sex and century (nf)

Figure 1 above is interesting for different reasons. For the variable of time, our previous study related to the discipline of History revealed a small difference in the Corpus of History English Texts (CHET), showing that the use of these type of adverbs in the nineteenth century had been slightly higher in comparison with the eighteenth-century. In CELiST, this tendency is not only confirmed, but presents a sharper difference. It has already been suggested that, with the appearance of more structured disciplines and the need for debate in the nineteenth century, there may have been a reason for these results, and here this is clearly visible. This fact is a very interesting finding that will deserve further discussion.

As for the sex variable, results in CHET showed women to be less assertive than men when using this type of linguistic resource. Results in CELiST show this breach even more clearly: no cases of the adverbs under study have been found in the only female author present in the eighteenth century in this subcorpus, but the results in male authors in the nineteenth century almost double the ones occurring in the texts written by female authors, according to the normalised figures. The role of female scientists in this period was a supporting one, concentrated on more practical aspects of life connected with their domestic duties, as Moskowich (2013: 469) explains: "Men at the time were credited as being more able to engage in abstract reasoning, with women considered to be suited to the more practical, immediate aspects of life". Following this idea, we see that for the eighteenth century we have a female author, Elizabeth Blackwell, writing *A Curious herbal* in 1737. In the same fashion, texts written by women are mostly related to herbs, plants and gardens. In the following century, where women are more visible in this corpus, the topic has not varied much. Maria Elizabeth Jacson wrote *Botanical Lectures. By a Lady*; Priscilla Wakefield authored *An Introduction to the Natural History and Classification of Insects*, and Almira Hart Phelps Lincoln published *Familiar lectures on Botany*. Anne Pratt, *Flowers and their Associations*, Elizabeth Agassiz, *A First Lesson in natural history*, Phebe Lankester, *Wild Flowers worth notice: a selection of some of our native plants* and Emily Lovira Gregory, *Elements of Plant Anatomy*.

Not all types of text bore the same function as to the transmission and dissemination of science. Moskowich explains that:

> Given that women themselves were aware of the difficulties of getting a formal education, they tended to concern themselves instead with the weakest group of the population (children, other women) for whom the obstacles to instruction were greatest. They were interested in the dissemination of knowledge, and this perhaps explains the fact that they tended to choose the kinds of genre/types that were aimed at providing either formal or non-formal education in simple terms.
> (Moskowich 2013: 469)

Herbals, Botany and plant description may not abound in theories to be discussed, but more clearly show descriptions and perhaps some medical advice which would justify in this subcorpus the low number of occurrences.

As to the age in which the material of the corpus was written, Figure 2 offers a vision of the data obtained.

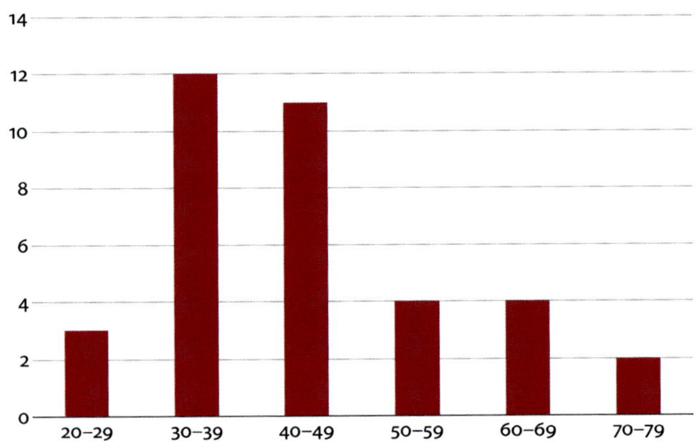

Figure 2. Distribution of certainty adverbs according to age of authors

The age of the authors when the texts were published is an interesting variable; also, how meaningful this may be in order to interpret the use of this type of adverbs in context, is what we will try to ascertain in the next lines. CHET results showed – contrary to our initial expectations – that older authors were trying to be more assertive and convincing. In CELiST, however, as we can see in Figure 2, occurrences are more in line to what might have been expected, and authors in their early mature years (already with some prestige and ready to debate and discuss theories and new approaches to knowledge) would be the ones using this type of adverbial elements in a more recurrent way, with the aim of enhancing discussion and making their positions clear.

Another aspect that the corpus is providing is the identification of the different types of texts in which the forms under study appear. This is important in this particular period, as each of them was written in order to reach different audiences, and their use represents an important way of understanding how scientific knowledge spread and was being built at the time. Figure 3 shows this information in normalised figures as per 10,000 words, and Table 2 includes the results per genre.

Table 2. Distribution according to genre

Text-type	Number of words	Tokens found	nf
Article	9,771	18	18.42
Lecture	20,099	20	9.95
Essay	20,020	19	9.49
Treatise	180,176	119	6.60
Textbook	50,074	21	4.19
Letter	30,052	12	3.99
Guide	60,068	20	3.33
Catalogue	30,045	7	2.33
Total	**400,305**	**236**	**5.89**

According to these results, Article is the genre showing the highest nf (18.42), followed by Lecture (9.95) and Essay (9.49). On the contrary, the lowest nf is shown by the category Catalogue (2.33) and Guide (3.33), more synthetic in their nature. Textbooks (4.19) are written by two men and two women, with tokens in male authors more than twice as many as in female ones (15 tokens vs. 6 tokens). It is also interesting to show that all 18 tokens occurring in the genre Article belong to only one author: Edward Jenner. The reason for this will be expanded in Section 4. The same happens with the category Catalogue: the 7 tokens only occur in the work of Sir James Edward Smith. In CHET, Treatise was the genre showing fewer results and Lecture was the one with highest number of occurrences. In this sense, the two subcorpora belong to different perspectives in science, and whereas in History the more humanistic and letters-oriented texts predominate, CELiST encompasses a combination of experimental material, more adequate for a treatise. Their audiences may be different as well, affecting the results obtained.

Also interesting is to observe whether there is any prevalence in the use of any specific adverb. In Figure 3 we can observe the number of examples obtained for each of them:

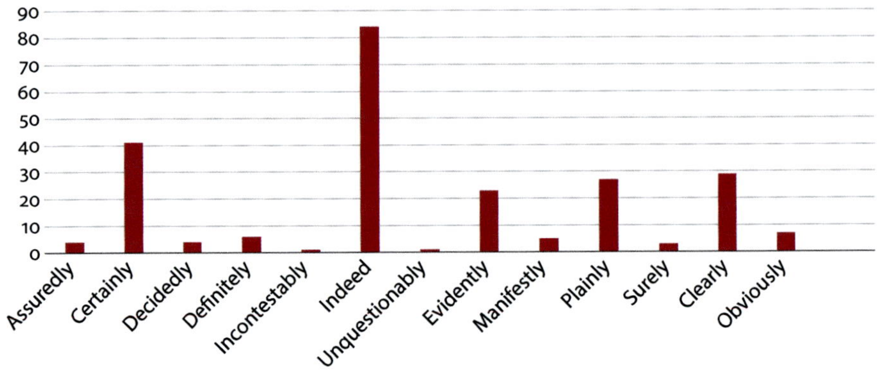

Figure 3. Distribution of certainty adverbs according to adverb-type

The most frequent type has been *indeed* as Figure 3 shows, coinciding with our previous research in CHET. Also, in both cases, *certainly* has been the second. The most frequent ones in CELiST are *indeed, certainly, clearly*, and also *plainly*. To explore further, we have focused on the analysis of specific examples in context, which will be carried out in the next section.

4. "My views amount to the following": A brief analysis of the examples

This subcorpus of the Coruña Corpus contains many further examples which, due to the limited extension of this chapter, we will not be able to discuss in detail. However, sufficient information has already been presented to show some distinctive features of eighteenth- and nineteenth-century scientific writing in English. Moreover, as we shall now see, the metadata files included in the corpus and available through the Coruña Corpus Tool (CCT) provide valuable insights for the interpretation of the results, in line with recent trends in historical pragmatics:

> Overall, nevertheless, the application of corpus linguistics to historical pragmatics has produced valuable results, as the possibility of making assumptions on the frequency and distribution of particular constructions over time and text-types provides sound, or at least, falsifiable, evidence for claims on linguistic change.
> (Rica and Visconti 2014: 133)

And two leading authors in the field have argued that:

> The availability of historical corpora and other electronic resources together with the increasing sophistication of corpus-linguistic tools has opened up a whole range of new research paradigms, and it has become possible for historical pragmaticists to probe into research questions that could not have been answered before.
> (Taavitsainen and Jucker 2015: 13)

As Smith (2020: 11) argues:

> Historical pragmatics is a mode of analysis that is not descriptive but also explanatory. Hitherto, most very valuable work in historical pragmatics has focused on corpus-analysis, especially of grammatical and lexical features.

Before focusing on the specific examples, a brief summary of the occurrences will help to contextualise them and give us further interesting insights on the results. As we will be having a look at individual authors and all the samples contain the same number of words, no normalising of frequencies is needed, and we will deal with raw numbers instead. Thus, we can see that the highest number of occurrences in the eighteenth-century group appears in the text by Keill, an essay entitled *Essays on several parts of animal oeconomy*, with a total of 10 examples. Keill is followed by Dodd, with 9 examples in his *An essay towards a Natural History of the Herring* and Borlase, also with 9 items appearing in his work *The Natural History of Cornwall*. With no uses of certainty adverbs (which is something meaningful as well) we have four works: Thomas Boreman (*A description of three hundred animals*); Elizabeth Blackwell (*A Curious Herbal, containing five hundred cuts of the moſt uſeful plants, which are now uſed in the practice of Physick*); John Brickell (*The Natural History of North Carolina*) and James Bolton (*An History of Fungusses, growing about Halifax*), being those classified as textbook, treatise and two catalogues. The last ones coincide with the results included in Table 2 in which Catalogue is the genre at the bottom of the scale.

As we move on to the nineteenth century we can observe, as set out above in Figure 2, that the increase of occurrences is relevant. All the texts published between 1800 and 1900 produced results, which goes in line with the increase of items, and marks a shift that can be interpreted as a tendency which is most remarkable and characteristic of this century, as opposed to the previous one. The author that has shown more occurrences in this century is Jenner, with his work *Some observation on the migration of Birds*. In order to understand this result, the metadata, where information on the work and life of the author are available, is key to understand the number of occurrences, as the title may have indicated a descriptive text and the presence of so many adverbs of this type would not have been easy to explain. According to the information in the metadata, Jenner was famous for being a pioneer of small vaccination, working on disease prevention and experimenting on immunity. In times of the coronavirus, he would have been in his element, experimenting on a vaccine which is still on its way. He had a passion for Natural History that led him to study Ornithology. In 1824, in his posthumous publication, he wrote about bird migrations and more interestingly, he helped to prove this phenomenon, contradicting the current theory at the time that believed that birds hibernated. He could prove this through dissection, and demonstrated that migratory behaviour had relation with changes in the reproductive system. This information contextualises clearly why this author was resorting to this type of adverbs in such a frequent manner.

Darwin's hugely influential work *On the Origin of Species by means of Natural Selection, or the Preservation of favoured Races in the Struggle for Life* comes up with 17 occurrences. This century is flooded with new and debatable theories, accounting for the number of examples in Table 1, and in which theories on the evolution of men and other species were at the centre of this scientific burst. Also, Huxley shows an interesting number of results, 16, with his work *On the Origin of Species, or the Causes of the Phenomena of organic nature: a course of six lectures to working men*. Both Darwin and Huxley were close at the time, both belonging to a circle of secularist and meritocratic scientists in the city of London which had the support of advanced liberals. Huxley was also a remarkable figure of its time, in the swirl of all these new theories related to the natural world. About his writing style, we find in the metadata that he was declared to be "perhaps the greatest virtuoso of plain English who has ever lived." His use of certainty adverbs may have contributed to this impression.

Let us now move on to look at some of the examples in detail. By using the CCT, we can access not just the basic searches which show the node and a few words collocating left and right, but we do also see the sample where these results are taken from. In a study such as this, having the token in its broader context proves to be key for the interpretation of the data.

In Section 2 above selected examples of each type of adverb were included. In number (1), we find additional markers that help to attain this sense of addressing the readership, with the intention of persuading them of the truth:

(1) but it is most *assuredly* the fact (…) – I *say* there is good reason to *believe* that (Huxley 1863: 34)

The verbs *say* and *believe* support this idea. Example number (6) refers directly to the word truth:

(6) This ingenious Hypothefis hath a great Appearance of *Truth*, and is *indeed* the beft (…). (Hughes 1750: 74)

On the contrary, the same resource is used for the denial of the Truth:

(7) (…) they are *unqueftionably* wrong. (Smith 1795: 246)

Example (8), again, refers to the truth of the arguments:

(8) (…) that *they are the greateft Arguments that could poffibly be urged for the truth of it*. Nothing does more *evidently* demonftrate the Intentions of Nature (…). (Keill 1717: 132)

More interestingly, in Example (9) we do find an element of punctuation in the same sentence as the adverb, fulfilling a complementary rhetorical function:

(9) If they are both like the same thing, they ought *manifestly* to be like each other!. (Huxley 1863: 109)

In number (11), we find the author addressing her audience directly:

(11) *I know you will approve this proposal;* for, though some inattentive observers have despised the science as trifling and useless, they have exposed their ignorance by such an opinion. *Surely*, nothing can be more suited (…). (Wakefield 1816: 2)

Finally, we think it is interesting to refer to the five examples resulting in the search for *evidently* in the text written by Keill. On the right context just after the adverb we find the verbs: *demonſtrate, appear, demonſtrate, prove, ſhews.* These verbs intensify and indicate a clear scientific ethos, related to a discourse that aims at proving, ascertaining what is being said in an attempt to show the reader rigour and objectivity.

Among all the information we have gathered, we would like to focus particular attention in one author. To this purpose we have produced further searches, examples of which are not numbered as they were not listed in the samples in Section 2. Darwin and his publications *On the Origin of Species* were, no doubt, a milestone in the history of scientific thought, responsible for bringing in a new epistemological dimension in the field of Natural Sciences. Our interest and expectations would be, therefore, addressed to attest how adverbs of certainty could contribute to the language of persuasion he may have resorted to. Interesting examples confirm that there was a necessity to address his audience. The instances using the adverb *clearly* can attest this. It is also noteworthy that he uses verbs such as *see* or *understand* as a direct reference to the readers:

(14) We shall hereafter, I think, *clearly* <u>see</u> why embryological characters are of such high classificatory importance. (Darwin 1859: 426)

(15) We can *clearly* <u>see</u> how it is that all living and extinct forms can be grouped together in one great system; and how the several members of each class are connected together by the most complex and radiating lines of affinities. (Darwin 1859: 433)

(16) On my view of characters being of real importance for classification, only in so far as they reveal descent, we can *clearly* <u>understand</u> why analogical or adaptive character, although of the utmost importance to the welfare of the being, are almost valueless to the systematist. (Darwin 1859: 427)

And it continues again to use this same adverb in another example, where the pursuit of truth is directly mentioned:

(17) As the embryonic state of each species and group of species partially shows us the structure of their less modified ancient progenitors, we can clearly see why ancient and extinct forms of life should resemble the embryos of their descendants, – our existing species. Agassiz believes this to be a law of nature; but I am bound to confess that I only hope to see the law hereafter proved true. (Darwin 1859: 449)

In Example (18) we find a direct reference to the audience:

(18) I have now given the leading facts with respect to rudimentary organs. In reflecting on them, *every one must be struck with astonishment:* for the same reasoning power which *tells* us *plainly* that most parts and organs are exquisitely adapted for certain purposes, *tells* us with equal plainness that these rudimentary or atrophied organs, are imperfect and useless. (Darwin 1859: 453)

As we have pointed out in our previous study, it is clear that persuasion and the – sometimes direct – appeal to the reader's attention, may not be achieved solely by using this type of adverbs, but these elements expressing conviction powerfully signal this intention on the writer's side. In some examples, the expression of the adverb and the combination with other verbs and elements (such as the exclamation mark) recreates the opening of a silent conversation with the reader. As Smith (2020) writes:

> After all, a spoken or written text at any given point in time is a communicative event, the interface between speaking/writing and listening/reading. And unlike most forms of speaking before the period of speech-recordings, writing has communicative functions over space and time, with all the constraints – and openness to change – those functions imply.

In this study it is clear that we are bringing forward a dialogue between writer and reader, as the use of these adverbs of certainty could be directly addressing the reader with the ultimate aim of convincing him/her; they would have, thus, an almost performative function in some cases. One of the most promising results is found in the presentation of *The Origin of Species* by Darwin, which would deserve further research. The combination of the adverbs with some other elements such as performative verbs and those related to orality, such as *tell* or *speak*, show an author who is intentionally trying to bridge to his audience, in the way of a direct conversation with them, to convince them further of the arguments he is using; in this sense, I see it as a potential thread of interest that could be further developed.

5. Conclusion

After discussing the main aspects of the study, we can conclude that certainty adverbs in CELiST present thought-provoking results that confirm the initial hypothesis: certainty adverbs constitute a visible and frequent element used by the authors included in this subcorpus. As to the main conclusions regarding the different parameters analysed, we can say that in terms of age, mid-career authors in their early maturity presented to highest number of examples, coinciding with a moment in which they would be debating and discussing their proposals. The time variable shows that the nineteenth century clearly has a much higher number of occurrences, perhaps to the more structured division of disciplines and the need for debate. As for genre, women are more visible in the nineteenth century as well, but fewer examples appear, perhaps due to the fact that their texts were related to Botany and plant description. These topics used a more descriptive language and therefore had less need for the use of conviction adverbs. Finally, as regards genre, we find that articles, lectures and essays were the preferred genres. The reason for this would be that it was adequate as it could combine experimental material along with more descriptive passages.

In the light of the evidence shown above, further studies should be conducted in order to obtain more solid conclusions, but the material reflects the intentional use of adverbs when addressing the readers with the purpose of reaffirming the truth of what is being exposed; it establishes a connection with the audience, and seeks to contribute to the necessary belief in true facts. These were times of fierce scientific debate, and bold new theories were in swing – as the case of Darwin's *Origin of Species* – so language was necessarily the great, powerful and necessary medium for standing to them. It was the means where they could defend their beliefs, and different mechanisms such as certainty adverbs were used to enhance such sense of *auctoritas*.

Acknowledgements

I would like to thank Prof Jeremy J. Smith for advice and comments.

Works cited

Conrad, Susan and Biber, Douglas. 2000. "Adverbial marking of stance in speech and writing". In Hunston, Susan and Thompson, Geoff (eds.), *Evaluation in Text: authorial stance and the construction of discourse*. Oxford: Oxford University Press. 56–73.

Crespo, Begoña. 2011. Persuasion markers and ideology in eighteenth century philosophy texts (CEPhiT). *Revista de Lenguas para Fines Especificos*, 17: 199–228.

Esteve-Ramos, María José and Lareo, Inés. 2019. "A corpus-based study of some certainty adverbs in the *Corpus of History English Texts*". In Moskowich, Isabel; Crespo, Begoña; Puente-Castelo, Luis and Monaco, Leida María (eds.), *Writing History in Late Modern English*. Amsterdam: John Benjamins. 168–183. https://doi.org/10.1075/z.225.09est

Downing, Angela. 2002. "*Surely you knew!*" *Surely* as a marker of evidentiality and stance. *Functions of Language*, 8/2: 253–285.

Jucker, Andreas and Taavitsainen, Irma. 2010. *Historical Pragmatics*. Berlin: Mouton de Gruyter. https://doi.org/10.1515/9783110214284

Lareo, Inés; Monaco, Leida Maria; Esteve-Ramos, María José and Moskowich, Isabel (comps.). 2020. *The Corpus of English Life Sciences Texts (CELiST)*. A Coruña: Universidade da Coruña. https://doi.org/10.17979/spudc.9788497497848

Moskowich, Isabel. 2021. "The making of CELiST, a bunch of disciplines". In Moskowich, Isabel; Lareo, Inés and Camiña, Gonzalo (eds.), *"All families and genera": Exploring the Corpus of English Life Sciences Texts*. Amsterdam/Philadelphia: John Benjamins. 1–19. https://doi.org/10.1075/z.237

Moskowich, Isabel. 2013. Eighteenth-century female authors: women and science in the Coruña Corpus of English Scientific Writing. *Australian Journal of Linguistics*, 33/4: 467–487. https://doi.org/10.1080/07268602.2013.857570

Moskowich, Isabel; Lareo, Inés; Lojo Sandino, Paula and Sánchez-Barreiro, Estefanía (comps.). 2019. *Corpus of History English Texts*. A Coruña: Universidade da Coruña. https://ruc.udc.es/dspace/handle/2183/21849

Quirk, Randolph; Greenbaum, Sidney; Leech, Geoffrey and Svartvik, Jan. 1985. *A Comprehensive Grammar of the English Language*. New York: Longman.

Ricca, Davide and Visconti, Jacqueline. 2014. "On the development of the Italian truth adverbs davvero and veramente". In Taavitsainen, Irma; Jucker, Andreas and Tuominen, Jukka. 2014. *Diachronic Corpus Pragmatics*. Amsterdam: John Benjamins. 133–154. https://doi.org/10.1075/pbns.243.09ric

Simon-Vanderbergen, Anne-Marie and Aijmer, Karin. 2007. *The Semantic Field of Modal Certainty*. Berlin: Mouton de Gruyter. https://doi.org/10.1515/9783110198928

Smith, Jeremy J. 2020. *Transforming Early English. The Reinvention of Early English and Older Scots*. Cambridge: Cambridge University Press. https://doi.org/10.1017/9781108333474

Taavitsainen, Irma; Jucker, Andreas and Tuominen, Jukka (eds.). 2014. *Diachronic corpus pragmatics*. Amsterdam/Phildelphia: John Benjamins. https://doi.org/10.1075/pbns.243

Taavitsainen, Irma and Jucker, Andreas. 2015. Twenty years of historical pragmatics. *Journal of Historical Pragmatics*, 16/1: 1–27. https://doi.org/10.1075/jhp.16.1.01taa

CHAPTER 12

Evaluative *that* structures in the Corpus of English Life Sciences Texts

Francisco Alonso-Almeida and Francisco J. Álvarez-Gil
University of Las Palmas de Gran Canaria

1. Introduction

Evaluation seems to be a key term in applied linguistics, especially in the analysis of academic and technical discourse (Mauranen and Bondi 2003; Stotesbury 2003; Jalilifar, Hayati and Don 2018). Scholars write their findings and their estimations concerning their research objects using a variety of discourse devices, some of which contribute to revealing their own position with respect to their texts. These devices can be lexical and grammatical. Biber and Finegan (1989: 97ff), for instance, examine some categories of stance, namely positive and negative affect, and certainty and doubt in evidentiality. While affect involves the study of adjectives, verbs and adverbs primarily, evidentiality is evaluated by means of adjectives, verbs and modal verbs. Semantic items in the study of evidentiality included in Biber and Finegan's study include such categories as emphatics and hedges. In our paper, we focus on one language strategy, i.e. *that*-structures, whose evaluative dimension has been acknowledged earlier in the literature, e.g. Hyland and Tse (2005a, 2005b), Hyland and Jiang (2018b), and Kim and Crosthwaite (2019). These studies will be central to our methodology of research.

The use of *that*-clauses appears to be a rich source for designating authorial perspective (cf. Charles 2007). The range of perspectives, as we shall see in due course, varies according to the matrices preceding the *that*-clauses, and these may, therefore, include such notions as authority, tentativeness, attribution and affectivity, among others. In our study, we will focus on these aspects specifying and evaluating viewpoints in the discourse of Life Sciences in Late Modern English, as found in the texts in the Corpus of English Life Sciences Texts (CELiST) (Lareo *et al.* 2020). We will furthermore take advantage of the potential of this corpus to discriminate data according to metadata variables, and we will try to detect variation according to the variable of sex. In this context, our research questions are the following: (a) what the commonest stancetaking language forms are used in the

https://doi.org/10.1075/z.237.12alo

texts analysed, (b) what information sources appear to be the preferred ones, hence revealing also the degree of authorial involvement, (c) the level of commitment and authority towards the information presented the authors take in the presentation of knowledge, and (d) the main differences concerning the use of these in the texts of men and women.

This chapter is structured as follows. In Section 2, we present a definition of stance, and we describe evaluative *that*-complement clauses and their potential as a stancetaking device, mainly after Hyland and Tse (2005a, 2005b) and Hyland and Jiang (2018b). The categories explained in these studies will be used in our analysis and interpretation of the *that*-structures identified in CELiST. Section 3 includes a description of the methodology concerning the interrogation of our corpus and the analysis of data. The results will be presented in the subsequent section, along with a discussion of examples illustrating the different evaluative categories of the *that*-structures found in CELiST. Finally, Section 5 offers the conclusions drawn from our study.

2. Stance and evaluative-*that*

Stance in discourse is related to perspectivisation. Biber *et al.* (1999: 966) consider this phenomenon as the expression of the speakers' "personal feelings, attitudes, value judgements, or assessments". From here, we can see that stance is a multifaceted concept involving interpersonal linguistic cues, as pointed out in Clift (2006), which altogether disclose the position of speakers (or writers). The study of stance involves such notions as evaluation (Martin 2000; Mauranen and Bondi 2003), evidentiary justification (Chafe 1986; Marín-Arrese 2011; Estellés and Albelda Marco 2018), affectivity and social relations (Hyland 2005a; Abdollahzadeh 2011; Wetherell 2013), authority, including from a gender perspective (Kendall 2004; Fox 2008), and mitigation (Hyland 1998; Hyland 2005b; Caffi 2007; Alonso-Almeida 2015), among other aspects.

The idea of attitude is put forward in the notion of stance by Hyland (2005a: 176), while Johnstone (2009) stresses the interpersonal function of certain stancetaking devices: "Stance is generally understood to have to do with the methods, linguistic and other, by which interactants create and signal relationships with the propositions they utter and the people they interact with" (2009: 30–31). This definition frames adequately our study of evaluative *that* in CELiST, as it considers the relationship held between writers and their texts and writers and their audience. Developing from Halliday's interpersonal function aimed at "enacting our personal and social relationships with the other people around us" (Halliday and Matthiessen 2014: 30), Thompson (2001: 73) explains the significance of evaluation in the process of argumentation. The interpersonal perspective is also taken in Crismore and Farnsworth (1989); Vande Kopple (2002); Hyland (2005b); Abdollahzadeh (2011) and Rozumko (2019), among others.

This is the view in Hyland and Tse (2005a) in order to explain the use of evaluative *that*-structures in scientific and academic discourse. According to these authors, these structures are "[p]erhaps one of the least noticed of these interpersonal" (2005a: 40) although the rhetorical function of these structures is evident in discourse, especially in those genres with a clearly distinct persuasive target. From a formal perspective, these structures present a matrix to which the *that*-clause is attached. The form of these matrices is varied and may present either a copula or a lexical verb, e.g. *We believe…, I think…, Some people say…* In semantic terms, Hyland and Tse (2005a: 40) label the matrix as evaluation, also containing the source of evaluation, and the *that*-clause is referred to as the evaluated entity, as signalled in the following instance:

(1) [evaluation & source of evaluation →] *I will show* [evaluated entity →] *that the biofunctional account can give us more content specifity than Fodor supposes.* (Hyland and Jiang 2018b: 140)

The evaluation may also be realised by a noun or an adjective which provides the evaluation of the entity, as in the following instances in (2) and (3) from Hyland and Tse (2005b: 125) and Hyland and Tse (2005a: 42), respectively. The noun *evidence* in (2) and the adjective *possible* describe different degrees of epistemic meaning concerning the evaluated entity given in the *that*-clauses.

(2) …there is *evidence* that traders may reduce their cost from trading by splitting orders so as to dampen the pressure on inventory holding.

(3) …it is *possible* that mainstream teachers overcompensate and are especially lenient with NNSS.

These statements can be categorised according to the (a) entity evaluated, (b) the author's stance, (c) the source of evaluation and (d) the formal aspect in which the evaluation is given, according to Hyland and Tse (2005a: 46ff). These same categories are given in Hyland and Tse (2005b) and our description also includes additional subcategories incorporated in Hyland and Jiang (2018b: 145–146).

These authors provide subcategories for these groups of *that*-structures. In the case of the entity evaluated, they include "an interpretation of the writer's claim; of the content of previous studies; of research goals; and of the research methods, models, or theories that had been drawn on" (Hyland and Tse 2005a: 46). Another subclass in this group is "common or accepted knowledge" given in Hyland and Jiang (2018b: 146). In short, the function of these is to evaluate the propositional content framed by the matrix, e.g. *The authors confirmed that the method of enquiry was carefully selected in all cases*. In this case, the evaluation applies to an interpretation of the method of research. Authorial stance is divided into attitudinal and epistemic stance concerning the propositional content framed by the lexical matrix along with a neutral subcategory, e.g. *It means that…*, as found in Hyland and Jiang (2018b: 145–146). The former concerns aspects of affect and obligation and the latter inform on differing degrees of probability as to the likelihood of the proposition to happen, and Hyland and Tse (2005a: 46) consequently see aspects of truth and accuracy affecting the proposition modulated.

In the case of the source of evaluation, the categories included come from the reworked version in Hyland and Jiang (2018b) and innovation deals with a distinction concerning the human source of information provided in the *that*-structure. In this respect, the human attribution of source can be either the author (*I*) or a third party (*they*). We have added a third category, i.e. the author and colleagues (*we*) so that we can specify those cases in which the subject or subjects of conception may report on subjective or intersubjective meanings. The source of evaluation can also be abstract entities (e.g. *the method revealed…*) or even concealed sources, which denote opaque conceptualisers (e.g. *it is generally known…*). In the cases of unidentified sources of evaluation, the authors claim that this might be a deliberate means to avoid responsibility for the claim made (2005a: 46). The last category, i.e. evaluative expression, classifies the evaluation given in the matrix according to its form into non-verbal predicate, namely noun and adjectival predicates, and verbal predicates, namely research acts (e.g. *the data show that…*), discourse acts (*the authors say…*), and cognitive acts (*the author thinks…*). This threefold division shows similarities with Marín-Arrese's model for the analysis of epistemic stancetaking devices (Marín Arrese 2009; Marín-Arrese 2011). This model has been used for the analysis of evidentials in seventeenth-century texts by women to reveal evaluation of propositional content, and characterise, therefore, women's style in technical discourse (cf. Alonso-Almeida and Mele-Marrero 2014).

Kim and Crosthwaite (2019) follow this model for the study of evaluative *that*-structures in Hyland and Tse (2005b) with some minor modifications. Kim and Crosthwaite's model presents the subcategory of "anticipation of reader's claim" (2019: 5) in the category of evaluative entity. In addition, they divide Hyland and Tse's subcategory of "evaluation of research methods, models and theories (2005b) into (a) evaluation of methods and (b) evaluation of models, theories and hypothesis. This level of attention is surely an interesting option for Kim and Crosthwaite's analysis of academic texts, but we will follow the categories put forward in Hyland and Tse (2005a, 2005b) including the additional categories in Hyland and Jiang (2018a) in our examination of *that*-structures in Life Sciences texts in Late Modern English.

3. Method

We have used the Coruña Corpus Tool (CCT) (cf. Barsaglini-Castro and Valcarce (2020)) for our interrogation of *that*-structures in CELiST. We are aware that these expressions may appear with an elision of the item *that*, but, as pointed out in Hyland and Tse (2005b: 130), the tendency in formal writing is to keep the form *that* and variation in this respect is not significant. Our enquiry returned all cases of *that* in the corpus, including adjectival and pronominal uses. Computerised searches cannot accurately isolate exact instances of evaluative *that*-structures, especially in a corpus that has not been tagged for part-of-speech categories. The process has, therefore, had to be done manually in order to identify evaluative *that* samples, only. Our analysis of findings has also taken into consideration the variable of sex (see Moskowich (2013); Crespo and Moskowich (2015) and Crespo (2015) for analyses concerning the involvement of women in science matters in Modern English). Comparison of results considering this criterion has been made with a normalisation of data to 10,000 words. The number of words in CELiST amounts to 400,305 words. The group of those texts written by men contains 320,118 words, and the group of texts written by women presents 80,187 words.

4. Evaluative *that*-expressions: Analysis and results

We have identified 1,176 cases of evaluative *that*-structures in CELiST through the manual analysis of all cases of *that* retrieved from our corpus interrogation of this term. As expected from the uneven distribution of words per sex, the compilation of texts written by men appears to have more instances of this structure, i.e. 981 cases, than the compilation of texts written by women, i.e. 196 cases. The normalisation of data to 10,000 words reveals that the difference is apparently less marked, namely 30.6 cases in the texts written by men and 24.4 in the texts written by women. These last figures reflect clear variation in the use of evaluative *that*-structures with slightly higher frequency leaning on the side of the group of texts written by men. The calculation of the loglikelihood ratio (LL) evinces this difference with a ratio of 8.7 (significance level $p < 0.05$), with overuse in the group of men's writings. A more straightforward measure of variation is gained through the log-ratio difference analysis (cf. McEnery and Baker 2016: 23). Hardie (2014) explains that a log-ratio analysis shows how significant the difference existing between two groups of texts may be, as it is an "effect-size" statistic, as Hardie points out. In our particular case, the variation difference between the use of the evaluative *that*-structures in texts by women and texts by men is 0.3, which indicates low significance in this respect, as the binary log-ratio is below 1. This means that, being the log-ratio so close to zero, the use of evaluative *that*-structures between the set of texts written by men and the set of texts written by women is really very similar.

Figure 1 reflects the distribution of evaluative *that*-structures per subcategory:

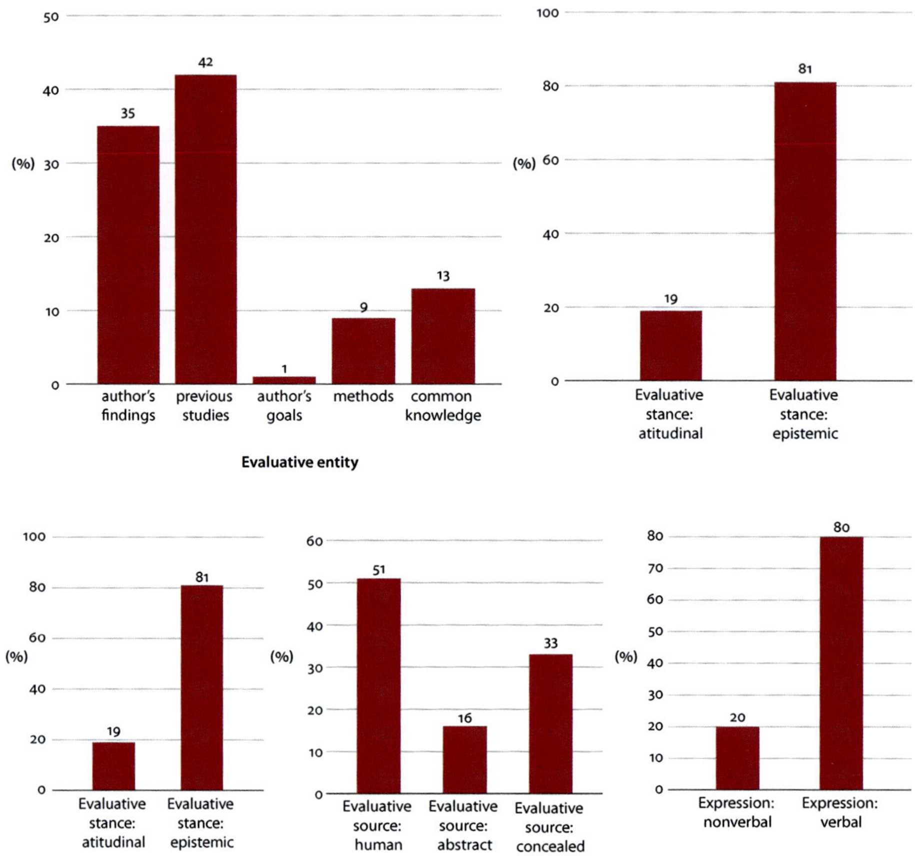

Figure 1. Distribution of subcategories of *that*-structures in CELiST

Epistemic devices, human attribution of authorship and verbal expression are the preferred choices in CELiST. In what follows, we include a description of these categories used for the analysis of *that*-structures, namely (a) evaluative entity, (b) evaluative stance, (c) evaluative source, and (d) expression. We also provide an indication of frequency and relative variance per sex along with a description of the language associated with these categories in our corpus.

4.1 Evaluative entity

The distribution of evaluation entity types in CELiST per sex category is given in Figure 2. As already stated, the subcategories included in this group refer to the entity evaluated, and so they mean to provide an evaluation of the writer's claim, an evaluation of the content of previous studies, an evaluation of research goals and of the research methods, models, or theories, as pointed out in Hyland and Tse (2005a: 46), along with an evaluation of common or accepted knowledge; this last included in Hyland and Jiang (2018b: 146). These subcategories were clearly designed by Hyland and Tse (2005a) to be applied to a corpus of scientific articles, and this means that some are genre-specific. For this reason, the number of *that*-structures identified to belong to any of these categories amount to 613. The texts written by men show more cases of all these subcategories, i.e. 550 cases (nf 17.1) vs. 63 (nf 7.8) in the case of the texts written by women. For the subcategory "interpretation of author's goals", no one single case has been identified in the set of texts written by women in CELiST.

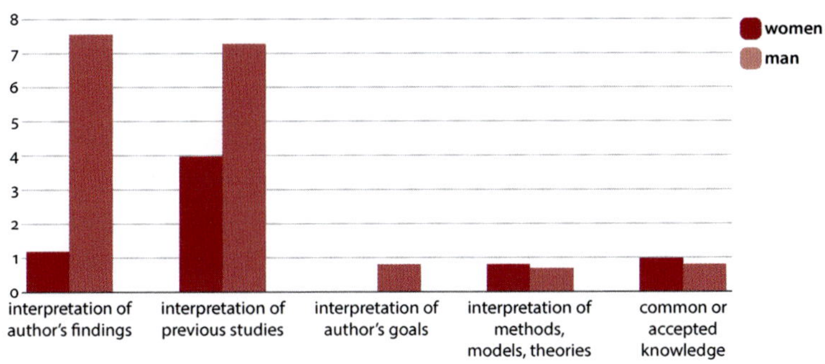

Figure 2. Distribution of evaluative entity per sex category, normalised data

The following are instances for each subcategory, excerpted from the corpus (underlining indicates the *that*-structure):

(4) As *evidence appears to me conclusive, that* the several domestic breeds of Pigeon have descended from one wild species (Darwin 1859; man); interpretation of author's findings.

(5) Some years since the celebrated [Mons]. De Reaumur [note] Memoires de l'Academie Royale des Sciences, & c. 1714. [endnote] communicated to the Royal Academy of Sciences at Paris, a Paper, in which he undertook to demonstrate that the shock of. The Torpedo was the effect of a stroke… *it will be no ways difficult to demonstrate, that* the whole of [M]. de Reaumur's pretended discovery is a perfect non-entity (Bancroft 1769; man); interpretation of previous studies.

(6) …although *we may perhaps learn something experimentally but that* at present we are an enormous distance from the goal I indicated (Huxley 1863; man); interpretation of the research goals.

(7) …and *it is only since the science of chemistry has proved that* the active principles of these vegetable productions are the same in many different plants and can be extracted from chemical processes (Lankester 1879; woman); interpretation of methods, models, theories.

(8) *It has even been stated on good authority that* rudiments of teeth can be detected in the beaks of certain embryonic birds (Darwin 1859; man); common or accepted knowledge.

4.2 Evaluative stance

Evaluative stance refers to authorial perspectivisation of knowledge. The analysis of cases found in CELiST shows the results given in Table 1. Attitudinal evaluative stance presents more cases in the set of texts written by women. Indeed, the log-likelihood ratio clearly indicates overuse in the case of this group of texts, and the log-ratio suggests little difference between texts written by men and by women. In the case of epistemic evaluative stance, men seem to show more cases of epistemic devices than women, as seen in the normalised data in this table. The log-likelihood ratios confirm these higher frequencies in the men's texts. In terms of how significant that difference is, the log-ratio values indicate that the use of epistemic neutral evaluative strategies appears to be more than 2 times more common in the case of those texts written by men.

Table 1. Distribution of evaluative stance per sex category, normalised figures; log-likelihood ratio

Evaluative stance				
		Man	**Woman**	**LL – log-ratio**
attitudinal	affect	1.5	3.7	0.09 –0.12
	obligation (deontic)	1.5	3.7	0.01 –0.02
epistemic	doubt	12.0	5.4	6.15 +0.56
	certainty	8.9	7.9	–25.97 –0.43
	neutral	20.9	13.3	54.39 +2.22

4.2.1 *Attitudinal stance*

The expression of affect to evaluate propositional content in CELiST is more likely to occur in texts written by women. The language normally associated with affect includes adjectives and participles, e.g. *reasonable, remarkable, surprising, striking, important*, and verbs indicating emotion or expectation, e.g. *feel, fear, hope*, as the following instances illustrate:

(9) *It is remarkable that* the jaws of insects do not meet, like those of quadruped and birds, crosswise, but lengthwise. (Wakefield 1816; woman)

(10) *I hope that* the sea-anemone has interested you so much, that you will like to hear about some other animals of the same kind, which live also in the sea, and of which I have a strange and wonderful story to tell you. (Agassiz 1859; woman)

(11) *It is not surprizing that* the milk of this animal is so salutary, as it brouzes only on the tops, tendrils and flowers of the mountain shrubs, and medicinal herbs; rejecting the grosser parts. (Pennant 1766; man)

The use of obligation markers is also more common in the texts by women, although the log-ratio value indicates that the difference between men and women is actually very low, really not significant, i.e. 0.02. Obligation is rare as such in technical terms unless it is understood in terms of necessity. For this, we can find such matrices containing either the words *necessity* or *necessary* as, for instance, in the template 'it BE necessary that…'. The commonest way to indicate deontic meaning in CELiST is by means of modal verbs, e.g. *shall, will, should* and *must*, as shown in the following examples:

(12) …yet if we reflect a moment *we shall see that* very many of our own wants are the result of habit. (Lincoln 1832; woman)

(13) …but *we should with a good magnifying-glass endeavour to show that* each one of the little white strap-shaped petals. (Lankester 1879; woman)

(14) …*I must confess that* I have never seen one of either that to my apprehension, could properly be so distinguished. (Speechly 1786; man)

(15) *I should tell you that* all pigeons have a tendency to do this at times, but in the Pouter it is carried to an enormous extent. (Huxley 1863; man)

4.2.2 *Epistemic stance*

Epistemic stance indicates the evaluation of the proposition in terms of the likelihood of the event described to happen. This is achieved through devices indicating doubt or complete assurance. In the case of those devices conveying doubt, they present

a. verbs indicating deduction and inferential processes, e.g. *suppose, seem (to/that), appear (to/that), estimate, believe, conjecture, imagine, perceive, assume*
b. adjectives and adverb in matrices, e.g. *improbable, probable, unlikely, perhaps, probably*, which may be accompanied by intensifiers reinforcing the author's perspective, e.g. *highly* and *very*, and
c. modal verbs, e.g. *may, might, would*, and *could.*

The function of these strategies is to hedge the information presented after the *that*-structure, as seen in the examples below:

(16) After all that has been said, *it is not improbable that* these creatures so essentially different from any other class of animated beings, possess senses of which we have no idea. (Wakefield 1816; woman)

(17) There being such quantities of recent shells on the coasts of Cornwall, *it might reasonably be expected that* extraneous and fossil-shells, such as pectunculi, conchites, ammonitæ, and other remains of marine testaceous animals should be found in proportionable plenty in stones, quarries, and mines. (Borlase 1758; man)

The expression of certainty in our corpus seeks to suggest confidence in the information provided. This is especially realised by means of matrices including adjectives and adverbs in the realm of clarity, truthfulness and obviousness, e.g. *obvious, certain, evident(ly), sure, true, undoubtedly*. Sometimes this is also manifested by the use of the verb *assure*. The following are instances in this group:

(18) *It is evident that* the shelter of trees is not necessary to it, since it may be occasionally be found in pastures. (Pratt 1840; woman)

(19) *It is most certain, that* they hide themselves in the most Secret Places. (Blackwell 1737; woman)

The neutral category includes those matrices containing verbs of communication and visual perception, which allows evaluating how the information has been obtained, as in these instances:

(20) ...*we find that* the rhizoids of the two plants are of equal rank, as well in respect to their place and manner of origin, as in their structure. (Gregory 1895; woman)

(21) *It often happens that* they perforate the cuticle and wound the true skin, in which case, large and thick warts are formed. (Godman 1828; man)

4.3 Evaluative source

This category provides information which may be indicative of authorial accountability and of the degree of reliability one object or abstract entity may be granted. The evaluative sources can be human, an abstract entity or a concealed subject. As shown in Table 2, human sources are preferred in relation to the other two, and women seem to use these referential devices more often than men. The log-ratio value, i.e. 0.10, indicates however that the difference according to the category of sex is not meaningful. Difference significance is certainly perceived in the set of texts written by men for the category of abstract or inanimate source with a score of 2.39. The log-likelihood ratio is also relevant, i.e. 48.72, in this case with overuse in the group of texts written by men. The difference in the use of concealed (not identified) sources with a log-ratio of 0.21 (more common in those texts by men) suggests very small variation between men and women texts. The fact that none of the works are co-authored makes significant the distinction between the first-person singular and first-person plural established here. In a general outline of evaluative *that*-structure, the distinction should be kept with a twofold perspective, one to implicate the authors and another to involve the author(s) and other people.

Table 2. Distribution of evaluative sources per sex category, normalised figures; log-likelihood ratio. Human subcategories are given in percentages

Evaluative source				
		Man	**Woman**	**LL – log-ratio**
human		24.6	26.1	0.42 –0.10
	author	26.8	15.7	
	other humans	42.3	58.3	
	we	30.9	26.1	
abstract entity – inanimate source		13.1	2.5	48.72 +2.39
concealed – source not identified		18.1	15.7	1.19 +0.21

4.3.1 *Human -subjective*

The human category represents attribution made by the author(s), alone or as a member of a group, or made by a third party. As seen in Table 2, texts written by men and by women tend to show third-person informants in their presentation of knowledge. This is followed by first-person plural references in the case of texts by men. Women have an equal preference for the first-person singular and first-person plural acknowledgement. In the case of the texts written by men, the first-person singular is the least frequent.

The following represents instances attributed to the authors, indicating thus a subjective perspective of the information provided. In these cases, the matrices accurately indicate the authors' involvement in the elaboration of information, although a side effect may also designate the authors' commitment to the truth of the propositional contents:

(22) *I observe, that* the Plant is strong, and the Stalk thick and marrowy. (Blair 1723; man)

(23) *I have sometimes thought that* the shape of our county might have contributed in a great measure to deprive us of these curiosities. (Borlase 1758; man)

(24) *I have been told that* there is a gold-striped Pine, [viz]. a pine with yellow stripes. (Speechly 1786; man)

(25) *I have said that* the coral animals grow in clusters. (Agassiz 1859; woman)

(26) *I may, however, say that* the leaves are entire, varying from narrow oblong to nearly oval. (Lankester 1879; woman)

The opposite also holds, and attribution to third parties may imply a lack of involvement in the information presented, and accountability is also external to the author. The language associated with the matrices preceding the *that*-structures is frequently verbs of perception and communication, e.g. *say, report* and *observe*, as in the following examples:

(27) *Pliny reporteth, that* when they encounter and fight with rats, they use Rue as a preservative against their Bite. (Brickell 1737; man)

(28) *Serapius, an Arabian Physician, says that* Spinage creates Wind; so that those who are troubled Chollic had better not eat it. (Blackwell 1737; woman)

(29) *Tournefort says, that* about paris they use it in Fomentations for Pains and Swellings of Haemorrhagies. (Blackwell 1737; woman)

(30) *Authors agree, that* juſt before their Approach to the North Seas the Air and Water give Tokens. (Dodd 1752; man)

(31) *[Dr]. Stokes justly remarks ([Bot]. [Arr].) that* Jacquin's Selinum sylvestre, figured in [Flo]. Austriaca, is this pant, and not the sylvestre of Linnæus. (Smith 1795; man)

(32) *Linnæus observed that* the wood anemone expanded in Sweden at the same time as the swallow returned from its migration. (Pratt 1840; woman)

Intersubjectivity is frequently deployed in technical and scientific writing to indicate shared responsibility as to the information presented. This is achieved by the form *we*, as in the instances in (33)-(36). As said above, this represents the second most preferred attribution human device in the set of texts written by women:

(33) *We know that* in many instances beasts and birds accommodate themselves to circumstances, which marks discrimination and superiority of intellect. (Wakefield 1816; woman)

(34) Although *we may suppose that* many things now necessary to our comfort and even our lives, are rendered so by nature; yet if we reflect a moment we shall see that very many of our own wants are the result of habit. (Lincoln 1832; woman)

(35) *Finally, we have seen that* natural selection which results from the struggle for existence, and which almost inevitably induces extinction and divergence of character in the many descendants from one dominant parent-species, explains that great and universe feature in the affinities of all organic beings. (Darwin 1859; man)

(36) *We see that* in insects ring distinctions have given way to regional distinctions. (Packard 1898; man)

4.3.2 *Abstract entity–inanimate source*

Abstract entities and inanimate sources in this category include terms frequently describing processes and means of research and communication as *observation, microscope, investigation, account, arguments, supposition, report, experience, evidence,* and *proof,* among others. The evaluative scope of these terms is undeniable, as they may make manifest the author's idea concerning the degree of reliability and factuality the described phenomenon represents. In the following instances, the degree of confidence concerning the information conveyed depends on the evaluation made of the source entities. It seems clear that the information collected using technological means, i.e. *the microscope* in (37), is more trustworthy than an impressionistic conclusion. The use of *experience* and *investigations* are also used with a mitigating force to support the information. Another conclusion following from these sources obviously corresponds to the notion of accountability in relation to the manifestation of source, as the author may thus avoid the responsibility for the claims, since a particular procedure can be subject to verification, as in the cases of (37) and (39). The use of *our experience* in (38) reports on an intersubjective positioning, hence reducing the amount of responsibility.

(37) …but *the microscope shows that* they are composed of very numerous close-set joints. (Wakefield 1816; woman)

(38) …*our experience shows that* this valuation has hitherto been arbitrary. (Darwin 1859; man)

(39) Regarding the development of the bundles in the embryo, *present investigations show that* in some cases the bundle of the cotyledon runs down in the axillary one of the root. (Gregory 1895; woman)

4.3.3 *Concealed–source not identified*

Concealing the subject of inception may mean the avoidance of responsibility for the information given. Unidentifiable sources frequently appear in the pattern 'it BE + ADJ (*obvious, necessary, curious, certain*)'. Another set of common structures in this category is communicative, experiential and cognitive evidential matrices given in the passive voice, and may contain embedded adverbials infusing the idea of an intersubjective perspective as to the claim made, e.g. *it is (generally) thought, it is (generally) supposed, it is well known, it seems (generally) allowed, it is seen, it is reported, it must be confessed, it is said, it may be said.* Another frequent structure is the use of *it* followed by evidential *seem to/that* or evidential *appear to/that*. The following examples illustrate all this:

(40) ...and *it is found that* the hog eats but 72 and refuses 171 plants. (Pennant 1766; man)

(41) Sweet William (dianthus barbatus) has its flowers in that species of head which is called a fascicle though *it seems that* the mode in which the flowers of sweet William are put together, would place it more properly under the term Cyme than Head. (Jacson 1804; woman)

(42) *It is evident that* the shelter of trees is not necessary to it, since it may occasionally be found in pastures. (Pratt 1840; woman)

4.4 Expression

The forms exhibited by *that*-structures are non-verbal and verbal. The former includes noun and adjective predicates, and the latter involves research, discourse and cognitive acts in Hyland and Tse's (2005a, 2005b) and Hyland and Jiang's model (2018b). As shown in the following table, verbal *that*-structures are more frequent than non-verbal ones. The distribution of these forms in CELiST is different per category, including some significance according to the variable of sex, as we shall show in what follows.

Table 3. Distribution of expression types per sex category, normalised figures; log-likelihood ratio

		Expression		
		Man	**Woman**	**LL – log-ratio**
non-verbal	noun predicate	2.2	0.9	22.61 +2.12
	adjective predicate	4.2	3.9	6.8 +0.88
verbal	research acts	10.7	7.0	9.71 +0.62
	discourse acts	5.6	6.1	0.21 –0.11
	cognitive acts	6.3	8.2	3.38 –0.38

4.4.1 *Non-verbal predicates*

Nonverbal predicates include noun or adjective complement clauses. These two types, which we shall explain in the following sections, present some variation in the set of texts written by men in relation to the set of texts written by women with a log-likelihood ratio of 22.61 with overuse in the set of texts by men. The log-ratio value, i.e. 2.12, indicates that noun predicates are more than 4 times more common in the texts written by men.

4.4.1.1 *Non-verbal noun predicates*

Non-verbal noun predicates refer to those nouns followed by *that*-structures which contribute to the presentation of information and of its epistemic status. As may follow from Jiang and Hyland's study on stance nouns in disciplinary writing (2015), noun complementation may help communicate the author's position concerning the reality described in terms of certainty, which to some extent may involve notions of tentativeness. This structure may also be helpful to reveal factual truth and to seemingly reveal an intention to communicate authority. In general, this structure has a strong potential to define and to present the current status of information, whether they represent actual truth (or partial truth), or they indicate relative conclusions from ongoing research. This has the obvious implication of allowing the evaluation of the contribution being made to the current state of affairs of the discipline, and so they enable the perspectivisation of information. The use of *his inference that, matter of fact that* and *this principle* indicates authority both from a deductive process in (43) and from a formalised description of reality in the case of (44) and (45).

(43) *I am more convinced of the fact by his inference, that,* since it is well known that many insects have the have the power of uttering sounds, as the large beetle, the bee, the wasp, the gnat, and the fly, we cannot suppose that this power is given without a corresponding capacity in their fellows to understand their meaning. (Wakefield 1816; woman)

(44) *This therefore is matter of fact that* the blood consists of a simple and limpid Fluid in, in which swim Corpuscles of various Figures and Magnitudes, and endued with different Degrees of an attractive force. (Keill 1717; man)

(45) From *this Principle that* the blood consists of corpuscles of various figures and magnitudes, and endued with various Degrees of an attractive Power, and that of such Particles the Fluids secerned by the Glands are composed. (Keill 1717; man)

4.4.1.2 *Non-verbal adjective predicates*

Non-verbal adjective predicates provide an open evaluation of the propositional content and its truth, as in the examples in (46) and (47). In the case of the latter, the adjective is additionally reinforced by the evidential adverb *undoubtedly*. Other adjectives with this function in CELiST report on epistemic uses, e.g. *(im)probable, (im)possible, likely, certain, true, apparent, evident*, deontic uses, e.g. *necessary*, and attitudinal uses, e.g. *remarkable, plain, strange, improper.*

(46) …it will be *obvious that* the sizes must be very various between these two extremes. (Jacson 1804; woman)

(47) we may be allowed to suppose what is *undoubtedly true that* the shock of the Torpedo and that of the Torporific Eel, are both communicated in a familiar manner, and by similar means, it will be no ways difficult to demonstrate, that the whole of [M]. de Reaumur's pretended discovery is a perfect non-entity. (Bancroft 1769; man)

4.4.2 *Verbal predicates*

Verbal predicates are extremely useful to indicate how information is elaborated, obtained or communicated, and most importantly, they make clear authorial involvement in the formulation of new or existing knowledge. Verbal predicates can be of three types, namely (a) research acts in the real world, (b) discourse acts, and (c) cognitive acts. The normalised figures in Table 3 above show that these three types of acts are more frequent in the case of the group of texts written by men. This is also supported by the log-likelihood values, all of them with a significance level $p < 0.05$. These values indicate that research acts are likely to appear in the texts written by men and discourse and cognitive acts are likely to appear in the texts written by women. The log-ratio values, however, reveal that the use of these verbal predicates is actually similar.

4.4.2.1 *Research acts in the real world*

Research acts, i.e. "experimental activities or actions carried out in the real world" (Hyland and Tse 2005b: 130), include verbs indicating achievement (*demonstrate, show*), procedure (*decide, apply*), finding (*find, happen*) and visual inspection (*see, observe*), among others, as exemplified in the following:

(48) *Later experiments have proved that* plants derive sustenance from the various ingredients which compose the soil, and also from the atmosphere. (Pratt 1840; woman)

(49) …thus *we see that* though the outer surface of the common Cowry, or Tide Shell, is of a pale colour with dark spots, when that is taken off, it is of a fine violet colour. (Donovan 1794; man)

(50) ...but *Böttcher has demonstrated that* a free nucleus or a nucleus removed from a cell contracts actively. (Macalister 1876; man)

4.4.2.2 *Discourse acts*

Discourse acts are characterised by the presence of verbs in the domain of speech, including verbs of argumentation and report. Examples in CELiST include cases of *say, tell, affirm, report, assert, claim, mention, answer,* and *inform*, among others. The following are excerpts from the corpus:

(51) ... *in his account of Ireland reports that* such was the richness of the pastures in that kingdom. (Pennant 1766; man)

(52) *Pliny asserts that* it never blooms except when the winds blow. (Pratt 1840; woman)

4.4.2.3 *Cognitive acts*

Cognitive acts include verbs designating mental activities, and these include such notions as thought, reasoning, perception, deduction, inference, awareness and understanding. The verbs in the matrices in this category in CELiST are *think, know, suppose, believe, imagine,* and *consider*. The evidential forms of *seem* and *appear* belong to this category, as in (54) and (55). This category of *that*-structure clearly reveals the author's intellectual involvement in the elaboration of meaning, especially deductive processes and learning, as seen in (53), (56) and (57):

(53) *Shall we suppose, too, that* by chance the prevailing colour of vegetation is green, that by chance, the prevailing colour of vegetation is green, that colour upon which, above all, the eye rests with the most agreeable sensations?. (Lincoln 1832; woman)

(54) *It appears that* the root absorbs them from the soil, by which it is nourished. (Lincoln 1832; woman)

(55) *It seems, however, that* the communication exists only through the puncta lachrymalia, and that that this opening is connected only with a glandular apparatus. (Jardine 1835; man)

(56) *I presume that* the "bastard-wing" in birds may be safely considered as a digit in a rudimentary state. (Darwin 1859; man)

(57) Yet *we know that* inventors often proceed by strides, their ideas originating in some sudden happy thought suggested by a chance occurrence. (Galton 1889; man)

5. Conclusion

The analysis of evaluative *that*-structures has shown that Life Sciences texts in the Modern English period exhibit a number of *that* complement clauses leading to provide different types of information concerning authorial communicative intention. This analysis has considered the variable of sex in order to highlight possible variation. The general conclusion in this respect is that even if some variation has been attested in texts written by men with respect to those written by women, there seems to be an intention to use similar stylistic choices in the scientific output of either sex, as evinced by log-ratio values in the majority of *that*-structure categories. Still, certain statistically significant differences have been found, as in the categories of evaluated entity, for instance.

Evaluative stance devices in the texts in CELiST are deployed to indicate perspectivisation of knowledge. The texts written by women are likely to present more cases of attitudinal affect and obligation devices than men texts, as follows from a variance analysis. It is not rare to find obligation markers in scientific texts, as these devices are used to indicate necessity as well as expert authority in the sense of signalling procedures and research paths, for instance, as pointed out in Álvarez-Gil (2021). Epistemic stance devices are, however, more common in the texts authored by men, mainly for hedging purposes and to signal the status of the information presented. In the case of certainty markers, these suggest confidence and conviction as to the truth of the propositional content framed by the epistemic matrices. As to the category evaluative source, our findings show a tendency to use human entities, either the author or third parties. Subjective claims are used in both sets of data, although they appear more frequently in the texts written by men. In the texts written by women, evidence supports an intention to rely more on intersubjective claims in order to gain greater epistemic support in their scientific community.

The use of abstract entities and concealed entities has outstanding evaluative potential. These devices are deployed to attribute knowledge to a source other than the author. This clearly suggests an intent to avoid authorial responsibility for the truth of the propositions presented. As to the form of the matrices, verbal predicates are preferred over non-verbal. The latter show overuse in the case of those texts by men. Women seem to prefer communicative and cognitive matrices in order to frame propositional content in their utterances. This way, they show their involvement in the construction, reception and dissemination of information. Conversely, texts written by men appear to show more cases of research acts to describe how information has been gained. This tendency clearly reflects a statistical difference between men and women writing in CELiST. Overall, the distinctions mentioned concerning the formal appearance of *that*-structures in CELiST may suggest a clear inclination to favour a cognitive versus a procedural way of making science and scientific claims, perhaps reflecting the contemporary situation of women in the scientific community. The claims here stated should be contrasted against the other subcorpora included in the Coruña Corpus.

Works cited

Abdollahzadeh, Esmaeel. 2011. Poring over the findings: Interpersonal authorial engagement in applied linguistics papers. *Journal of Pragmatics*, Elsevier B.V. 43/1: 288–297. https://doi.org/10.1016/j.pragma.2010.07.019

Alonso-Almeida, Francisco. 2015. On the mitigating function of modality and evidentiality. Evidence from English and Spanish medical research papers. *Intercultural Pragmatics*, 12/1: 33–57. https://doi.org/10.1515/ip-2015-0002

Alonso-Almeida, Francisco and Mele-Marrero, Margarita. 2014. Stancetaking in seventeenth-century prefaces on obstetrics. *Journal of Historical Pragmatics* 15/1: 1–35. https://doi.org/10.1075/jhp.15.1.01alo

Álvarez-Gil, Francisco J. 2021. "Authority and deontic modals in Late Modern English: Evidence from the Corpus of Life Sciences Texts". In Moskowich, Isabel; Lareo, Inés and Camiña, Gonzalo (eds.), *"All families and genera": Exploring the Corpus of English Life Sciences Texts*. Amsterdam/Philadelphia: John Benjamins. 249–264. https://doi.org/10.1075/z.237.13alv

Barsaglini-Castro, Anabella and Valcarce, Daniel. 2020. The Coruña Corpus Tool: Ten Years On. *Procesamiento del Lenguaje Natural*, 64: 13–19.

Biber, Douglas; Johansson, Stig; Leech, Geoffrey; Conrad, Susan and Finegan, Edward. 1999. *Longman Grammar of Spoken and Written English*. London: Longman.

Biber, Douglas and Finegan, Edward. 1989. Adverbial stance types in English. *Discourse Processes*, 11: 1–34. https://doi.org/10.1080/01638538809544689

Caffi, Claudia. 2007. *Mitigation* (Studies in Pragmatics). London: Elsevier.

Chafe, Wallace. 1986. "Evidentiality in English conversation and academic writing". In Chafe, William and Nichols, Johanna (eds.), *Evidentiality: The linguistic coding of epistemology*. Norwood, NJ: Ablex. 261–272.

Charles, Maggie. 2007. Argument or evidence? Disciplinary variation in the use of the Noun that pattern in stance construction. *English for Specific Purposes*, 23: 203–218. https://doi.org/10.1016/j.esp.2006.08.004

Clift, Rebecca. 2006. Indexing stance: Reported speech as an interactional evidential. *Journal of Sociolinguistics*, 10/5: 569–595. https://doi.org/10.1111/j.1467-9841.2006.00296.x

Crespo, Begoña. 2015. Women writing Science in the eighteenth century: A preliminary approach to their language in use. *Anglica. An International Journal of English Studies*, 24/2: 103–127.

Crespo, Begoña and Moskowich, Isabel. 2015. Involved In Writing Science: Nineteenth-Century Women in the Coruña Corpus. *International Journal of Language and Linguistics*, 2/5: 76–88.

Crismore, Avon and Farnsworth, Rodney. 1989. Mr. Darwin and his readers: Exploring interpersonal metadiscourse as a dimension of ethos. *Rhetoric Review*, 8/1: 91–112. https://doi.org/10.1080/07350198909388880

Estellés, Maria and Albelda Marco, Marta. 2018. "On the dynamicity of evidential scales". In Bates Figueras, Carolina and Cabedo Nebot, Adrián (eds.), *Perspectives on Evidentiality in Spanish. Explorations across genres*. Amsterdam, New York: John Benjamins. 25–48. https://doi.org/10.1075/pbns.290.02est

Fox, Barbara A. 2008. Evidentiality: Authority, Responsibility, and Entitlement in English Conversation. *Journal of Linguistic Anthropology*, 11: 167–92. https://doi.org/10.1525/jlin.2001.11.2.167

Halliday, Michael and Matthiessen, Christian. 2014. *An Introduction to Functional Grammar*. New York: Taylor & Francis. https://doi.org/10.4324/9780203783771

Hardie, Andrew. 2014. Log ratio – an informal introduction. *ESCR Centre for Corpus Approaches to Social Science (CASS)*. Retrieved May 17, 2020, from http://cass.lancs.ac.uk/log-ratio-an-informal-introduction/

Hyland, Ken. 1998. *Hedging in Scientific Research Articles* (Pragmatics & Beyond New Series). Vol. 54. Amsterdam: John Benjamins Publishing Company. https://doi.org/10.1075/pbns.54

Hyland, Ken. 2005a. Stance and engagement: A model of interaction in academic discourse. *Discourse Studies*, 7/2: 173–192. https://doi.org/10.1177/1461445605050365

Hyland, Ken. 2005b. *Metadiscourse. Exploring Interaction in Writing*. London, New York: Continuum.

Hyland, Ken and Jiang, Feng (Kevin). 2018a. "In this paper we suggest": Changing patterns of disciplinary metadiscourse. *English for Specific Purposes*, 51: 18–30. https://doi.org/10.1016/j.esp.2018.02.001

Hyland, Ken and Jiang, Feng (Kevin). 2018b. 'We Believe That … ': Changes in an Academic Stance Marker. *Australian Journal of Linguistics*, 38/2: 139–161. https://doi.org/10.1080/07268602.2018.1400498

Hyland, Ken and Tse, Polly. 2005a. Evaluative that constructions: Signalling stance in research abstracts. *Functions of Language*, 12/1: 39–63. https://doi.org/10.1075/fol.12.1.03hyl

Hyland, Ken and Tse, Polly. 2005b. Hooking the reader: A corpus study of evaluative that in abstracts. *English for Specific Purposes*, 24/2: 123–139. https://doi.org/10.1016/j.esp.2004.02.002

Jalilifar, Alireza; Hayati, Samira and Don, Alexanne. 2018. Investigating metadiscourse markers in book reviews and blurbs: A study of interested and disinterested genres. *Studies About Languages*, 2824/33: 90–107. https://doi.org/10.5755/j01.sal.33.0.19415

Jiang, Feng (Kevin) and Hyland, Ken. 2015. "The fact that": Stance nouns in disciplinary writing. *Discourse Studies*, 17/5: 529–550. https://doi.org/10.1177/1461445615590719

Johnstone, Barbara. 2009. "Stance, style, and the linguistic individual". In Jaffe, Alexandra (ed.), *Stance: sociolinguistic perspectives*. Oxford: Oxford University Press. 29–71. https://doi.org/10.1093/acprof:oso/9780195331646.003.0002

Kendall, Shari. 2004. Framing authority: Gender, face, and mitigation at a radio network. *Discourse and Society*, 15/1: 55–79. https://doi.org/10.1177/0957926504038946

Kim, Chanhee and Crosthwaite, Peter. 2019. Disciplinary differences in the use of evaluative that: Expression of stance via that-clauses in business and medicine. *Journal of English for Academic Purposes*, 100775: 1–14. https://doi.org/10.1016/j.jeap.2019.100775

Lareo, Inés; Monaco, Leida Maria; Esteve-Ramos, María José and Moskowich, Isabel (comps.). 2020. *The Corpus of English Life Sciences Texts (CELiST)*. https://doi.org/10.17979/spudc.9788497848

Marín-Arrese, Juana I. 2011. Epistemic legitimizing strategies, commitment and accountability in discourse. *Discourse Studies*, 13/6: 789–797. https://doi.org/10.1177/1461445611421360c

Marín Arrese, Juana I. 2009. "Effective vs. epistemic stance, and subjectivity/ intersubjectivity in political discourse. A case study". In Tsangalidis, Anastasios and Facchinetti, Roberta (eds.), *Studies on English modality. In honour of Frank R. Palmer.* Bern/ New York: Peter Lang. 23–52.

Martin, James. 2000. "Beyond exchange: APPRAISAL systems in English". In Hunston, Susan and Thompson, Geoffrey (eds.), *Evaluation in text: authorial stance and the construction of discourse.* Oxford: Oxford University Press. 142–175.

Mauranen, Anna and Bondi, Marina. 2003. Evaluative language use in academic discourse. *Journal of English for Academic Purposes,* 2/4: 269–271. https://doi.org/10.1016/S1475-1585(03)00045-6

McEnery, Anthony and Baker, Helen. 2016. *Corpus Linguistics and 17th-Century Prostitution: Computational Linguistics and History.* London, New York: Bloomsbury Publishing.

Moskowich, Isabel. 2013. Eighteenth Century Female Authors: Women and Science in the Coruña Corpus of English Scientific Writing. *Australian Journal of Linguistics* 33/4: 467–487. https://doi.org/10.1080/07268602.2013.857570

Rozumko, Agata. 2019. Between acknowledgement and countering: Interpersonal functions of English reportative adverbs. *Journal of Pragmatics,* 140: 1–11. https://doi.org/10.1016/j.pragma.2018.11.011

Stotesbury, Hilkka. 2003. Evaluation in research article abstracts in the narrative and hard sciences. *Journal of English for Academic Purposes,* 2/4: 327–341. https://doi.org/10.1016/S1475-1585(03)00049-3

Thompson, Geoff. 2001. Interaction in academic writing: learning to argue with the reader. *Applied Linguistics,* 22/1: 58–78. https://doi.org/10.1093/applin/22.1.58

Vande Kopple, William J. 2002. "Metadiscourse, discourse and issues in composition and rhetoric". In Barton, Ellen and Stygall, Gail. (eds.), *Discourse Studies in Composition.* Cresskill, NJ: Hampton Press. 91–113.

Wetherell, Margaret. 2013. Affect and discourse –What's the problem? From affect as excess to affective/discursive practice. *Subjectivity,* 6/4: 349–368. https://doi.org/10.1057/sub.2013.13

CHAPTER 13

Authority and deontic modals in Late Modern English

Evidence from the Corpus of Life Sciences Texts

Francisco J. Álvarez-Gil
University of Las Palmas de Gran Canaria

1. Introduction

Modal verbs in Modern English are still a rich source of research, if the area is still well underrepresented in the literature, not only from a semantic perspective. In this chapter, I include information on a specific modal meaning, i.e. deontic, to see the way in which authors use this modal meaning to indicate authority, among other aspects. Even if there is a general tendency to evaluate mitigation in scientific discourse (cf. Caffi 1999; Caffi 2007; Czerwionka 2012; Alonso-Almeida 2015; among others), commitment to the truth of the utterance along with certain degrees of imposition appear essential to convey scientific thought. The strategic use of deontic modality and epistemic modality has been found to conveniently boost interest in good reputation and eventually this may represent an increase in the book sales, as pointed out in Alonso-Almeida and Mele-Marrero (2014).

In other words, mitigation is a community need in order to avoid potential face-threatening acts, but lack of determinacy may equally result in a lack of reliance by potential expert readers. On top of this, expressing necessity in the scientific domain is very much linked to argumentative reasoning, and this means that the use of deontic modals cannot be avoided.

Life Sciences is an interesting area to see the frequency of occurrence of deontic modals, as observation may force the indication of obligatory phenomena in certain biological processes in order to meet an expected result. This would represent an unintentional use of this modal meaning and would eventually cancel any interpretation from a politeness perspective (Culpeper 2011: 22). The corpus of analysis is a compilation of texts in the domain of Life Sciences (Lareo *et al.* 2020), as found in the Coruña Corpus of English Scientific Writing (henceforth Coruña Corpus). This corpus, as has been noted elsewhere in chapter one (Moskowich 2021), allows data segregation in terms of sex distinction, and we will take advantage of

https://doi.org/10.1075/z.237.13alv

this feature to see whether this criterion causes variation. The analysis of the data to detect deontic meaning is mainly led by the information in Collins (2009). My notion of modality relies on Palmer (1990, 2001) and Hoye (1997), among others, as I will explain later on in this chapter. Studies solely focusing on modal verbs in the Coruña Corpus include Crespo (2015); Puente-Castelo and Monaco (2016); Alonso-Almeida and Álvarez-Gil (2019); Moskowich and Crespo (2019), and Alonso-Almeida (forthcoming).

The chapter is structured as follows: Section 2 presents definitions in the domain of modality that will guide our analysis and interpretation of deontic modals in CELiST. This section is followed by a short description of the methodology procedure and a presentation of the results obtained through the interrogation of the texts. A discussion of data is offered in Section 5 and, after this, the conclusions drawn from this study are given.

2. Deontic modality

Modality is a term in linguistics that is often associated with the expression of possibility, probability, obligation and necessity: "Modality can be defined more broadly in terms of qualification of a proposition or of the content of an utterance" (Ormelius-Sandblom 1996: 102). In other words, modality has scope over the proposition and modifies its contents to convey the speaker's communicative intentions (cf. Lyons 1977: 452). There are different language devices able to express modality, but I will focus on modal verbs in this chapter. There are several classifications of modal meanings in the literature if only the tripartite division among epistemic, dynamic and deontic modality seems to enjoy general acceptance. Epistemic modality is "concerned with matters of knowledge or belief on which basis speakers express their judgements about state of affairs, events or actions" (Hoye 1997: 42). Dynamic modality, which is a subcategory of Palmer's event modality (2001), involves senses of the speaker's willingness and ability to perform an activity. Alonso-Almeida and Álvarez-Gil (2020: 64) state that "dynamic modality is, we argue, an important language feature in technical texts, as uses of modals can be only justified according to the notions of dispositions and potentiality."

Deontic modality is my focus of research here, and I shall explain it in what follows. One definition of deontic modality relates this type with the "necessity of acts in terms of which the speaker gives permission or lays an obligation for the performance of actions at some time in the future" (Hoye 1997: 43). These notions of permission and obligation are part of the definitions for this term in the literature (cf. Cann 1993: 279). Auwera and Plungian (1998: 81) states that deontic modality "identifies the enabling or compelling circumstances external to the participant as some person(s), often the speaker, and/or as some social or ethical norm(s) permitting or obliging the participant to engage in the state of affairs". These ethical or moral nuances are further explained in Mortelmans *et al.* who suggest that "deontic modality offers an indication of the degree of moral desirability of the state of affairs" (2009: 16), as in *We should pay the bill this time* (my example).

The idea of authority is explicitly given in Collins (2009: 22), and he states that deontic meaning "occurs when the factors impinging on the actualization of the situation referred to in the utterance involve some type of authority – as when a person or a set of rules or a social convention is responsible for the imposition of an obligation or a granting of permission". Authority is, therefore, considered one key aspect in the notion of deontic modality, and this implies different perspectives. Actually, for Portner (2009: 2), "deontic modality has to do with right and wrong according to some system of rules", and this obviously implies an evaluative dimension of this modal type. In this sense, the advantages of deontic modality for rhetorical purposes are straightforward, as pointed out in Charlow and Chrisman (2016: 3): "Deontic modality, we said, is a phenomenon that encompasses normative concepts and thought (e.g. practical reasoning)". In this context, a distinction should be made between the notions of deontic possibility (or permission), including meanings related to what is permitted, and deontic necessity (or obligation), involving such meanings as 'owe' duty, 'belong', be good/proper' (Auwera and Plungian 1998: 82, 98). These two deontic types are exemplified in the following instances (my examples):

(1) You may open your exam book, now!. (permission)

(2) Nobody can smoke in public places in Spain. (prohibition)

As already pointed out, there are several ways in which modality can be linguistically manifested, namely adverbials, modal verbs, clitics and lexical matrices. I have selected modal verbs in this study in order to characterise the texts in CELiST to be compared with other registers found in the other subcorpora in the Coruña Corpus in future research. Biber *et al.* (1999: 483) indicate that there are nine modal verbs in present-day English and another set of peripheral or marginal modals. The former group includes *can, could, may, might, shall, should, will, would* and *must*, and the latter group includes *need (to), ought to, dare (to)* and *used to*. There are such other modulating periphrastic idiomatic expressions as *have to* and *be supposed to* (Biber *et al.* 1999: 484); these along with marginal modal verbs are not considered here. My interest is on central modal verbs, exclusively.

Modal verbs are classified according to specific morphological, syntactic and semantic criteria, as pointed out in Denison (1993: 292). These features are: (1) modal verbs do not present non-finite forms, (2) tense-distinction takes place in the majority of these verbs, (3) modal verbs do not show third-person singular present indicative suffixes, (4) most modals can show their contracted version to form the negative and they may also be given as clitics (e.g. *will* and *'ll*), (5) modal verbs do not have imperative forms, (6) they are followed by the bare infinitive, (7) modal verbs have scope over the propositional content, (8) more than one modal verb can co-occur in some dialects, and (9) as operators, they may share a same set of the so-called NICE properties: direct negation with *not, n't*; inversion of subject and verb; coda functionality, i.e. the modal used alone can retrieve the meaning of the lexical verb through context as in (3) (my example), and emphatic use.

(3) A: *May I use this pen?* B: *Of course, you may.*

3. Method

CELiST has been fully interrogated for the occurrences of central modal verbs. This analysis has been carried out using the updated version of the Coruña Corpus Tool (CCT 2019) (cf. Barsaglini-Castro and Valcarce 2020). Headwords, as well as contracted forms, were obtained, each of them given in a concordance line. In order to isolate deontic meanings, all modal verbs have been studied in context, as there is no computerised way to differentiate meaning automatically. This procedure has been done in an Excel spreadsheet where each concordance line representing one modal verb case has been conveniently tagged following the classification of modal meanings in Section 2. The analysis performed this way allows me to obtain statistics as to the frequency of each modal verb as well as each deontic modal meaning attested in the corpus. My analysis has also considered the variable of sex (see Moskowich (2013) and Crespo and Moskowich (2015) for the role and the language of women in science in Modern England) and whenever an attempt to compare results according to this criterion has been made, data are normalised to 10,000 words. Otherwise, raw data are offered. Although the number of words in CELiST has been noted elsewhere in this volume, I think convenient to quote them here again for ease of reference. The entire corpus contains 400,305 words: 320,118 words belong to the set of men texts and 80,187 words to the set of women texts.

4. Deontic modals in CELiST

The computerised search of modal verbs in CELiST returned a grand total of 4,005 modal verbs, of which 1,672 present deontic nuances. Using the variable of sex, 1,382 verb tokens were identified in texts written by men, while 290 were in texts written by women. In normalised figures (nf/10,000), these are 43.2 in the samples by men and 36.2 in the samples by women. In what follows, I indicate the deontic modal verbs deployed in our corpus, their exact meanings and their functions in detail.

4.1 The form of deontic modals

Deontic modals in CELiST are distributed, as shown in Figure 1:

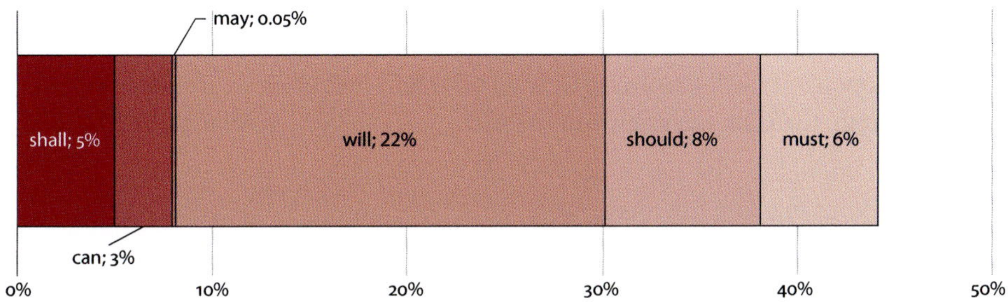

Figure 1. Distribution of deontic modal verbs in CELiST (percentages based on total modal verb meanings in the corpus)

As seen in this figure, the number of deontic modal verbs in the corpus amounts to ca. 44%. The rest of the cases, i.e. 56%, is shared by epistemic and dynamic modals, but these are beyond the scope of the present paper. The verbs identified to have a deontic meaning in CELiST are *can, may, will, shall, must* and *should.* Concerning the variable of sex, Table 1 offers both the raw data and the normalised figures of these forms.

Table 1. Deontic modal verbs per sex variable in CELiST, raw data and nf/10,000 words

	Deontic			
	Man		Woman	
	Raw data	**nf**	**Raw data**	**nf**
can	83	2.59/2.6	31	3.86/3.9
may	6	0.19/0.2		
will	701	21.89/21.9	135	16.83/16.8
shall	159	4.97/5	27	3.37/3.4
should	245	7.65/7.6	46	5.74/5.7
must	188	5.87/5.9	51	6.36/6.4
	1382	**43.2**	**290**	**36.2**

The form *will* is clearly the most frequent modal verb used in the corpus for the expression of deontic meaning. Using log-likelihood (LL) statistics (cf. Rayson 2008), we can see that the difference between men and women writers is meaningful with a LL ratio of 8.30 (significance level $p < 0.05$), with overuse in texts written by men. Second to this verb in frequency is the form *should* with a LL ratio of 3.43. Examples of *will* and *should* with deontic nuances are given below:

(4) And I *will* shew in the second Part of this Treatise how the Lympha may be drawn off, by Glands appointed to separate other Fluids; so that for this Operation she makes no Part, is at no Expence of Blood. (Keill 1717; man)

(5) Many plants *will* grow only in water; we find here such as are peculiar to the marsh, the lake, the river, and the sea. (Lincoln 1832; woman)

(6) The mould on the top of the pots *should* be taken off as deep as can be done without injuring the roots. (Speechly 1786; man)

(7) Children *should* be induced to show tenderness to every thing that breathes. (Wakefield 1816; woman)

The forms *must* and *shall* follow in frequency. The modal verb *must* occurs more frequently in texts written by women than in those written by men with a LL ratio of 0.25. In the case of *shall*, this form appears more often in the documents by men with a LL ratio of 3.82. The following excerpts present examples of these two modal verbs:

(8) And therefore to know the whole Pressure of the Air, we *must* determine the internal Surfaces of the Lungs. (Keill 1717; man)

(9) I *must* tell you something about the mouth, and about the inside of our little Sea-Anemone. (Agassiz 1859; woman)

(10) And likewise with the engraved figures, we *shall* soon attain a pretty good idea of the different parts of a Flower. (Withering 1776; man)

(11) The dairymaid, as she follows her cows to their pasture, drives them onward with a branch of this tree, persuaded that by this precaution she *shall* preserve them from danger. (Pratt 1840; woman)

The forms *can* and *may* are a less frequent group of deontic modals in CELiST. The form *can* has been attested with a normalised frequency of 2.6 in the texts by men and 3.9 in the texts written by women. This deontic meaning of *can* has a LL ratio of 3.38 with overuse in the case of the set of texts written by women. In the case of *may*, this form has been only attested in those texts written by men. The normalised figure for this modal verb is 0.24. Examples of these deontic modal verbs are provided, below:

(12) Therefore it *cannot* be called Hepatica according to the Opinions of the Antients. (Sloane 1707; man)

(13) There is also a viscous juice surrounding the stems of some plants, which effectually defends them from the depredations of insects, as they *cannot* extricate themselves from this glutinous material. (Jacson 1804; woman)

(14) In Certainty of which, and that my Readers *may* never be deficient of Methods of preparing them for Food. (Dodd 1752; man)

The meanings of deontic possibility and deontic necessity described in Section 2 have been attested in our corpus with a distribution of 9.1% and 90.9%, respectively. The following instances exemplify possibility and necessity, in this order:

(15) Therefore the sublimest Genius must, with Humility, own, that we know but in Part; and that by searching we *cannot* (till God is pleased farther to unveil his now secret, tho' always wise Ends) find out the Causes and Effects of several Phænomena, which future Ages may possibly discover. (Hughes 1750; man)

(16) We *must* remember that it is not only by budding that they [little coral animals] increase, but also by eggs, which when hatched are the little pear-shaped free Corals which float about for a while, so that they not only multiply by dividing and branching, but also by the addition of all the little animals that are born from their eggs. (Agassiz 1859; woman)

The use of *cannot* makes reference to the impossibility to gain knowledge in the current state of science in the eighteenth century. The author acknowledges that there is no way in which researchers may learn the reasons for some phenomena and the consequences they may have on certain aspects of nature. Science has to wait until technological advances may allow for more fine-grained analysis of facts. In the case of (16), the modal verb *must* is used to imply an essential aspect to be borne in mind in the author's opinion as to the origin of coral animals. From here, she adds further information in her description of the way in which these animals reproduce and grow.

4.2 The function of deontic modals

The meanings of deontic possibility and necessity underlie the functions of deontic modal verbs in CELiST, which I have identified into these groups: (a) unsuitability, (b) prescription, (c) resultative use, (d) reference to procedure in research, (e) reference to observed phenomena, (f) expectation or estimation based on actual knowledge (g) failure to perform an action as based on opinion or evidence; (g) prediction based on lack of attested knowledge, and (h) appeal to the reader or colleague. I explain these in what follows.

4.2.1 *Prescription*

The function of prescription seeks to provide readers with information regarding what is necessary to do in some biological and natural processes, as shown in the following instances with *must*, and this function, therefore, refers mainly to the meaning of deontic necessity:

(17) Crow foot is caustic, and may be used to draw Blisters, where Cantharides cannot be had; but they *must* not ly on too long, for fear of ulcerating the Part. (Blackwell 1737; woman)

(18) If you have a mind to pot Herrings after they are pickled, you *must* lay them all Night in Milk, and parboil them before they are put into the Pot, adding more Vinegar. (Dodd 1752; man)

(19) You *must* be extremely careful when you press on the thorax not to crush it more than you can possibly avoid. (Donovan 1794; man)

In the instances in (17) and (18), the modal verb *must* appears in what seems stretches of instructive text-types (see Werlich 1976), and *must* unambiguously indicate authority in the presentation of directions in a procedure. The example in (19) is a directive and gives a clear indication of how a particular technical process needs to be done. The form *must* in this instance modulates the propositional content to entail this necessity meaning.

This function also includes the idea of advisable recommendation and suggestion to carry out a particular event, as shown in (20) and (21):

(20) …and therefore there was an absolute necessity that they *should* be broken and divided in the Lungs by the Force of Respiration. (Keill 1717; man)

(21) A full statement of these defects will be found in [Mr]. Milne's Botanical Dictionary, a book which *should* be in the hands of all young botanists, as much information may be derived from it. (Jacson 1804; woman)

The modal verb *should,* in these instances indicates desirability in different ways. The first case given in (20) seems to emerge from a standard technical procedure. The second use in (21) follows from a more subjective view, even if the expression of desire follows from the author's own professional expertise.

4.2.2 *Resultative use*

Deontic modal verbs with a resultative function are used to indicate necessity in CELiST in argumentative or expositive discourse, as evinced in the examples below:

(22) And consequently when in ordinary Expiration the Pressure upon the Larynx is 2 Ounces, the Pressure upon the whole internal Surfaces of the Lungs *will* be 14412 [lib]. Weight, and the utmost Force of the Air in breathing when the Pressure upon the Larynx is 7 Ounces *will* be 50443. (Keill 1717; man)

(23) When all these days are stuck into the flesh of animals, sometimes one after another, and sometimes all at once, the blood and humours of the adjacent parts *must* unavoidably be extravasated upon which a tumour *must* consequently ensue, the little orifice whereof is closed up by the compression of the external air. (Goldsmith 1774; man)

(24) The egg-cell which it produces, on being fertilized, gives rise to a plant of extremely low morphological character; we *must* therefore pass to the next class of plants to reach a root of different character, or a true root. (Gregory 1895; woman)

(25) The groove will receive the water that falls from the rook, and if worked in in the above manner, the water will descend to one end of the stove if small; but if the Hot-house be large, it will be more convenient to descend from both ends to the middle, where a semicircular stone, one foot in diameter, worked in the form of a bason, *should* receive the waste. (Speechly 1786; man)

These instances present cases of necessity deontic modals, namely *will* in (22), *must* in (23) and (24) and *should* in (25) to indicate seemingly objective results obtained from a logical sequence of facts, e.g. calculations in (22). This use contrasts with the use of *should* in (25), which represents a more subjective stance and so the propositional content modulated by this modal verb seems to indicate expected results from the actions described earlier in the text and framed by *will*. An interesting discursive aspect of modal verbs with a resultative function is the fact that they are often accompanied by result or inference adverbs (Biber *et al.* 1999: 559), e.g. *therefore* and *consequently*, thus strengthening the idea of the logical outcome. This function of deontic modals does not include inferential processes, which I categorise as epistemic necessity modality, as described in Auwera and Plungian (1998), which often implies some evidentiary qualifications. As I said, this is not the case here.

4.2.3 *Reference to procedure in research*

Another use of deontic modals is found in references to procedure in research in CELiST, as in the instances below:

(26) I *shall* first examine with what Velocity the Blood in the Liver would have moved, if it had been carried there by such Arteries as are every where else. (Keill 1717; man)

(27) I have told you that not only are islands built up by Corals, but parts of continents also; and I *will* show you how the whole peninsula of Florida has been patiently added to the continent of North America on which you live. (Agassiz 1859; woman)

(28) The use of the Anchovy and their Cure *shall* be seen hereafter. (Dodd 1752; man)

(29) The characters of the Orders, are most frequently taken from the number of the Pointals; but sometimes from some other circumstances either of the Chives or Pointals which *will* be noticed hereafter. (Withering 1776; man)

In these examples, the authors deploy such modal verbs as *shall* and *will* with the intention to signal necessary steps in their scientific accounts and to signal position of information in their texts. Expressions in this category include *I shall first examine* in (26), *I will show you* in (27), *The use of the Anchovy... shall be seen hereafter* in (28) and *which will be noticed hereafter* in (29). The idea of chronological position in discourse is strengthened with the time adverbials *first* and *hereafter*. The particle *and* in (28) very likely functions as a sequencer. The combined use of these modal and adverbial devices has an outstanding textual function, as described in Hyland (2005: 41ff) and Halliday and Matthiessen (2014). Some devices in this category, e.g. *I will show you* and *I shall first examine*, also serve as engagement markers, as defined in Hyland and Paltridge (2011: 182), and the function of these is therefore to show how "writers acknowledge and connect to others, recognizing the presence of their readers by focusing their attention, acknowledging their uncertainties and including them as discourse participants."

4.2.4 *Appeal to the reader or colleague*

The appeal to the reader or colleague function is strongly related to the function described in the previous section in the sense that this can be categorised as an engagement marker. The modal verbs used in these expressions show a deontic necessity meaning to indicate desirability and a mild requirement to readers in order to enhance their understanding, as shown in the following examples:

(30) It *must* be observed that the Agasœus of [Dr]. Caius, is a very different species from the Agasseus of Oppian; which he describes as being a small kind of dog. (Pennant 1766; man)

(31) But when we see them at home, as they are brought from foreign lands, we *must* remember that all the soft and moving parts, the tentacles or fringes that wave so gracefully in the water, are gone. (Agassiz 1859; woman)

These instances present cases of *must* reporting necessity, although its use is different in each case. In (30), this modal verb modulates a passive construction, which seems to mitigate an otherwise direct command introduced by an imperative form, i.e. *Observe that* + proposition. The illocutionary force is evidently milder with the modal form. The same modal verb *must* in (31) is used in the context of an inclusive *we*, clearly addressing her readers. This use is noticeably connected with the textual genre in which it is given, i.e. Lecture, and so the use of *we must...* is a directive whose deontic force is obviously attenuated, as this is clearly identified as characteristic of the language of instruction. Relying on the notion of face and politeness (Brown and Levinson 1987; Leech 2014; Culpeper *et al.* 2018), these expressions do not appear to represent lack of polite behaviour in discourse, even if they maximise the notion of academic authority in the pedagogic design of the lesson. Directives (Searle 1969) work well in teaching, as pointed out in Moore (2017: 36): "When a person produces a directive, he or she establishes a social relationship with the addressee of the directive. In classrooms, directives used by teachers "provide an index of a teacher's social control" (Erickson and Mohatt 1982: 150)".

4.2.5 *Observed phenomena*

This category refers to examples of modal verbs indicating deontic necessity in descriptive passages, as the ones given below:

(32) The fur is very soft and fine: the general color of these animals is of a yellowish white, mixed with a deep grey: these colors, tho' they appear at first sight confusedly blended together, yet on a close inspection *will* be found to be disposed like the streaks on the skin of the tiger, pointing from the back downwards, rising from a black list that runs from the head along the middle of the back to the tail. (Pennant 1766; man)

(33) The hawthorn belongs to the rose order Rsaceæ, and any one who examines the flower of the briar rose, *will* see that there is a great family likeness in all the plants composing it. (Pratt 1840; woman)

The modals in this category normally frame perception verbs. In the examples above, the form *will* precedes the lexical verb *find* in the passive in (32), which indicates visual scrutiny. The idea of objectivity seems to underlie this combination of *will* with the perception verb. The same procedure is shown in (33), and the verb *see* follows the modal *will* in this descriptive fragment. The degree of subjectivity is minimised in these passages, as the deontic necessity meaning of the modal verb along with the meaning of the perception verbs enhance the idea of factuality.

4.2.6 *Expectations based on actual knowledge*

The description of phenomena may rely on expectations which are based on the authors' actual knowledge, as in the instance in (34) with the modal verb *will*. In this instance, choice allows for three options. All of them are presented as feasible by means of the modal verb *will* which suggests deontic necessity in accordance with the author's expertise:

(34) ...it will be seen that the expense of growth becomes relatively greater with each increment of size; and that hence, of two similar creatures commencing life with different sizes, the larger one in reaching its superior adult bulk, will do this at a more than proportionate expense; and so *will* either be delayed in commencing its reproduction, or *will* have a diminished reserve for reproduction, or both. (Spencer 1867; man)

4.2.7 *Impossibility to perform an action as based on opinions or evidence or lack of means*

This category indicates the use of modal verbs exhibiting deontic possibility and reports on the impossibility to carry out a particular activity due to different reasons. These can include regulations, human or natural laws, lack of technological means, the individual's own lack of knowledge or adequate physical conditions, including the conditions of the species being described. The following examples illustrate these aspects:

(35) The Havanna Pine is much praised by many, but I *cannot* commend it from my own knowledge. (Speechly 1786; man)

(36) For, inasmuch as the thing propagated proceeds from two organisms of different sexes and different makes and temperaments, and as the offspring is to be either of one sex or the other, it is quite clear that it *cannot* be an exact diagonal of the two. (Huxley 1863; man)

(37) Now the Sea-Anemone has upon these fringes or tentacles, as I will call them, because that is their true name, numbers of what are called lasso-cells. They are so small that you *cannot* see them with your naked eye. (Agassiz 1859; woman)

The modal verb indicating deontic possibility is prototypically *can* followed by *not* or any other negative adverb, e.g. *never*. All the examples given above present cases of *cannot*. In (35), this modal verb indicates the author's impossibility to recommend the product mentioned because he does not have the exact information to be able to do so. The example in (36) indicates that the two options described are mutually exclusive, apparently due to natural laws. Finally, the third example illustrates a case in which the author clarifies readers about the impossibility to see the species described without the help of some technological appliance to magnify its size, probably a microscope.

5. Conclusions

The present chapter has examined the meanings and functions of deontic modal verbs in CELiST, whose distribution represents more than 40% of the modal verbs in this compilation, second after epistemic verbs. The inspection of these modals has revealed that *will* is distinctly the commonest form in this corpus regardless of the variable of sex in relation to authorship. Concerning this variable, normalisation of data has also shown that men are more likely to use deontic modals than women in their technical texts. Within the traditional division of deontic modals between possibility and necessity modals, the latter are attested to occur in a higher frequency in the entire corpus. This presence of necessity modals reinforces the idea of futurity underlying some functions of deontic modal verbs.

The functions of modal verbs driven by these notions of necessity and possibility contribute patently to the elaboration and presentation of knowledge. This includes the idea of academic authority, which stands vital in technical accounts of natural phenomena. Authority, however, is more positively on a par with expertise than with imposition in these texts. Deontic modals are contextually justified to indicate perceptions of mild obligation, desirability, volition, advisability and recommendation in the form of prescribed directions to achieve a particular goal. Precisely, the use of directives modulated by means of modal verbs, indicating necessity slightly attenuates a directive that could have been construed with an imperative form.

A similar pragmatic effect of mitigation should follow from the use of deontic necessity modals in argumentative discourse. The combination of these modals with result and inference adverbs and a carefully laid logical sequencing suggests the resultative function of the deontic modals. The modals used in this context have not been found to have an evidentiary qualification. A logical marking as to the elaboration of information strengthens the idea of authority in the texts analysed. In the same vein, authority and scientific adequacy are also gained by a careful indication of the procedure of research. This indication is clearly signalled by necessity deontic modals in combination with adverbial devices acquiring thus a textual function, which is, in turn, an indication of the author's expertise in scientific contents and methodology of research.

Expertise and authority are also revealed in the use of deontic necessity modals to elaborate logical information in those situations in which phenomena cannot be observed directly, even if this use may imply subjective reasoning. I have seen, however, that the degree of subjectivity is partially attenuated by the use of necessity modal verbs deployed jointly with perception verbs that emphasise the notion of factuality. In turn, evincing means of elaborating information seems to pursue a softening of possible face-threatening situations leading to lack of politeness behaviour and collegial cooperation. This social function of deontic modality is confirmed in the use of such deontic expressions as *I will show you* and *we must remember*, for instance. Their function as engagement markers shows the relationship between the writers and their audience while they maximise their authoritative voice in the context of logical argumentation, which apparently seeks avoidance of straightforward imposition. As I have noted earlier in this paper, engagement markers may also originate in CELiST as a consequence of the textual genres some texts exemplify. In the case of a lecture, for instance, I have shown that the instructive purpose of this genre benefits the presence of engagement markers embedding necessity deontic modals both to show authority as well as to show control over the pedagogic situation. In terms of politeness, these deontic expressions are particularly tolerated as they are felt like a generic feature, which includes an indexical reference, e.g. *we*, rather than as an act of imposition. Impossibility is also presented as a biological, physiological, legal or technological means to explain certain natural phenomena as well as to justify impracticalities concerning certain procedures.

Works cited

Alonso-Almeida, Francisco. Forthcoming. Los modales dinámicos en textos de historia en lengua inglesa (1700–1900). *Signos. Revista de lingüística.*

Alonso-Almeida, Francisco. 2015. On the mitigating function of modality and evidentiality. Evidence from English and Spanish medical research papers. *Intercultural Pragmatics*, 12/1: 33–57. https://doi.org/10.1515/ip-2015-0002

Alonso-Almeida, Francisco and Álvarez-Gil, Francisco J. 2020. 'so that it may reach to the jugular'. Modal Verbs in Early Modern English Recipes. *Studia Neofilologiczne*, XVI: 61–88. https://doi.org/10.16926/sn.2020.16.04

Alonso-Almeida, Francisco and Álvarez-Gil, Francisco J. 2019. "Modal verb categories in CHET". In Moskowich, Isabel; Crespo, Begoña; Puente-Castelo, Luis and Monaco, Leida-Maria (eds.), *Writing history in Late Modern English: Explorations of the Coruña Corpus.* Amsterdam/Philadelphia: John Benjamins. 150–161. https://doi.org/10.1075/z.225

Alonso-Almeida, Francisco and Mele-Marrero, Margarita. 2014. Stancetaking in seventeenth-century prefaces on obstetrics. *Journal of Historical Pragmatics*, 15/1: 1–35. https://doi.org/10.1075/jhp.15.1.01alo

Auwera, Johan Van der and Plungian, Vladimir A. 1998. Modality's semantic map. *Linguistic Typology*, 2:1: 79–124. https://doi.org/10.1515/lity.1998.2.1.79

Barsaglini-Castro, Anabella and Valcarce, Daniel. 2020. The Coruña Corpus Tool: Ten Years On. *Procesamiento del Lenguaje Natural*, 64: 13–19.

Biber, Douglas; Johansson, Stig; Leech, Geoffrey; Conrad, Susan and Finegan, Edward. 1999. *Longman Grammar of Spoken and Written English.* London: Longman.

Brown, Penelope and Levinson, Stephen. 1987. *Politeness: Some Universals in Language Usage.* Cambridge: Cambridge University Press. https://doi.org/10.1017/CBO9780511813085

Caffi, Claudia. 1999. On mitigation. *Journal of Pragmatics*, 31/7: 881–909. https://doi.org/10.1016/S0378-2166(98)00098-8
Caffi, Claudia. 2007. *Mitigation* (Studies in Pragmatics). London: Elsevier.
Cann, Ronnie. 1993. *Formal Semantics. An Introduction*. Cambridge: Cambridge University Press. https://doi.org/10.1017/CBO9781139166317
Charlow, Nate and Chrisman, Matthew. 2016. *Deontic Modality*. Oxford: Oxford University Press. https://doi.org/10.1093/acprof:oso/9780198717928.001.0001
Collins, Peter. 2009. *Modals and Quasi-modals in English*. Amsterdam, New York: Rodopi. https://doi.org/10.1163/9789042029095
Crespo, Begoña. 2015. Women Writing Science in the Eighteenth Century: A Preliminary Approach to Their Language in Use. *Anglica. An International Journal of English Studies*, 24/2: 103–127.
Crespo, Begoña and Moskowich, Isabel. 2015. Involved In Writing Science: Nineteenth-Century Women in the Coruña Corpus. *International Journal of Language and Linguistics*, 2/5: 76–88.
Culpeper, Jonathan. 2011. *Impoliteness: Using Language to Cause Offence*. Cambridge: Cambridge University Press. https://doi.org/10.1017/CBO9780511975752
Culpeper, Jonathan, Paul Kerswill, Ruth Wodak, Anthony McEnery, Francis Katamba, Jonathan Culpeper and Claire Hardaker. 2018. *Politeness in Interaction. English Language, Description, Variation and Context*. London: Red Globe Press.
Czerwionka, Lori. 2012. Mitigation: The combined effects of imposition and certitude. *Journal of Pragmatics*, 44/10: 1163–1182. https://doi.org/10.1016/j.pragma.2012.05.002
Denison, David. 1993. *English Historical Syntax: Verbal Constructions (Longman Linguistics Library)*. London, New York: Longman.
Erickson, Frederick and Mohatt, Gerald. 1982. "Cultural organization of participation structures in two classrooms of Indian students". In Spindler, George (ed.), *Doing the ethnography of schooling: Educational anthropology in action*. Prospects Heights, IL: Waveland Press, Inc. 132–174.
Halliday, M. A. K. and Matthiessen, C. M. I. M. 2014. *An Introduction to Functional Grammar*. New York: Taylor & Francis. https://doi.org/10.4324/9780203783771
Hoye, Leo Francis. 1997. *Adverbs and Modality in English*. New York: Routledge.
Hyland, Ken. 2005. *Metadiscourse. Exploring Interaction in Writing*. London, New York: Continuum.
Hyland, Ken and Paltridge, Brian. 2011. *The Continuum Companion to Discourse Analysis*. London, New York: Continuum.
Lareo, Inés; Monaco, Leida Maria; Esteve-Ramos, María José and Moskowich, Isabel (comps.). 2020. *The Corpus of English Life Sciences Texts (CELiST)*. A Coruña: Universidade da Coruña. https://doi.org/10.17979/spudc.9788497497848
Leech, Geoffrey. 2014. *The Pragmatics of Politeness*. Oxford: Oxford University Press. https://doi.org/10.1093/acprof:oso/9780195341386.001.0001
Lyons, John. 1977. *Semantics* (Semantics). Cambridge: Cambridge University Press.
Moore, Ekaterina. 2017. Managing classroom transgressions: Use of directives in a reading practice. *Linguistics and Education*, 41: 35–46. https://doi.org/10.1016/j.linged.2017.07.006
Mortelmans, Tanja, Boye, Kasper and Auwera, Johan van der. 2009. "Modals in the Germanic Languages". In Hansen, Björn and Haan, Ferdinand de (eds.), *Modals in the Languages of Europe. A Reference Work*. Berlin, New York: Mouton de Gruyter. 1–58.

Moskowich, Isabel. 2021. "The making of CELiST, a bunch of disciplines". In Moskowich, Isabel; Lareo, Inés and Camiña, Gonzalo (eds.), *"All families and genera": Exploring the Corpus of English Life Sciences Texts*. Amsterdam/Philadelphia: John Benjamins. 1–19. https://doi.org/10.1075/z.237

Moskowich, Isabel. 2013. Eighteenth Century Female Authors: Women and Science in the Coruña Corpus of English Scientific Writing. *Australian Journal of Linguistics*, 33/4: 467–487. https://doi.org/10.1080/07268602.2013.857570

Moskowich, Isabel and Crespo, Begoña. 2019. "Arguments That Could Possibly Be Urged": Modal Verbs and Tentativeness in the Coruña Corpus. *Languages*, 4/57: 1–12. https://doi.org/10.3390/languages4030057

Ormelius-Sandblom, Elisabet. 1996. "The modal particle schon: Its syntax, semantics, and pragmatics". In Swan, Toril and Westvik, Olaf Jansen (eds.), *Modality in Germanic Languages. Historical and Comparative Perspectives*. Berlin, New York: Mouton de Gruyter. 75–131.

Palmer, Frank R. 1990. *Modality and the English Modals*. 2nd edn. New York: Longman.

Palmer, Frank R. 2001. *Mood and Modality*. Cambridge: Cambridge University Press. https://doi.org/10.1017/CBO9781139167178

Portner, Paul. 2009. *Modality*. Oxford: Oxford University Press.

Rayson, Paul. 2008. From key words to key semantic domains. *International Journal of Corpus Linguistics*, 13/4: 519–549. https://doi.org/10.1075/ijcl.13.4.06ray

Puente-Castelo, Luis and Monaco, Maria. 2016. "The might of might": A mitigating strategy in eighteenth and nineteenth century female scientific discourse. *Revista Canaria de Estudios Ingleses*, 72: 143–168.

Searle, John R. 1969. *Speech Acts: An Essay in the Philosophy of Language*. Cambridge: Cambridge University Press. https://doi.org/10.1017/CBO9781139173438

Werlich, Egon. 1976. *Genre Analysis*. Heidelberg: Quelle and Meyer.

CHAPTER 14

A study of coherence relations in the English scientific register

Conjunctions in the Corpus of English Life Sciences Texts

Iria Bello Viruega and Elisa Narváez García
University of the Balearic Islands / Heidelberg University

1. Introduction

Discourse is more than a succession of utterances; it is an assembly of meaningful connections between information units, where every link in the chain presupposes others and the interpretation of the whole depends on that of each other (Halliday and Hasan 1974: 4). These units of information are tied in different kinds of semantic or pragmatic relations, which are vital in discourse understanding and production, since discourse cannot be understood in its isolated parts but in the fusion of their meanings. Thus, (1):

(1) [A]nd the Glands of the Inteſtines, receive a leſs Quantity in Proportion than any others, *becauſe* they cannot be ſo much dilated by the greater Force of the Blood, as others, which are not ſo much ſtimulated by the Medicine.
(Keill 1717: 168; emphasis added)

does not simply account for the intestines' reception of purgative and their inability to be dilated, but for the more complex relation between these two facts connected in a causal, coherence relation. In other words, what is meant in (1) is that the intestines' inability to be dilated, caused their reception of a smaller quantity of purgative. In this example, the coherence relation is marked by the conjunction *because*. Conjunctions are one of the five means of coherence considered by Halliday and Hasan (1974). These devices correspond with what has been studied as *connectives*, "one-word items or fixed word combinations that express the relation between clauses, sentences, or utterances in the discourse of a particular speaker" (Sanders and Pander Maat 2006: 591). They cast explicit relations, as in (1). Coherence relations, however, can also be cast in the absence of a connective, that is the case of the implicit relations.

https://doi.org/10.1075/z.237.14bel

(2) [A]nd the Glands of the Inteſtines, receive a leſs Quantity in Proportion than any others. They cannot be ſo much dilated by the greater Force of the Blood, as others, which are not ſo much ſtimulated by the Medicine.

In (2), although there is no linguistic marking of the relation, the causal meaning remains as one of the possible interpretations of the utterance. It is to be said that even though the causal interpretation is perfectly possible in the absence of the connective, there is a wider margin of interpretation. This feature, at times, turns implicit relations into more ambiguous ones in comparison to the explicit relations, where the presence of a connective restraints the possible interpretations. The present study focuses on explicit relations. This decision was taken based on practical issues: the presence of linguistic marks to look for in the analysed corpora facilitates the retrieval of cases.

Different taxonomies have been proposed to group semantic relations. Due to its clearness and simplicity, this study will adopt Halliday and Hasan's (1974) four-category model. It constitutes a middle point between the overly detailed twenty-three-relation system outlined in the Rhetorical Structure Theory (Mann and Thompson 1987; Mann, Matthiessen and Thompson 1992) and Sanders, Spooren and Noordman's (1992) minimalist, yet complex, taxonomy based on what they called *relational criterion*. Halliday and Hasan (1974) distinguished between additive, adversative, causal, and temporal relations.

Additive relations imply the conjunction of certain elements that are considered to be narrowly linked to each other (3). Adversative relations, conversely, imply the conjunction of elements that would normally be considered to be disjoint (4). Causal relations, as has been shown in (1), tie a causing event/phenomenon to a caused event/phenomenon (5). Finally, temporal relations order certain elements in time (6).

(3) *Beſides* its Uſe in dividing the Breaſt, it preſerves the Heart from being hurt in its Motion by the bony Sides of the Cheſt. (Gibson 1720: 27; emphasis added)

(4) The bite of this spider is deemed fatal *but* I never knew of any animal who experienced its effects. (Bancroft 1769: 241; emphasis added)

(5) By the by, we may obſerve how that Steel, being an elaſtick Body, heats the Blood more than any other Mineral; and how by its Elaſticity, the Force of its own Particles in removing Obstructions, as well as thoſe of the Blood, increaſe; and *therefore* it is a better Deobſtruent, than ſome other Minerals, which have a greater Gravity. (Keill 1717: 129; emphasis added)

(6) Not all the nuclei undergo the above changes; but some of them stop short in their development, undergo atrophy, and appear *finally* to be absorbed as pabulum by the protoplasm of the nest. (Balfour 1880: 46; emphasis added)

In this study, we observe the frequency and evolution in the use of additive, adversative, causal and temporal conjunctions during the two centuries covered by the Corpus of English Life Sciences Texts (CELiST). Our analysis also considers two diatopic and diaphasic variables, together with other two extralinguistic parameters (sex and age of authors). In this chapter, we begin with a brief description of the corpus analysed and the variables This is followed by the presentation and discussion of the results. The main findings of this study are summarised in the conclusions.

2. Methodology

2.1 Corpus description

The corpus material for this study was taken from the Corpus of English Life Sciences Texts (CELiST) (Lareo *et al.* 2020), one of the six subcorpora of the Coruña Corpus of English Scientific Writing (CC) (Moskowich and Crespo 2007; Moskowich and Parapar López 2008; Crespo and Moskowich 2010). As stated in Moskowich (2021), CELiST is compliant with all the specifications for corpus design stated by McEnery and Wilson (1996) and was designed to study the variation of the scientific register in English across the eighteenth and nineteenth centuries. This version is a closed corpus with a finite size of 400,305 words (Table 1) and it is machine-accessible, thanks to a search tool (Coruña Corpus Tool, CCT, Parapar López and Moskowich 2007; Barsaglini-Castro and Valcarce 2020) that was designed for its joint use.

Table 1. Number of words in CELiST

Century	Word count	%
18th Century	200,220	50.02
19th Century	200,085	49.98
Total	**400,305**	

All CC subcorpora contain biographical information about the author's gender, age and place of education, which was taken as extralinguistic variables in this study. In CELiST there are eight texts written by women. All of them, with the sole exception of Elizabeth Blackwell's (1737) work, were published in the nineteenth century.

Age is the second biographical variable listed in CC. In Table 2, the total number of words is divided into four age groups. Younger writers are more numerous in CELiST. More specifically, authors aged less than 35 constitute 25% of the total and those between 35 and 44 years of age make up 35%. There is a fifth category ("not given" in Table 3) that includes the work of Thomas Boreman, an Englishman that lived in the first part of the eighteenth century but whose date of birth remains unknown. Given that this group is formed only by one text, it was excluded from the analysis.

Table 2. Number of words according to age of author

Age	Word count	%
25–34	100,125	25.01
35–44	140,189	35.02
45–54	80,147	20.02
55+	69,824	17.44
not given	10,017	2.50
Total	**400,305**	

Table 3 shows the distribution of words according to the place of education of authors. As in other CC subcorpora, more than half the texts (54%) have an English author. This can be seen as a consequence of the political and economic power of the country as well as part of the effect of the establishment of the London Royal Society in 1660, which became a bastion of the Scientific Revolution and created, together with the neighboring Gresham College, Cambridge and Oxford universities, a network of scholars. The development and institutionalisation of science took a slower pace in other more peripheral areas. In the case of Scotland, the "Act of Union" (1707) and the Scottish Enlightenment granted an early development of science. Thus, the eight texts written by Scottish authors in the corpus are evenly distributed across the two centuries. The situation of U.S. authors, conversely, is different, as five of the six texts written by American scholars were published in the nineteenth century. The effects of colonialism can be alluded to explain the underrepresentation of both U.S. and Irish authors in the corpus.

Table 3. Number of words according to place of education of authors

Place	Word count	%
England	210,041	52.47
Ireland	50,090	12.51
Scotland	80,085	20.01
U.S.	60,089	15.01
Total	**400,305**	

The compilers of the corpus provided information about text typology. Their decisions were based on the notion of paragenre, that is, "a genre belonging to one professional community" (Moskowich 2012: 37) as well as functional constraints in the form of "communicative format" (Moskowich 2017: 97). There are eight different text typologies in the collection of texts used for this study: Article, Catalogue, Essay, Guide, Lecture, Letter, Textbook, and Treatise. Their distribution is shown in Figure 1.

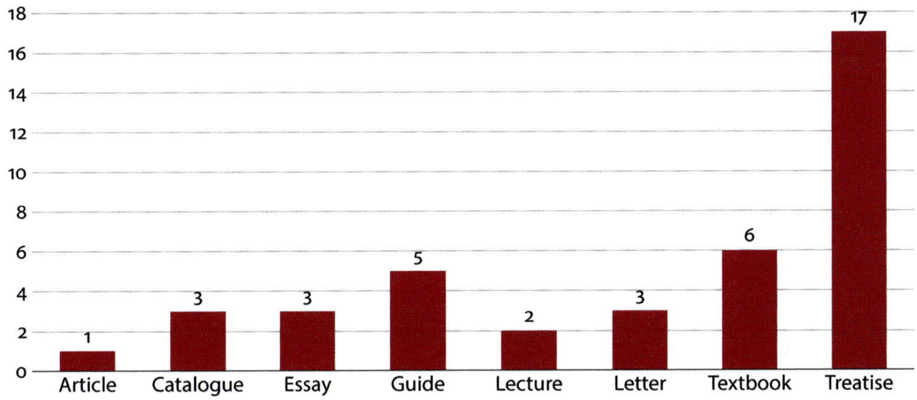

Figure 1. Genres in CELiST

For this analysis, genres were grouped in two categories according to the level of education of their intended addressee. On the one hand, specialised texts (Articles, Essays, Letters, and Treatises) are aimed at the scientific community. They are expected to follow, to some extent, the vocabulary and lexicogrammar peculiarities adopted by the community of scholars publishing their work in the *Philosophical Transactions*. On the other hand, non-specialised texts (Catalogues, Guides, Lectures, and Textbooks) are targeted at students. They are used in educational contexts to facilitate access to specialised knowledge (Swales 1990, 2004; Hyland 2000, 2011; Bhatia 2002; Parodi, Burdiles, Moreno de León and Julio 2018). According to Bolívar and Parodi (2015), the differences between professional and academic discourse can be found in several domains, namely, production and circulation contexts, participants and their interactions, communicative purposes and linguistic choices. With our categorisation of texts in this study, we aim to shed light on the last domain. The number of words for both groups is shown in Table 4:

Table 4. Number of words in CELiST according to level of specialisation

Level of specialisation	Word count	%
non-specialised	160,286	40.04
specialised	240,019	59.96
Total	**400,305**	

2.2 Variables and procedure

In addition to the four variables discussed (sex, age, place of education and genre/level of specialisation), data were analysed according to the classification of conjunctions proposed by Halliday and Hasan (1974), which consists of four large groups: additive, causal, adversative and temporal. Table 5 provides an account of the conjunctions considered in this study.[1]

Table 5. Conjunctions analysed

Additive	*accordingly, after all, and, as for, as well, besides, by the way, concerning, considering, especially, equally, for example, for instance, further, in addition, in fact, in the same way, in other respects, in other words, in particular, including, indeed, likewise, moreover, namely, notably, nor, of course, on that account, on this account, on the one hand, on the other hand, on the subject of, or, or else, particularly, regarding, specifically, that is, that is to say, the fact that, to illustrate, similarly*
Causal	*as a result, as long as, because, consequently, due to, for, for the purpose of, for this purpose, for this reason, given that, hence, if, in case, in that case, in consequence, in order that, in order to, in this respect, inasmuch, it follows, on account of this, otherwise, owing to, seeing that, so, so as to, so that, then, therefore, thus, to the end that, unless*
Adversative	*actually, although, and yet, at any rate, at least, but, conversely, even if, however, in any case, instead, meanwhile, nevertheless, notwithstanding, on the contrary, on the other hand, rather, regardless, though, whereas, while, whilst*
Temporal	*after a time, after that, afterwards, at first, at last, at once, at the same time, at this point, as yet, briefly, earlier, eventually, finally, first, formerly, henceforth, hereafter, hitherto, in conclusion, in short, lastly, next, next day, next time, originally, previously, secondly, sequential, so far, soon, subsequently, thirdly, to conclude, to the point, to resume, to return to, to sum up*

These conjunctions were searched with the help of the CCT. After a disambiguation process that discarded instances like the adjectives *first* and *next* or the adverb *so*. The results of the analysis are presented in the following section.

3. Data analysis

A total of 25,245 conjunctions were found in the corpus. However, two additive conjunctions, the polyfunctional unit *and* (Culpepper and Kytö 2000) together with *or* gather alone 64.79% of cases.[2] To observe better the behaviour and interactions

1. Appendix 1 provides a list of the conjunctions that were searched in the corpus but had no presence in it.

2. An account of the frequency of each conjunction is offered in Appendix 2.

of the remaining conjunctions, these two words were excluded from the analysis. Thus, in this chapter a total of 8,890 conjunctions were studied.

The evolution in the use of conjunctions in the two centuries analysed is shown in Figure 2. In general terms, a steady increase of explicit coherence markers is observed.

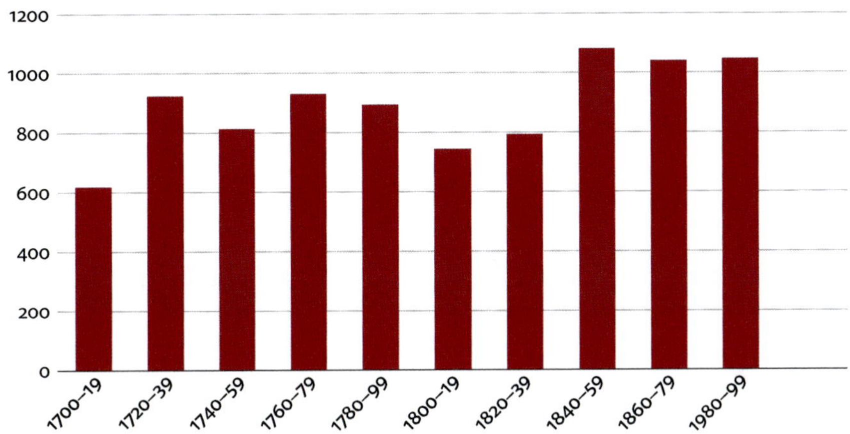

Figure 2. Evolution in the use of conjunctions in CELiST

Three phases can be identified: an initial phase, where the mean frequency per 10,000 words is 154.4. An increase of 49.5% marks the beginning of the following phase. In the period 1720–1839, the mean frequency of conjunctions per 10,000 words raised to 212.5. Finally, another sharp increase (36.1%) between the 1820–39 period and the subsequent time group set the tone for the last years of the nineteenth century, when the mean frequency rose to 264.

3.1 Type of relation

Halliday and Hasan's (1974) study of coherence relations was taken as a model for this variable. Table 6 offers the frequency of the typology discussed by them.

Table 6. Frequency of conjunctions according to type of relation

	Additive	Causal	Adversative	Temporal
No.	1,482	2,953	3,473	982
%	16.68	33.22	39.07	11.03

Data corroborate the difference between simple and complex coherence relations. Cognitively, additive relations are the simplest. The juxtaposition of two arguments is frequently understood as an additive relation (Kleijn, Pander Maat and Sanders 2019) and the inclusion of an additive conjunction can be seen as a reinforcement of the informative and argumentative structures (De Cesare 2017). In other words, they are not necessary. This explains their low frequency in the corpus (16.68%). Temporal relations add a temporal trait and are slightly more difficult but less frequent, according to corpus results.

Causal and adversative relations are more complex because, in some way, they also convey addition and temporality. The difference between them lies in the argumentative path established between the two segments they connect. Segments holding a causal relation follow the same argumentative path (Ducrot 1983). They are in a way similar to additive relations because, even if they are more complex, they can be understood without a linguistic mark. It has been shown, however, that the inclusion of a causal conjunction facilitates the processing and comprehension of the relation (Kleijn, Pander Maat and Sanders 2019), which, in terms of use, makes them frequent connectives.

On the other hand, segments follow opposite argumentative paths in adversative relations. In order to fully understand this relation, the reader needs to reach a conclusion that cancels or weakens the inference that could have been obtained from the first segment (Anscombre and Ducrot 1976; Blakemore 1987). This makes their presence essential for comprehension (Rudolph 1996) and hence very frequent in texts, as evinced by the 39.07% of adversative conjunctions found in the corpus. In some way, the cognitive complexity of the relation makes adversative conjunctions more indispensable than the other types.

The evolution of all four categories (see Figure 3) also depicts this two-level distinction in terms of cognitive complexity. With very short lapses of time in which the situation was reversed, additive relations were more frequent than temporal

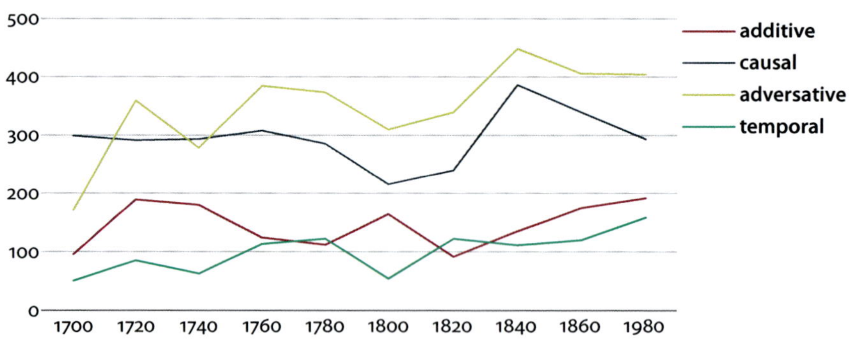

Figure 3. Evolution in the type of relation

ones and adversative relations prevailed over causal ones. What is common in all groups is the progressive augmentation along the two centuries. This is especially evident in additive and temporal relations in the last part of the nineteenth century. All in all, temporal relations have the highest growth rate (1.8% annually), whereas the frequency of additive conjunctions remained stable and shows an annual growth rate of 0.07%.

3.2 Sex of author

The first thing to be observed in Table 7 is that the number of conjunctions is higher in male-authored texts. These texts have a normalised frequency of 230.6 conjunctions per 10,000 words, which translates in a difference of 18.4%.

Table 7. Frequency of conjunctions according to sex of the author

Sex	No.	nf (10,000)
female	1,509	188.2
male	7,381	230.6

A closer look at the four types of conjunction considered reveals a balanced distribution. This information is provided in Table 8, where it can be seen that in texts written by both men and women, adversative conjunctions constitute the most numerous group (42.68% and 38.33% in male and female texts, respectively).

Table 8. Type of conjunction and sex of author

Sex		Additive	Causal	Adversative	Temporal
female	No.	220	521	644	124
	%	14.58	34.53	42.68	8.22
male	No.	1,262	2,432	2,829	858
	%	17.10	32.95	38.33	11.62

They are followed by causal, additive, and temporal relations. Texts written by men have a higher percentage of additive (17.10% vs. 14.58%) and temporal (11.62% vs. 8.22%) conjunctions. In contrast, women used more causal (34.53% vs. 32.95%) and adversative (42.7% vs 38.3%) conjunctions. This slight difference may be put in context considering the situation of women writers in the Late Modern period. In this age, women did not serve as linguistic models to others. However, their speech was scrutinised and there were clear expectations on how a lady should write. As Tieken-Boon van Ostade (2010) discussed, there were defenders of the alleged 'female inaccuracy' but, on the other side, there were also fathers, brothers and husbands – their immediate social context – who cared about their speech and invested their time coaching them. In their texts, well-instructed female writers were expected to show some knowledge of classical languages, eminently Latin, awareness of norms of correctness concerning grammar and punctuation, as well as some formality in their style (Tieken-Boon van Ostade 2010).

Concerning scientific writing, the pervasive negotiation between the assertion of the author's role in the advancement of science and the mitigating acknowledgment of the work of the scientific community was especially troublesome for female scholars (Puente-Castelo and Monaco 2016). Their constraints in a group that they accessed as outsiders were bigger (Moskowich 2016). Pragmatic strategies, such as persuasion (Ducrot 1983; Allen, Qin and Lancaster 1994; Bazerman 1998; Moessner 2009; Crespo 2015; Bello 2016) may serve, on the one side, to overcome this gender gap. In this light, the pervasive use of adversative conjunctions may also reflect a subtle persuasive strategy to indicate the disparity of ideas, as in (7).

(7) [A] full ſtatement of theſe defects will be found in Mr. Milne's *Botanical Dictionary*, a book which ſhould be in the hands of all young botaniſts, as much information may be derived from it; *but* it is to be lamented, that the author, inſtead of pointing out the errors of the Linnean ſyſtem with the candour due to a work of ſuch great ability, has marked the ſmalleſt failings with a moſt ungenerous acrimony. (Jacson 1804: 15; emphasis added)

On the other side, female scholars in the eighteenth and nineteenth centuries showed evidence of detachment (Crespo and Moskowich 2015), objectivity, and scrupulousness (Allen, Qin and Lancaster 1994). After all, this may be a result of the expectations of formality laid on them as women and of the need female authors felt to include causal and adversative conjunctions as explicit markers of logical thinking.

3.3 Age of author

Age is generally regarded as an important independent variable in the study of language change and variation (Wagner 2012). In Table 9, the frequencies of conjunctions according to the age of the author are revealed.

Table 9. Frequency of conjunctions according to age of author

Age	No.	nf (10,000)
25–34	2,005	200.2
35–44	3,314	236.4
45–54	1,655	206.5
55+	1,716	245.8

Senior authors had the highest normalised frequencies of use (235.4). Young scholars are positioned on the other side of the balance: they used conjunctions 18.5% less. Age groups 55+ and 25–34 are situated on the far ends of the scale. However, middle-aged authors reverse the situation and make it impossible to create a direct correlation between age and conjunction use, as authors aged 35–44 used 15.3% (236.4) more conjunctions than the youngest group and 12.6% more than those aged 45–55.

The typological breakdown offered in Table 10 shows a steady distribution. All four conjunction categories have a very similar presence in the four groups analysed.

Table 10. Type of conjunction and age of author

Age		Additive	Causal	Adversative	Temporal
25–34	No.	274	716	778	237
	%	13.67	35.71	38.80	11.82
35–44	No.	584	1,174	1,222	344
	%	17.62	35.43	36.87	10.08
45–54	No.	272	495	712	168
	%	16.44	29.91	43.02	10.15
55+	No.	312	512	660	232
	%	18.18	29.84	38.46	13.52

Whereas temporal conjunctions represent 11% to 14% of the total in all four age groups, additive conjunctions range from 13% to 18%. In the case of adversative conjunctions, they represent around 37% of the total in age groups 1, 2 and 4, while in group 3 (45–54), the percentage rises to 43%. Perhaps the most remarkable finding is that younger writers seem to favor explicit causal relations, as evinced by the fact that this typology represents around 35% of the total in the two youngest groups, a figure that drops to approximately 30% in age groups 3 and 4. Although there do not seem to be plausible explanations to account for these trends yet, it may be still useful to delve deeper into these findings.

3.4 Place of education

This variable informs about where authors acquired their scientific writing habits, that is, where they learned to write rather than where they initially learned to speak. It is therefore a direct way to gain information about the production context of writers, and it may also serve to study diatopic variation. Table 11 shows the absolute and normalised frequencies for the four places of education considered in this study.

Table 11. Frequency of conjunctions according to place of education

Place	No.	nf (10,000)
England	4,800	228.5
Ireland	1,134	226.4
Scotland	1,544	192.8
U.S.	1,412	234.9

U.S. writers were the most prolific in the use of conjunctions. Their texts had 234.9 conjunctions per 10,000 words. Their European counterparts used less explicit coherence markers: English and Irish writers used 228.5 and 226.4 conjunctions, respectively, whereas Scottish scholars fell behind with a normalised frequency of 192.8 conjunctions. These results may timidly corroborate the estrangement of the American variety that has been already pointed out (Puente-Castelo and Monaco 2016; Bello 2019). Fear of language fragmentation and a desire for uniformity to arrest language change in the eighteenth century (Bailey 2010) may have held European varieties together, let alone actual political and economic ties.

The distribution of types of conjunction and places of education is offered in Table 12. As in other variables in this study, the four types of conjunctions analysed have a very similar distribution in all four groups.

Table 12. Type of conjunction and place of education

Place		Additive	Causal	Adversative	Temporal
England	No.	898	1,556	1,840	506
	%	18.71	32.42	38.33	10.54
Ireland	No.	137	434	440	123
	%	12.08	38.27	38.80	10.85
Scotland	No.	252	474	597	221
	%	16.32	30.70	38.67	14.31
US	No.	195	489	596	132
	%	13.81	34.63	42.21	9.35

Adversative conjunctions are the most numerous group, with about 38.5% of the total of European texts to 42.2% of American texts. Explicit causal relations constitute more than one third in the case of authors educated in Ireland (38.27%) and the US (34.63%). Their presence in English (32.42%) and Scottish (30.70%) texts is not lagging far behind. These two nations seem to have favored the use of additive conjunctions instead, as 18.71% and 16.32% of these markers fall into this category. Temporal conjunctions close the list. In this case, Scottish authors favored their use (14.31%) in detriment of causal conjunctions, compared to other geographical groups. On the other hand, American authors used fewer temporal conjunctions (9.35%), which in their case resulted in an increase in the use of explicit adversative relations.

3.5 Genre and level of specialisation

The variable of text type informs about authors' purposes. Genres, understood as dynamic mental frames (Taavitsainen 2001) that belong in the cognitive dimension (Parodi 2015), work as horizons of expectation for both writers and readers. As for conjunctions in CELiST, Figure 4 shows that Treatise gathers half the number of total occurrences (4,169 tokens). Guide and Textbook, two academic genres, follow up, with 1,253 and 1,025 occurrences, respectively.

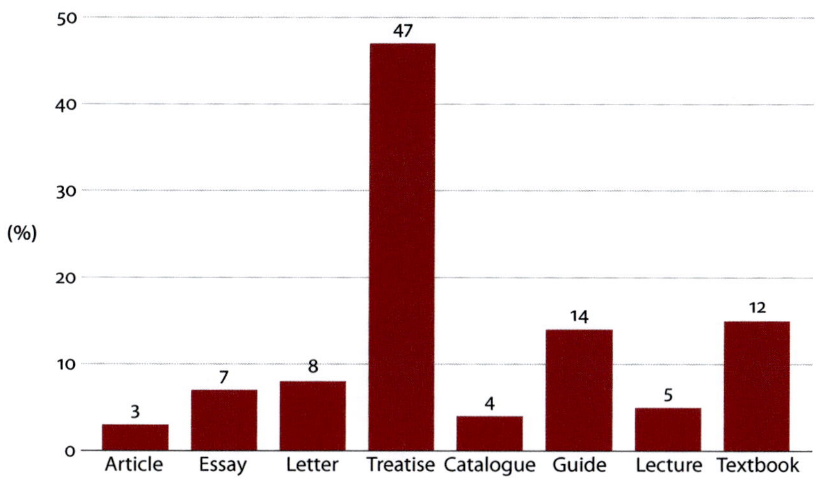

Figure 4. Conjunctions in CELiST according to genre

As the normalised frequencies show in Table 13, non-specialised texts displayed a higher frequency of conjunctions (239.9). This can be related to their function and the dichotomy transmitting / accessing knowledge (Parodi 2014).

Table 13. Frequency of conjunctions according to level of specialisation

Level of specialisation	No.	nf (10,000)
non-specialised	3,130	239.9
specialised	5,760	195.2

Specialised texts show the need to establish links within a group of professionals. Thus, "knowledge is not organised with pedagogical purposes of teaching new concepts, but [that of communicating] contemporary scientific information" (Parodi 2015: 481). The main goal is the transmission of knowledge within a community of practice (Gotti 2013). The interaction between participants establishes an expert relationship where pragmatic assumptions regarding shared knowledge and linguistic practices are expected. The expression of coherence relations implicitly may be one of these expectations.

In non-specialised texts, the emphasis is drawn to the interaction between teacher and students. More specifically, instructors need to provide a textual trail to ease the access and acquisition of new knowledge. New information is carefully organised in "separated units and presented gradually" (Parodi 2015: 481). If devices like nominalisations may ensure the progressive assimilation of new contents (Bello 2016), the expression of explicit relations through conjunctions can be considered proof of this careful organisation of new ideas. The argumentative organisation of texts will help readers create a mental representation of shared assumptions (Kintsch 1998; Van Dijk and Kintsch 1983) and the acquisition of specific reading and writing practices will eventually grant their access to the specialised community (Bhatia 2002).

More detailed information about the distribution of different conjunctions in the levels of specialisation under analysis can be seen Table 14.

Table 14. Type of conjunction and level of specialisation

Level of specialisation		Additive	Causal	Adversative	Temporal
non-specialised	No.	494	996	1,292	348
	%	15.78	31.82	41.28	11.12
specialised	No.	988	1,957	2,181	634
	%	17.15	33.98	37.86	11.01

Explicit adversative relations represent 41.28% of the total number of conjunctions found in non-specialised texts. With 37.86% of the total, their role in specialised texts is also prominent but a bit more limited, as in these texts, causal (33.98%) and additive (17.15%) conjunctions are more numerous than in academic samples. Finally, temporal conjunctions are equally represented in both types of texts, with around 11% of the total.

4. Conclusions

The present study investigated the role of conjunctions in CELiST. The frequency of use and evolution of the four-category model proposed by Halliday and Hasan (1974) were taken as independent variables, together with other four variables, namely sex and age of authors, place of education and level of specialisation. Perhaps the two more interesting findings of the study revolve around the distribution of the different types of relation in the corpus and their use according to the specialisation variable.

Data in Table 6 confirmed that there is a correlation between cognitive complexity and the expression of coherence relations. More basic operations like addition and temporality are less frequently expressed by means of a conjunction. Adversative relations, the most complex ones, are unsurprisingly the most frequently marked through a connective. This is closely related to Asr and Demberg's (2012: 2682) finding that "discourse relations which are discontinuous or temporally non-linear are much more likely to be expressed with an explicit discourse marker than those which are continuous". Their study confirmed the validity of the continuity hypothesis (Segal, Duchan and Scott 1991; Murray 1997) and the causality-by-default hypothesis (Sanders 2005) by addressing the expression of both explicit and implicit coherence relations in an annotated corpus of Contemporary English. The applicability of their results to English scientific register still needs validation. The results obtained in this study may be a good starting point for such an endeavor.

The second finding shows the connection between the level of specialisation of the text and the frequency of explicit coherence marks. Conjunctions are procedural units that guide comprehension processes (Blakemore 1987; Carston 2002). They ease disambiguation, which, in the case of conjunctions, is applied to the establishment of coherence relations between discourse segments. This facilitative role is especially important in instructional texts and, consequently, the results showed a more prominent use of conjunctions in non-specialised texts. Although this might shed some light on the communicative strategies used by English native speakers writing science in the Late Modern period, again, this study can only be seen as a suitable starting point for further research that might result in a more precise description of the English scientific register across the eighteenth and nineteenth centuries.

Works cited

Allen, Bryce; Qin, Jian and Lancaster, Frederik Wilfrid. 1994. Persuasive Communities: A Longitudinal Analysis of References in the Philosophical Transactions of the Royal Society, 1665–1990. *Social Studies of Science*, 24/2: 279–310.

Anscombre, Jean-Claude, and Oswald Ducrot. 1976. L'argumentation dans la langue. *Langages*, 42: 5–27. https://doi.org/10.3406/lgge.1976.2306

Asr, Fatemeh Torabi and Demberg, Vera. 2012. "Implicitness of Discourse Relations". In Kay, Martin and Boitet, Christian (eds.), *Proceedings of COLING 2012*. Mumbai: The COLING 2012 Organizing Committee. 2669–2684.

Bailey, Richard. 2010. "Variation and Change in Eighteenth-Century English". In Hickey, Raymond (ed.), *Eighteenth Century English: Ideology and Change*. Cambridge: Cambridge University Press. 182–199. https://doi.org/10.1017/CBO9780511781643.011

Barsaglini-Castro, Anabella and Valcarce, Daniel. 2020. The Coruña Corpus Tool: Ten Years on. *Procesamiento del Lenguaje Natural*, 64: 13–19.

Bazerman, Charles. 1988. *Shaping Written Knowledge: The Genre and Activity of the Experimental Article in Science*. Madison: The University of Wisconsin Press.

Bhatia, Vijay. 2002. "A Generic View of Academic Discourse". In Flowerdew, John (ed.), *Academic Discourse*. London: Pearson Education. 21–39.

Bello, Iria. 2016. Cognitive Implications of Nominalizations in the Advancement of Scientific Discourse. *International Journal of English Studies*, 16/2: 1–23. https://doi.org/10.6018/ijes/2016/2/262921

Bello, Iria. 2019. "On Cognitive Complexity in Scientific Discourse: A Corpus-Based Study on Additive Coherence Relations". In Moskowich, Isabel; Crespo, Begoña; Puente-Castelo, Luis and Monaco, Leida Maria (eds.), *Writing history in Late Modern English: Explorations of the Coruña Corpus*. Amsterdam: John Benjamins. 260–276. https://doi.org/10.1075/z.225.13vir

Blakemore, Diane. 1987. *Semantic Constraints on Relevance*. Oxford: Blackwell.

Bolívar, Adriana and Parodi, Giovanni. 2015. "Academic and Professional Discourse." In Lacorte, Manel (ed.), *The Routledge Handbook of Hispanic Applied Linguistics*. London: Routledge. 459–476.

Carston, Robyn. 2002. *Thoughts and Utterances: The Pragmatics of Explicit Communication*. Oxford: Blackwell. https://doi.org/10.1002/9780470754603

Crespo, Begoña. 2015. Women Writing Science in the Eighteenth Century: Some Hints about Their Language Use. *Anglica. An International Journal of English Studies*, 24/2: 103–128.

Crespo, Begoña, and Moskowich, Isabel. 2015. Involved in Writing Science: Nineteenth-Century Women in the *Coruña Corpus*. *International Journal of Language and Linguistics*, 2/5: 76–88.

Crespo García, Begoña, and Moskowich, Isabel. 2010. *CETA* in the Context of the *Coruña Corpus*. *Literary and Linguistic Computing*, 25/2: 153–164. https://doi.org/10.1093/llc/fqp038

Culpeper, Jonathan and Kytö, Merja. 2000. "The Conjunction AND in Early Modern English: Frequencies and Uses in Speech-related Writing and other Texts". In Bermúdez-Otero, Ricardo; Denison, David; Hogg, Richard M. and McCully, Christopher B. (eds.), *Generative Theory and Corpus Studies: A Dialogue from 10ICEHL*, (Topics in English Linguistics, 31). Berlin and New York: Mouton de Gruyter. 299–326.

De Cesare, Anna Maria. 2017. "Introduction: on 'Additivity' as a Multidisciplinary Research Field". In De Cesare, Anna Maria and Andorno, Cecilia (eds.), *Focus on Additivity. Adverbial Modifiers in Romance, Germanic and Slavic languages*. Amsterdam: John Benjamins. 1–22. https://doi.org/10.1075/pbns.278.00dec

Ducrot, Oswald. 1983. Operateurs argumentatifs et visée argumentative. *Cahiers de Linguistique Francaise*, 5: 7–36.

Gotti, Maurizio. 2013. "The Formation of The Royal Society as a Community of Practice and Discourse". In Kopaczyk, Joanna and Jucker, Andreas (eds.), *Communities of Practice in the History of English*. Amsterdam: John Benjamins. 269–285. https://doi.org/10.1075/pbns.235.16got

Halliday, Michael and Hasan, Ruqaiya. 1974. *Cohesion in English*. London: Longman.

Hyland, Ken. 2000. *Disciplinary Discourses. Social Interaction In Academic Writing*. London: Longman.

Hyland, Ken. 2011. "Academic Discourse". In Hyland, Ken and Paltridge, Brian (eds.), *The Continuum Companion to Discourse Analysis*. London: Continuum. 171–184.

Kintsch, Walter. 1998. *Comprehension: A Paradigm for Cognition*. Cambridge: Cambridge University Press.

Kleijn, Suzanne; Pander Maat, Henk L. and Sanders, Ted J. M. 2019. Comprehension Effects of Connectives Across Texts, Readers, and Coherence Relations. *Discourse Processes*, 56/5–6: 447–464. https://doi.org/10.1080/0163853X.2019.1605257

Lareo, Inés; Monaco, Leida Maria; Esteve-Ramos, María José and Moskowich, Isabel (comps.). 2020. *The Corpus of English Life Sciences Texts (CELiST)*. A Coruña: Universidade da Coruña. https://doi.org/10.17979/spudc.9788497497848

Mann, William; Matthiessen, Christian and Thompson, Sandra. 1992. "Rhetorical structure theory and text analysis". In Mann, William and Thompson, Sandra (eds.), *Discourse Description: Diverse Linguistic Analyses of a Fund-Raising Text*. Amsterdam: John Benjamins. 39–78. https://doi.org/10.1075/pbns.16.04man

Mann, William and Thompson, Sandra. 1987. Rhetorical Structure Theory: A Theory of Text Organization. *ISI/RS Report*, 87/1: 2–82.

McEnery, Tony and Wilson, Andrew. 1996. *Corpus Linguistics*. Edinburgh: Edinburgh University Press.

Moessner, Lilo. 2009. The Influence of the Royal Society on 17th-Century Scientific Writing. *ICAME Journal*, 33: 65–87.

Moskowich, Isabel. 2012. CETA as a tool for the study of modern astronomy in English. In Moskowich, Isabel and Crespo, Begoña (eds.), *Astronomy 'playne and simple': the writing of Science between 1700 and 1900*. Amsterdam/Philadelphia: John Benjamins. 35–57. https://doi.org/10.1075/z.173.03mos

Moskowich, Isabel. 2016. Lexical Richness in Modern Women Writers: Evidence from the Corpus of History English Texts. *Revista Canaria de Estudies Ingleses, Monographic Issue: Women Scientists, Women Travellers, Women Translators: their Language and their History*, 72: 110–128.

Moskowich, Isabel. 2017. Genre and Change in the Corpus of History English Texts. *Nordic Journal of English Studies*, 16: 84–106. https://doi.org/10.35360/njes.413

Moskowich, Isabel. 2021. "The making of CELiST, a bunch of disciplines". In Moskowich, Isabel; Lareo, Inés and Camiña, Gonzalo (eds.), *"All families and genera": Exploring the Corpus of English Life Sciences Texts*. Amsterdam/Philadelphia: John Benjamins. 1–19. https://doi.org/10.1075/z.237

Moskowich, Isabel and Crespo García, Begoña. 2007. "Presenting the *Coruña Corpus: A Collection of Samples for the Historical Study of English Scientific Writing*." In Pérez Guerra, Javier et al. (eds.), *'Of Varying Language and Opposing Creed': New Insights into Late Modern English*. Bern: Peter Lang. 341–357.

Moskowich, Isabel and Parapar López, José. 2008. "Writing Science, Compiling Science. The *Coruña Corpus of English Scientific Writing*." In Lorenzo Modia, María Jesús (ed.), *Proceedings from the 31st AEDEAN Conference*. A Coruña: Universidade da Coruña, Servizo de Publicacións. 531–544.

Murray, John. 1997. Connectives and Narrative Text: The role of Continuity. *Memory and Cognition*, 25/2: 227–236. https://doi.org/10.3758/BF03201114

Parapar López, Javier and Moskowich, Isabel. 2007. The Coruña Corpus Tool. *Revista de Procesamiento del Lenguaje Natural*, 39: 289–290.

Parodi, Giovanni. 2014. Genre Organisation in Specialised Discourse: Disciplinary Variation across University Textbooks. *Discourse Studies*, 16/1: 65–87. https://doi.org/10.1177/1461445613496355

Parodi, Giovanni. 2015. Variation across University Genres in Seven Disciplines. A Corpus-Based Study on Academic Written Spanish. *International Journal of Corpus Linguistics*, 20/4: 469–499. https://doi.org/10.1075/ijcl.20.4.03par

Parodi, Giovanni; Burdiles, Gina; Moreno de León, Tomás and Julio, Cristóbal. 2018. Hábitos lectores y géneros del discurso en filosofía y en economía y negocios: del discurso académico al discurso profesional. *Revista de lingüística teórica y aplicada*, 56/2: 117–152. https://doi.org/10.4067/S0718-48832018000200117

Puente-Castelo, Luis and Monaco, Leida María. 2016. The might of 'might': A Mitigating Strategy in Eighteenth and Nineteenth Century Female Scientific Discourse. *Revista Canaria de Estudios Ingleses*, 72: 143–168.

Rudolph, Elisabeth. 1996. *Contrast: Adversative and Concessive Relations and their Expressions in English, German, Spanish, Portuguese on Sentence and Text level*. New York / Berlin: de Gruyter. https://doi.org/10.1515/9783110815856

Sanders, Ted. 2005. "Coherence, Causality and Cognitive Complexity in Discourse". In Aurnague, Michel, Bras, Myriam, Le Draoulec Anne and Vieu, Laure (eds.), *Proceedings/Actes SEM-05, First International Symposium on the Exploration and Modelling of Meaning*. Toulouse: Université de Toulouse – Le Mirail. 105–114.

Sanders, Ted and Pander Maat, Henk. 2006. "Cohesion and Coherence: Linguistic Approaches". In Brown, Keith (ed.), *Encyclopedia of Language and Linguistics* (Volume 2). London: Elsevier. 591–595. https://doi.org/10.1016/B0-08-044854-2/00497-1

Sanders, Ted; Spooren, Wilbert and Noordman, Leo. 1992. Toward a Taxonomy of Coherence Relations. *Discourse Processes*, 15: 1–35. https://doi.org/10.1080/01638539209544800

Segal, Erwin; Duchan, Judith and Scott, Paula. 1991. The Role of Interclausal Connectives in Narrative Structuring: Evidence from Adults' Interpretations of Simple Stories. *Discourse Processes*, 14/1: 27–54. https://doi.org/10.1080/01638539109544773

Swales, John. 1990. *Genre Analysis. English in Academic and Research Settings*. Cambridge: Cambridge University Press.

Swales, John. 2004. *Research Genres: Explorations and Applications*. Cambridge: Cambridge University Press. https://doi.org/10.1017/CBO9781139524827

Taatvitsainen, Irma. 2001. Changing Conventions of Writing: The Dynamics of Genres, Text Types, and Text Traditions. *European Journal of English Studies*, 5/2: 139–150. https://doi.org/10.1076/ejes.5.2.139.7309

Tieken-Boon van Ostade, Ingrid. 2010. "Eighteenth-Century Women and the Norms of Correctness". In Hickey, Raymond (ed.), *Eighteenth-Century English: Ideology and Change*. Cambridge: Cambridge University Press. 59–72. https://doi.org/10.1017/CBO9780511781643.005

Van Dijk, Teun and Kintsch, Walter. 1983. *Strategies of Discourse Comprehension*. New York: Academic Press.
Wagner, Suzanne Evans. 2012. Age Grading in Sociolinguistic Theory. *Language and Linguistics Compass*, 6/6: 371–382. https://doi.org/10.1002/lnc3.343

Appendix 1. Conjunctions not found in the corpus

Additive	*add to that, additionally, along the same lines, also, alternatively, as a matter of fact, by extension, by the same token, either (… or), furthermore, I mean, in a like manner, into the bargain, let alone, much less, neither (… nor), not only (this)… but also (that), not to mention (this), on top of that, or, over and above that, (to) put (it) another way, to say nothing of, to tell you the truth, too, specifically, thus, what is more*
Causal	*apart from this, arising out of this, arising out of this, as, as a consequence, aside from this, being that, even if, for fear that, for the (simple) reason that, forasmuch as, granted (that), if not, if so, in other respects, in that case, in the hope that, in such an event, in that, in the event that, in view of (the fact that), insofar as, on (the) condition (that), on this/that account, on this basis, only if, otherwise, provided that, providing that, since, so much (so) that, thanks to (the fact that), that being the case/so, to avoid the risk of, to the extent that, to this end, under the (these/those/other) circumstances, with this in mind, with this intention*
Adversative	*admittedly, albeit, all the same, anyhow, as a matter of fact, at the same time, be that as it may, but even so, by (way of) contrast, despite (this/the fact that), either way, even supposing, even though, for all that, granted (this), granted that, in contrast (to), in either (any) case, in either (any) event, in spite of (this), just the same, nonetheless, nor yet, notwithstanding (the fact) that, still, when in fact, whatever happens, whichever happens, whichever way it is, yet*
Temporal	*an hour later, at this moment, before that, here from now, just then, meanwhile, then, thereupon, until then, up to now*

Appendix 2. Conjunctions found in the corpus

Additive conjunctions

Conjunction	No.	%	% (without *and*, *or*)
and	12,924	72.45	
or	3,432	19.24	
as well	166	0.93	11.20
nor	157	0.88	10.59
especially / efpecially	134	0.75	9.04
likewise	122	0.68	8.23
such as / fuch as	108	0.61	7.29

Conjunction	No.	%	% (without *and, or*)
besides	95	0.53	6.41
indeed	84	0.47	5.67
further	75	0.42	5.06
particularly	64	0.36	4.32
equally	48	0.27	3.24
concerning	44	0.25	2.97
for inſtance	30	0.17	2.02
that is	27	0.15	1.82
for example	26	0.15	1.75
namely	25	0.14	1.69
of course /of courſe	24	0.13	1.62
including	20	0.11	1.35
the fact that	18	0.10	1.21
moreover	17	0.10	1.15
on the other hand	17	0.10	1.15
similarly	15	0.08	1.01
in fact / in fact	14	0.08	0.94
as for	13	0.07	0.88
conſidering	13	0.07	0.88
that is to say	12	0.07	0.81
regarding	11	0.06	0.74
in other words	10	0.06	0.67
in particular	10	0.06	0.67
on the one hand	10	0.06	0.67
in the ſame way	9	0.05	0.61
accordingly	8	0.04	0.54
in addition	8	0.04	0.54
to illustrate	7	0.04	0.47
by the way	5	0.03	0.34
in other respects / in other reſpects	5	0.03	0.34
ſpecifically	5	0.03	0.34
after all	4	0.02	0.27
on the ſubject of	4	0.02	0.27
or else /or elſe	4	0.02	0.27
on that account	3	0.02	0.20
surely	3	0.02	0.20
what is more	3	0.02	0.20
I mean	2	0.01	0.13
on this account	2	0.01	0.13
notably	1	0.01	0.07
Total	**17,838**		
Total without *and, or*	**1,482**		

Causal conjunctions

Conjunction	No.	%
then	760	25.70
if	726	24.55
ſo that/so that	288	9.74
thus	249	8.42
therefore	244	8.25
because	159	5.38
ſo/so	64	2.16
hence	60	2.03
for	58	1.96
conſequently /consequently	52	1.76
otherwiſe / otherwise	40	1.35
in order to	39	1.32
ſo as to/ so as to	37	1.25
unleſs/ unless	27	0.91
due to	26	0.88
owing to	20	0.68
for the purpoſe of / for the purpose of	19	0.64
in conſequence / in consequence	13	0.44
for this purpoſe / for this purpose	12	0.41
in this respect	11	0.37
for this reaſon / for this reason	9	0.30
as long as	8	0.27
in caſe / in case	8	0.27
in order that	6	0.20
in that caſe/ in that case	5	0.17
inasmuch	4	0.14
ſeeing that	4	0.14
as a result of	1	0.03
given that	1	0.03
it follows	1	0.03
on account of this	1	0.03
to the end that	1	0.03
Total	**2,953**	

Adversative conjunctions

Conjunction	**No.**	**%**
but	2,029	58.69
while	274	7.93
rather	265	7.67
though	264	7.64
however	245	7.09
although	78	2.26
inſtead/instead	74	2.14
at least	57	1.65
and yet	30	0.87
notwithstanding	23	0.67
whereas	23	0.67
neverthele∫s / nevertheless	21	0.61
on the contrary	21	0.60
whilst	20	0.58
actually / actually	19	0.55
conversely	7	0.20
at any rate	6	0.17
in any case	4	0.12
in either case	4	0.12
meanwhile	3	0.09
on the other hand	3	0.09
even if	2	0.06
regardless	1	0.03
Total	**3,473**	

Temporal conjunctions

Conjunction	**No.**	**%**
firſt / first	137	13.95
ſoon / son	115	11.71
at the ſame time / at the same time	80	8.15
at firſt / at first	70	7.13
afterwards	66	6.72
hitherto	61	6.21
formerly	41	4.18
at once	40	4.07
earlier	40	4.07
ſo far / so far	38	3.87
originally	33	3.36
eventually	28	2.85

Conjunction	No.	%
hereafter	26	2.65
next	25	2.55
finally	24	2.44
at laſt / at last	17	1.73
previouſly / previously	16	1.63
as yet	15	1.53
secondly	15	1.53
sequential	14	1.43
subsequently	14	1.43
laſtly / lastly	13	1.32
briefly	8	0.81
thirdly	8	0.81
to return to	8	0.81
in ſhort / in short	5	0.51
after that	4	0.41
to conclude	4	0.41
to the point	4	0.41
after a time	3	0.31
henceforth	2	0.20
next day	2	0.20
to reſume	2	0.20
at this point	1	0.10
in conclusion	1	0.10
next time	1	0.10
to sum up	1	0.10
Total	**982**	

CHAPTER 15

Spotting register-internal variation in eighteenth- and nineteenth-century life sciences

Descriptiveness and argumentation in the Corpus of English Life Sciences Texts

Leida Maria Monaco
University of Oviedo

1. Introduction

Although it is still one of the most popular current trends in corpus linguistics, language variation has been an object of historical, sociological and anthropological interest for almost seventy years, with pioneering studies by Hymes (1962), Labov (1963, 1965), or Weinreich *et al.* (1968), among others, analysing language in the light of a number of social factors, such as the speakers' religion, ethnicity, education, or gender, and the impact they have on the speech community. The compilation of large electronic corpora from the 1980s onwards was crucial for the spread of register variation studies on a large scale, making it possible to compare different registers in different languages, often from a diachronic perspective.

Although the corpus described in this volume may seem relatively homogeneous in that it represents only one English scientific register – that of Life Sciences– in a particular timespan (1700–1900), it has been shown that variation found within a register may reveal unexpected patterns (Biber and Gray 2013) that could be helpful in comparing and contrasting different subregisters. While eighteenth- and nineteenth-century Life Sciences has been demonstrated to be a largely descriptive discipline on the map of Late Modern English science (Lareo and Esteve-Ramos 2008; Zea 2015; Moskowich and Monaco 2016; Monaco 2017), a closer look at variation within the register yields a more accurate picture. The aim of this chapter is to analyse CELiST from a genre and sex-of-the-author perspective with respect to a dimension of variation measuring descriptiveness (and/or argumentation on its opposite pole), which, in turn, is the result of a multidimensional analysis of register variation in Late Modern scientific English (Monaco 2017), following Biber's (1988) method.

https://doi.org/10.1075/z.237.15mon

Section 2 offers a brief background on variation studies in historical English, focusing on scientific and academic registers in particular. A description of the methodology followed in the present study – Biber's (1988) Multidimensional Analysis and its application on the Coruña Corpus (Monaco 2017) – and CELiST in particular– are offered in Section 3, results are analysed in Section 4 and, finally, some tentative conclusions are drawn in Section 5.

2. Register variation and change in historical and scientific English

If we are going to talk about register variation, we need first to define the somewhat polemical term *register*, which, along with *genre*, has been broadly debated in the literature (see Ferguson 1983; Hymes 1984; Biber 1988; Swales 1990, 2004; Conrad 2001; Bhatia 2002; Biber and Conrad 2009, among others). For this study I will be using the definition adopted in Biber and Conrad (2009: 6), whereby *register* stands for "a variety associated with a particular situation of use (including particular communicative purposes)" and can be described with regard to "three major components: the situational context, the linguistic features, and the functional relationship between the first two components". This definition allows any particular register – for instance, scientific prose, fiction or legalese– to be characterised in function of its pervasive lexical and grammatical features, rather than textual organisation, and, therefore, seems to be particularly suitable for corpus studies, which often analyse text samples, rather than full texts.

Genre, by contrast, will be used here as a type of text responding to a particular rhetorical organisation, its characteristic features being of a structural, rather than lexical and/or grammatical character (Biber and Conrad 2009: 7, 16–17). According to this definition, genres must therefore be considered in their totality for their description. For instance, in order to describe a scientific research article, we would need to take into account its structure (typically known as IMRAD, thanks to the Introduction-Methods-Results-And-Discussion sequence, generally accepted nowadays as standard). Another example would be a letter, which, regardless of its subject matter, would usually contain a greeting, followed by a few introductory lines, the body dealing with the matter discussed, and a closure, with a signature at the end. This said, Biber and Conrad's (2009) framework allows any given text to be analysed either as a genre (with a focus on its structure when considered as a whole) or as a register, or subregister of a broader register. An illustration of this could be, precisely, a scientific research article, a scientific letter, or a scientific essay, each of which may be formally considered as three different genres with a particular structural organisation, and yet the three of them being also subregisters of a broader register category, the scientific prose. According to the latter, each of these subregisters may be described in terms of their pervasive lexical and grammatical features fulfilling a particular situational function. This has been a recurring topic in variation studies following the multidimensional methodology and those stemming from it (see below).

Scientific and academic, as well as otherwise specialised English appear, indeed, to be a focal point of both synchronic and diachronic linguistics. Specialised historical corpora, such as the Corpus of Early English Correspondence (CEEC; see Nevalainen and Raumolin-Brunberg 1994), the Corpus of Early English Medical Writing (CEEM, as in Taavitsainen *et al.* 2010, and Taavitsainen and Pahta 2010), as well as Davies' (2007) sample of twentieth-century news reportage articles, published in the *Time* Magazine, allow a wide range of register variation research which permits to trace the evolution of politeness formulas (Nevalainen and Raumolin-Brunberg 1995) and code-switching structures (Pahta 2003, 2004), or to uncover subtle patterns of variation and change within a register, thus revealing a number of different subregisters, each with their own tendencies and behaviour (Biber and Gray 2013). On a larger scale, a virtually inexhaustible mine for diachronic research is, of course, ARCHER (A Representative Corpus of Historical English Registers; Biber *et al.* 1994), an on-going project which inspired, already in its early stages, several studies on register variation and change which are nowadays considered a reference in the field of historical linguistics (Biber and Finegan 1989, 2001; Atkinson 1999), and continues to inspire many others (see, for instance, Rissanen 2011; Pérez-Guerra 2012; Núñez-Pertejo 2013; or Rodríguez-Puente 2013, 2016).

In this line, the Coruña Corpus of English Scientific Writing (hereafter CC) – the "tree" on which CELiST (Lareo *et al.* 2020), the corpus on Life Sciences described in this volume of studies, is a branch– contains a growing number of scientific English subregisters sampled through texts published between 1700 and 1900 (see Moskowich and Crespo 2007; Moskowich and Parapar 2008, and Moskowich 2021), and has therefore contributed to a wide range of inquiries into the nature and peculiarities of Late Modern English scientific discourse. Among these, studies dealing with particular aspects in the evolution of scientific English during that period, such as morphology and lexis (Camiña-Rioboo 2012, 2013; Bello 2014), syntax (Lareo and Esteve-Ramos 2008; Lareo 2012; Gray and Biber 2012), semantics (Crespo 2011, 2012; Dossena 2016) or pragmatics (Alonso-Almeida and Lareo 2016; Puente-Castelo 2016, 2017, 2019), are complemented with others offering a more macroscopic view on register variation and change (Moskowich and Monaco 2014; Monaco 2017, 2019; Puente-Castelo 2021). Rather than focusing on a given lexical or grammatical category, the latter resort to multivariate statistics as used in Biber's (1988) Multidimensional Analysis to obtain a bigger picture of variation, both within and across the different subregisters represented in the CC. The method applied for this purpose and which is also the one we are using in the present study is explained in what follows.

3. Methodology: The Multidimensional Analysis applied

The Multidimensional (hereafter MD) Analysis was the result of Biber's (1988) attempt to explain the fundamental differences between speech and writing, as seen through a large number of spoken and written English registers, represented through different corpora. For this purpose, he used a multivariate statistical technique of data reduction known as factor analysis, which transforms a very large number of variables into a much smaller set. In this case, factor analysis groups several linguistic (lexical and grammatical) features into factors, each factor containing features that tend to co-occur in some texts because they share an underlying communicative function. In addition, each factor will normally yield two sets of linguistic features, rather than just one, called "positive" and "negative" because, pragmatically, they are in binary opposition, conveying opposite communicative functions (e.g. involved vs. informational; elaborate vs. spontaneous, etc.). For this reason each factor is regarded as a "dimension" of variation with respect to the communicative function interpreted through its features, whereby a given text has a position on this factor – or factor score– which responds to the individual co-occurrence patterns of that text, and may therefore be characterised as more or less informational, more or less elaborate, and so on. This, in turn, allowed Biber to also characterise individual registers with regard to each factor, or each communicative dimension. His pioneering five-dimensional model was the first one to compare and contrast a large number of contemporary English registers with regard to the following dimensions of variation: (1) Informational vs. involved production; (2) Narrative vs. non-narrative concerns; (3) Explicit vs. situation-dependent reference; (4) Overt expression of persuasion, and (5) Abstract vs. non-abstract style.

To date there have been many studies which replicated this technique on a wide range of corpora. While some apply Biber's (1988) original model on a specific register in order to compare it with respect to the general variability found among the different English registers – including diachronic studies, such as Atkinson (1999), mentioned earlier, as well as Biber and Finegan (2001) – others run a factor analysis of their own that yields different results, specific to each corpus (Lee (1999) on the BNC; Conrad (2001) on biology and history textbooks; Gray (2011) on a number of contemporary scientific English registers; Grieve (2016) on samples showing variation in American English, and so on). In these cases, it is usual for the registers analysed to reveal their own variation patterns, or dimensions. However, many of such studies have repeatedly yielded certain dimensions that appear to be common to different registers, and even to certain languages. As explained by Biber (2016: 16), "two linguistically similar parameters of variation have emerged repeatedly across MD studies: a basic oral/literate parameter of variation, and a narrative/non-narrative dimension", something that seems to demonstrate that certain functions in language appear to be, indeed, universal. This was likewise the case with Monaco's (2017) MD study of three registers in the CC: CETA (the Corpus of English Texts on Astronomy; see Moskowich and Crespo 2012), CEPhiT (the Corpus of English Philosophy Texts; see Moskowich *et al.* 2016), and CELiST (in its beta version at that time; see Lareo 2011), which revealed four dimensions of variation in this representation of Late Modern scientific English: (1) Involved/persuasive vs. impersonal style; (2) Argumentative vs. descriptive focus; (3) Elaborate vs. non-elaborate discourse, and (4) Narrative vs. non-narrative discourse, where dimensions 1 and 4 correspond to the aforementioned oral/literate and narrative/non-narrative parameters.

In the present study we are going to look at CELiST with respect to Dimension 2: Argumentative vs. descriptive focus. This dimension "isolates Life Sciences to some extent as a fundamentally descriptive discipline" (Monaco 2017: 227) with regard to Philosophy and Astronomy, presumably, due to the fact that a large percentage of eighteenth- and nineteenth-century Life Sciences was devoted to the cataloguing of different animal and vegetal species and to a detailed description of their external appearance and their internal organisation. Dimension 2 contains two opposing sets of lexical and grammatical features,[1] illustrated below in Examples (1) to (10), all of them taken from the CC. Positive features, or those presenting positive (i.e. above zero)[2] factor loadings are conjuncts, predictive modals, causative adverbs, certain (labelled as 'other') adverbial subordinators, and conditional subordinators, all of which tend to co-occur in some texts, functioning as markers of argumentation:

1. The linguistic features searched for in Monaco (2017) correspond to those analysed in Biber (1988). Closed lists of conjuncts, predictive modals, causative adverbs, other adverbial subordinators, conditional subordinators, hedges, downtoners and amplifiers can be found in Biber (1988: 235–239).

2. The number on a factor loading is not a measure of quantity, but an indicator of co-occurrence; the higher the factor loading, the stronger the co-occurrence. Above or below zero, in turn, indicate opposite polarities in terms of communicative function.

- Conjuncts:

  (1) Liberty, *therefore*, or a power to act or not to act, to do this or another thing under the same causes, is an Impossibility and Atheitical. (CEPhiT, Collins 1717)

- Predictive modals:

  (2) but, if it is before or behind the thread, it *will* be seen on it only from a direct front view. (CETA, Chauvenet 1871)

- Causative adverbs:

  (3) The second expression marked in italics can not be taken literally; *because* inequality is essential to preponderance. (CELiST, Lawrence 1819)

- Other adverbial subordinators:

  (4) and I say, that, *since* there must have been something from eternity, because there is something now, the eternal Being must be an intelligent Being. (CEPhiT, Bolingbroke 1754)

- Conditional subordinators:

  (5) therefore *if* we move from the Equator, towards either Pole, that Pole will accordingly rise above our Horizon. (CETA, Charlton 1735)

Negative features, or features which presented factor loadings below zero on this dimension, are, in turn, those generally involved in description: attributive adjectives and predicative adjectives (descriptors par excellence), as well as hedges, downtoners and amplifiers, most of which are adverbs and, as such, accompany adjectives so as to modulate the latter in one way or another:

- Attributive adjectives:

  (6) In some cases an *external* less *granular* layer of the ovum itself has been described as a *special* membrane. (CELiST, Balfour 1880)

- Predicative adjectives:

  (7) Those of the European Species of Coneys or Rabbits are very *scarce* in this Province, and are to be met with but in few places. (CELiST, Brickell 1737)

- Hedges:

  (8) They all move in elliptic orbits; for as to parabolic orbits, we may always substitute elliptic ones *more or less* elongated, which will satisfy observations equally as well. (CETA, Garland 1838)

- Downtoners:

  (9) Let us, therefore, use a little Freedom, and call them IMPRESSIONS, employing that Word in a Sense *somewhat* different from the usual. (CEPhiT, Hume 1748)

- Amplifiers:

(10) Its own dimensions, vast as they are when expressed in terms of any linear unit with which we are familiar, are *utterly* insignificant when compared with its distance from the fixed stars. (CETA, Bartlett 1855)

The co-occurrence of some of these linguistic features, either positive or negative, as can be particularly appreciated through Example (5), where three negative features – conjuncts, predictive modals, and conditional subordination– appear together, or Examples (8) to (10), where a hedge, a downtoner and an amplifier appear next to an adjective, corroborates here the mutual dependency of co-occurring features in a multidimensional model.

Before running the factor analysis with R, the linguistic features were searched for in the CC with the Coruña Corpus Tool (CCT; see Moskowich 2021; Camiña and Lareo 2016) and the CQPweb (https://cqweb.lancs.ac.uk/coruñav1/index.php; see Hardie 2012, 2016). In order to find out the position any given text occupies with regard to a particular dimension, factor scores were calculated for each text, after which mean scores were calculated for each subregister (i.e. corresponding to a particular scientific discipline, genre or sex of the author). Results are analysed in what follows.

4. Analysis of data

4.1 Variation across time and disciplines

Figure 1 plots Dimension 2 scores for CELiST and two other subcorpora of the CC – CETA and CEPhiT– in the eighteenth and nineteenth centuries in order to describe the register of Life Sciences with regard to that of Astronomy and Philosophy in the same period. Positive scores (on the right) indicate the presence of features

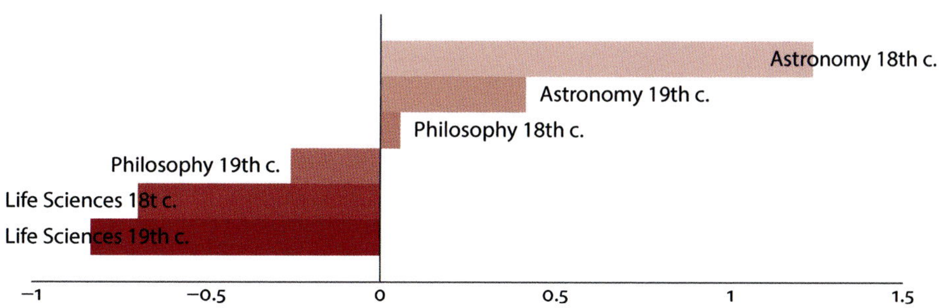

Figure 1. Relations among three scientific disciplines from the Coruña Corpus across two centuries along Dimension 2 "Argumentative vs. descriptive focus" (descending order)

conveying argumentative style, as well as little to no presence of descriptive features, whereas negative scores (on the left) indicate precisely the opposite picture – a much higher frequency of descriptive features and a relatively lower one of argumentative ones. The closer a discipline stands to the mean, the more balanced its distribution of positive and negative features.

As shown on Figure 1, it is eighteenth-century Astronomy that contains the highest proportion of linguistic features expressing logical relationships, even though decreasing considerably in the nineteenth century, but still staying above the mean. Life Sciences, by contrast, clearly appears as a descriptive discipline, both in the eighteenth and nineteenth centuries. Examples (11) and (12), which belong to eighteenth-century Astronomy and Life Sciences, respectively, illustrate this notable difference between two scientific disciplines which, at that time, could be both considered observational:

(11) Again, because the square of AYE is to the square of TA as TA to TG; therefore AYE will be to TA in the subduplicate ratio of TA to TG; therefore the rectangle contained by AYE, Tx, will be to the rectangle contained by TA, Tx, in the subduplicate ratio of TA to TG: and because the rectangle contained by TA, Tx, is to the square of TA as Tx to TA; that is, in the subduplicate ratio of TP to TA. (CETA, Stewart 1761)

(12) THE Sea Aster, or Sea Starwort, is a pretty plant, found growing in salt marshes, near the coast, and belong to the family Composit. It is seldom above a foot high, erect or decumbent at the base; slightly branched, the leaves linear, entire, and somewhat succulent. The flower-heads are in a compact corymb; the involucral bracts few and oblong. The centre of the disk of the flower is bright yellow, the circumference blue or purple. (CELiST, Lankester 1879)

Example (11) is a clear illustration of astronomical observations of a largely mathematical character which, as such, need all kinds of logical connectors to establish relationships of causality and proportional dependences. Example (12), in turn, shows a picture of an aquatic plant, 'painted' by means of all kinds of adjectives–which in turn are mitigated with downtoners such as *slightly* or *somewhat*–in order to provide as detailed a description as possible.

Back on Figure 1, Philosophy occupies a more centred position, closer to the average distribution of positive and negative features in the whole of the corpus analysed in the eighteenth century, and tending towards descriptiveness in the nineteenth, as shown through Examples (13) and (14) below, respectively:

(13) Since these simple ideas therefore are the foundations of human knowledge, this knowledge can neither be extended wider, nor elevated higher, than in a certain proportion to them. If we consider these ideas like foundations, they are extremely narrow, and shallow, neither reaching to many things, nor laid deep in the nature of any. (CEPhiT, Bolingbroke 1754)

(14) It has thus been shown incidentally that even Hume is not perfectly faithful to his philosophical principles. The Ego the central principle of connection is found to resist his efforts. Time and Space are also principles of connection which he first tries to ignore or explain away, and then unwarrantably assumes. The four relations of ideas which he assumes as intuitively perceived, are likewise inconsistent additions to a theory which professes to found on pure particulars. (CEPhiT, Seth 1885)

In contrast to Example (13), an extract from the mid-eighteenth century philosophical essay, which deals essentially with logical reasoning aided by adverbial subordinators such as *since*, conjuncts such as *therefore*, or the conditional clause starting with *If…*, Example (14), taken from a late nineteenth-century lecture, appears to contain fewer logical connectors (see below), and more attributive adjectives used here for the description of concepts (*philosophical principles*), a philosopher's attitude towards the latter (*not perfectly faithful*), or the limitations of his philosophical theory (*inconsistent additions, pure particulars*). Still, in both examples a more balanced distribution of argumentative and descriptive features can also be perceived through the presence of a few predicative adjectives (*extremely narrow and shallow*) in Example (13), and some logical connectors (*thus, likewise*) in Example (14).

All in all, although this analysis of variation across disciplines characterises Life Sciences as a clearly descriptive one, in contrast to Astronomy and Philosophy, a look at the variation found across the different genres in the former, offered in the next subsection, shows a more detailed picture.

4.2 Variation across genres

As explained earlier in this volume (see Moskowich 2021), the texts in CELiST are classified according to six genres: Treatise, Textbook, Essay, Lecture, Article, and Letter. Most of these genres are represented by samples both from the eighteenth and the nineteenth centuries, except for Essay, of which we only have samples from the 1700s, and Article and Lecture, present only in the 1800s half of the corpus.

Figure 2 below plots the Dimension 2 scores for all six genres in CELiST, ranging from moderately argumentative to clearly descriptive:

As we can see on Figure 2, not all the genres represented in CELiST are equally descriptive, Essay being clearly on the argumentative end of the scale, while the remaining genres are not descriptive to a similar extent. Rather, while some genres show a very strong presence of descriptive features, such as eighteenth-century Letter and Textbook, or, to a slightly lower degree, nineteenth-century Treatise and Textbook, others such as Article and Lecture, both from the 1800s, show a more moderate descriptive score, while the nineteenth-century Letter loses its strongly

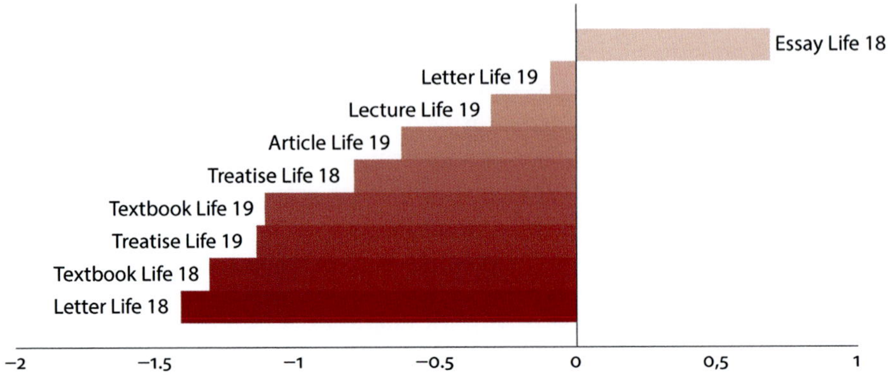

Figure 2. Scores for six genres in eighteenth- and nineteenth-century CELiST along Dimension 2 "Argumentative vs. descriptive focus" (descending order)

descriptive character and shows a much more balanced distribution of positive and negative features, as shown in examples (15) and (16):

(15) It is somewhat larger than a common House Sparrow, and has a conical, straight, sharp bill, of a light carnation colour. Its feathers are a confused assemblage of all the most lively and beautiful colours in nature: among these, yellow, scarlet, green, and a blackish purple, or indigo colour, have the greatest share: besides these, there are white, black, and blue. All these colours are mixed with such beautiful disorder, that it is impossible to convey an idea of their disposition. (CELiST, Bancroft 1769)

(16) In another kind, the individuals do not divide and widen as they grow higher, and can not therefore, by spreading, fill up the spaces between, which enlarge with their increasing height; but in those spaces the new buds form, thus filling all the intervals, and making a coral mass covered all over with thousands of closely packed pits, which mark the spots occupied each by a little animal. (CELiST, Agassiz 1859)

Extracts (15) and (16) above are both letters on Life Sciences, the first one from the eighteenth century, and the second one from the nineteenth. The former gives a very detailed and particularly colourful description of a sparrow, with an abundance of mostly attributive adjectives (*conical, straight, sharp bill; greatest share; beautiful disorder*). The description in this passage is entirely physical, without any reference to the behaviour or any movements whatsoever of the bird. The latter, in turn, describes a coral, which, being a marine invertebrate, resembles a plant very much. One would expect such a description to be more static than that of a bird, yet here, besides a few adjectives (*increasing height, closely packed pits, little animals*) we find verbs conveying actions (*divide, widen, grow, form, fill, occupy*) and which, for that purpose, require certain connectors of causality and logic (*therefore, thus*).

This difference might be owing to the fact that in the 1700s Life Sciences was more concerned with cataloguing and thus offering a more thorough yet more static description of the different species, both animal and vegetal, as well as of their constitution, whereas in the nineteenth century their behaviour gradually started to occupy a more important position.

Examples (17) and (18), in turn, are both instances of eighteenth-century Essay, which on Figure 2 shows a fairly strong positive score, indicating a high frequency of argumentative features:

(17) If the Sphere consists of Particles that attract in a reciprocal quadruplicate Proportion of their distances, the Force by which a Corpuscle will be drawn to the Sphere will be as [formula]. Now when the Corpuscle comes to touch the Sphere, PI becomes =0, and consequently whatever is divided by it, becomes infinite, and therefore the attractive Force of the Sphere, at the Contact being proportional to [formula] will be infinite. (CELiST, Keill 1717)

(18) And since Names are only distinguishing Marks of Ideas, and as we are near certain the Herring hath subsisted as long as the Creation, we need not cavil at particular Words, since they must be expressed by some. It is evident the Sprats are young Herrings, since they appear immediately after the Herrings are gone, and seem to be the Spawn just vivified (if I may use that Expression) and a more undeniable Proof of their being so, is their Anatomy, since on the closest search no Difference, but Size can be found between them. (CELiST, Dodd 1752)

While Example (17) appears argumentative *par excellence*, showing a strong co-occurrence of several features expressing logical relationships (conditional subordination, predictive modals (*will*) and conjuncts such as *consequently* or *therefore*), with few adjectives (*reciprocal quadruplicate Proportion*), Example (18) also shows a relative density of argumentative features, though with a lesser variety in their categories (with the logical connector *since* being used a few times, and one conditional structure expressing politeness). However, the latter also contains a number of adjectives (*distinguishing Marks, near certain, particular Words, evident, young Herrings, undeniable Proof*), which add a proportion of descriptive features to this extract. As suggested elsewhere (Crespo 2016; Moskowich 2016), the argumentative density of Essays might be determined by the characteristics of the genre, which often gives shape to a highly formal, elaborate kind of discourse, and which in turn is closely related to the subject matter dealt with through this genre (here, as it appears, physics in the first example and the anatomy of a herring in the second one).

We have seen, thus, how some of the subregisters found within a register may show different trends, revealing new variation patterns in the latter. A glimpse at male/female author variation is offered in what follows.

4.3 Variation across male and female scientific discourse

Figure 3 plots mean dimension scores for texts written by male and female authors in CELiST, CETA and CEPhiT in the eighteenth and nineteenth centuries:

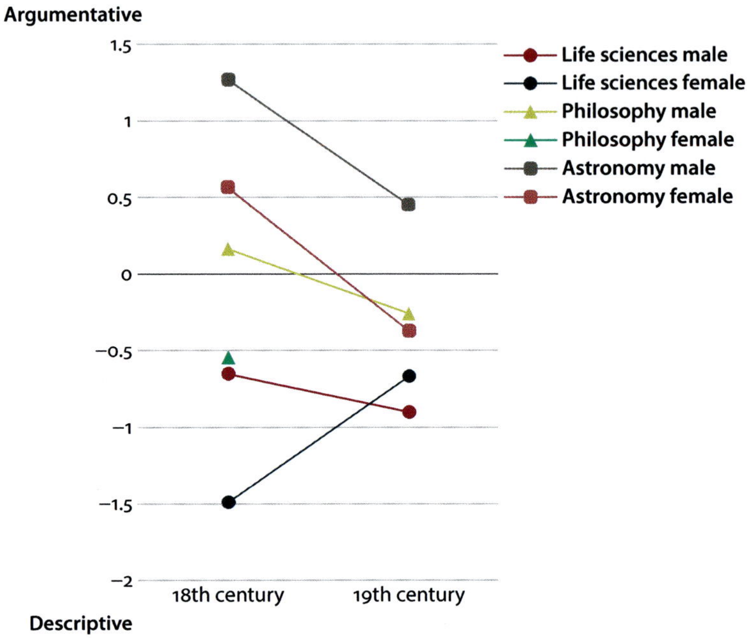

Figure 3. Distribution of male and female authors in three scientific disciplines across two centuries along Dimension 2 "Argumentative vs. descriptive focus"

All three subcorpora contain texts written by women scientists, except nineteenth-century Philosophy. As shown on Figure 3, texts by male authors present higher scores with regard to those written by women in the eighteenth century, which entails a stronger presence of argumentative style, with Astronomy and Philosophy, as well as the male part of Philosophy and Life Sciences, showing a trend towards a more descriptive style in the nineteenth century. The general drift of Astronomy and Philosophy away from argumentation, which could also be appreciated on Figure 1 (see Section 4.1), appears to mirror the progressive abandonment of a rhetoric which had been common before the scientific method had started to gradually replace logical reasoning in scientific discourse (Puente-Castelo 2017). The female subregister in CELiST, however, appears to take the contrary direction and, instead, gains argumentative style in the 1800s, to the point that their mean dimension score is slightly higher than the one for Life Sciences written by male authors.

Examples (19) and (20), belonging to male and female eighteenth-century CELiST, respectively, illustrate the differences between these two subregisters, while Example (21) offers a glimpse at the case with Life Sciences written by women in the nineteenth century:

(19) They have about five or six young ones at a time, which remain sporting in and out of this false Belly, till they are able to fend for themselves. They have long Tails without Hair, like a Rat, but as thick as a Man's Thumb; and appear as if they were Scaly, which they will readily twine about your Finger or Cane, in which Posture you may carry them where you please. (CELiST, Brickell 1737)

(20) The Leaves are restringent and vulnerary, good to stop all kind of Fluxes & preternatural Evacuations; to dissolve coagulated Blood, to help those who are bruised by Falls, outwardly it is used as a Cosmetic to take off Freckles, Sun-burn and Morphew and is good in restringent Gargarisms. (CELiST, Blackwell 1737)

(21) One kind produces the hard protuberances on trees of different kinds, which are called gall-nuts, or nut-galls; others which are softer and more spongy are called apple-galls or berry-galls. Another kind of insect, called cochineal, attaches itself to the bark of trees, and preys upon the juices. (CELiST, Lincoln 1832)

Although all three examples above contain descriptions of either animals or plants, they do present some differences. While Example (19), written by a male eighteenth-century zoologist, contains a number of adjectives (*false belly, long Tail, thick, Scaly*), this description of a possum is not limited to the little animal's physical appearance, but goes beyond that to focus on its reproductive habits and the movements of their tails. Example (20), in turn, belongs to an eighteenth-century female botanist, who focuses on the characterisation of a plant and its medicinal properties through a more frequent use of adjectives (*restringent and vulnerary, preternatural evacuations, coagulated blood, good, restringent Gargarisms*).

Example (21), conversely, written by a female scientist almost a hundred years later, describes, just like Example (19), an animal – in this case, an insect– and compares its action on the different trees (which can be either *hard protuberances* of *different kinds*, as well as *softer* or more *spongy*) with those caused by another insect, this time through verbs, rather than adjectives. Still, this general shift towards a less descriptive style in nineteenth-century women scientists seems somewhat difficult to describe at this stage, considering that most female authors in CELiST, both in the 1700s and the 1800s, are botanists, rather than zoologists – something which, perhaps, might be owing to the social restrictions of the time, according to which a woman was supposed to stay in her garden, rather than explore the forests and fields (see Moskowich 2021)– and, thus, the higher or lower relative frequency of descriptive features may be due to other factors which should be further investigated.

5. Concluding remarks

In this chapter we have analysed CELiST, both externally (i.e. with regard to other registers in the CC) and internally (i.e. by looking at some of the different subregisters represented in it), with respect to a dimension of variation that measures descriptive and argumentative style, a result of a previous study (Monaco 2017, after Biber 1988). The analysis of CELiST with regard to CETA and CEPhiT characterised Life Sciences as a fundamentally descriptive discipline both in the eighteenth and nineteenth centuries, as opposed to Astronomy, which stood out as a densely argumentative scientific register, especially in the eighteenth century. Philosophy, in turn, presented a more balanced distribution of argumentative and descriptive linguistic features, but showed, along with Astronomy, a shift away from argumentation in the 1800s, which seems to parallel the progressive abandonment of logical reasoning in Western science.

In order to find register variation within CELiST, two different variables have been used: genre and sex of the author. The former revealed that, despite the general characterisation of Late Modern English Life Sciences as a descriptive discipline, each subregister occupies its own position along the dimension, some of which stand much closer to the mean than others, whereas the subregister represented by the genre Essay appears isolated on the argumentative side of the scale. This seems to suggest that certain genres not only give shape to a text, but also determine its communicative style as shown through its pervasive linguistic features. The sex-of-the-author variable, in turn, uncovered other interesting patterns, both in the three subcorpora, and within CELiST in particular as compared to CETA and CEPhiT. First of all, it has been found that, in the 1700s, texts written by male authors present higher argumentative scores than those written by women scientists in all three scientific registers. Furthermore, it has also been shown that Life Sciences female authors do not follow the general drift towards a more descriptive style in the nineteenth century, shifting instead in the opposite direction and losing descriptive features. Although the examples we have seen suggest differences in subject matter as one of the possible reasons for such a shift, a further study is needed to investigate this tendency, as well as the differences found between male- and female-authored texts in all three scientific disciplines.

Works cited

Alonso-Almeida, Francisco and Lareo, Inés. 2016. "The status of seem in the nineteenth-century Corpus of English Philosophy Texts (CEPhiT)". In Moskowich, Isabel; Camiña, Gonzalo; Lareo, Inés and Crespo, Begoña (eds.), *'The Conditioned and the Unconditioned': Late Modern English texts on Philosophy*. Amsterdam: John Benjamins. 145–166. https://doi.org/10.1075/z.198.08alo

Atkinson, Dwight. 1999. *Scientific discourse in sociohistorical context: The Philosophical Transactions of the Royal Society of London, 1675–1975*. Mahwah, NJ: Erlbaum.

Bello, Iria. 2014. On how the 'Motion of the Stars' changed the language of science: A corpus-based study of deverbal nominalizations in Astronomy texts from 1700 to 1900. Unpublished PhD dissertation. Universidade da Coruña.

Bhatia, Vijay. 2002. Applied genre analysis: A multi-perspective model. *Ibérica*, 4: 3–19.

Biber, Douglas. 1988. *Variation across speech and writing*. Cambridge: Cambridge University Press. https://doi.org/10.1017/CBO9780511621024

Biber, Douglas. 2016. "Using multi-dimensional analysis to explore cross-linguistic universals of register variation". In Lefer, Marie-Aude and Vogeleer, Svetlana (eds.), *Genre- and register-related features in contrast*. Amsterdam: John Benjamins. 7–14. https://doi.org/10.1075/bct.87.02bib

Biber, Douglas and Conrad, Susan. 2009. *Register, genre, and style*. Cambridge: Cambridge University Press. https://doi.org/10.1017/CBO9780511814358

Biber, Douglas and Finegan, Edward. 1989. Drift and the evolution of English style: A history of three genres. *Language*, 65/3: 487–517. https://doi.org/10.2307/415220

Biber, Douglas and Finegan, Edward. 2001. "Diachronic relations among speech-based and written registers in English". In Conrad, Susan and Biber, Douglas (eds.), *Variation in English: Multi-Dimensional Studies*. Harlow: Longman. 66–83.

Biber, Douglas and Bethany Gray. 2013. Being specific about historical change: The influence of sub-register. *Journal of English Linguistics*, 41:104–134. https://doi.org/10.1177/0075424212472509

Biber, Douglas; Finegan, Edward and Atkinson, Dwight. 1994. "ARCHER and its challenges: Compiling and exploring A Representative Corpus of Historical English Registers". In Fries, Udo; Schneider, Peter and Tottie, Gunnel. (eds.), *Creating and using English language corpora. Papers from the 14th International Conference on English Language Research on Computerized Corpora, Zurich 1993*. Amsterdam: Rodopi. 1–13.

Camiña-Rioboo, Gonzalo. 2012. "Accounting for observations in the heavens in the 18th-century: New nouns to explain old phenomena". In Moskowich, Isabel and Crespo, Begoña (eds.), *Astronomy 'playne and simple': The writing of science between 1700 and 1900*. Amsterdam: John Benjamins. 93–121. https://doi.org/10.1075/z.173.06cam

Camiña-Rioboo, Gonzalo. 2013. Noun formation in the scientific register of Late Modern English: A corpus-based approach. Unpublished PhD dissertation. Universidade da Coruña.

Camiña, Gonzalo and Lareo, Inés. 2016. "Editorial policy in the Corpus of English Philosophy Texts: Criteria, conventions, encoding and other marks". In Moskowich, Isabel; Camiña, Gonzalo; Lareo, Inés and Crespo, Begoña (eds.), *'The Conditioned and the Unconditioned': Late Modern English texts on Philosophy*. Amsterdam: John Benjamins. 45–59. https://doi.org/10.1075/z.198.03cam

Conrad, Susan. 2001. "Variation among disciplinary texts: A comparison of textbooks and journal articles in biology and history". In Conrad, Susan and Biber, Douglas (eds.), *Variation in English: Multi-Dimensional Studies*. Harlow: Longman. 94–107.

Crespo, Begoña. 2011. Persuasion markers and ideology in eighteenth century philosophy texts. *Revista de Lenguas para Fines Específicos*, 17: 199–228.

Crespo, Begoña. 2012. "Astronomy as scientific knowledge in Modern England". In Moskowich, Isabel and Crespo, Begoña (eds.), *Astronomy 'playne and simple': The writing of science between 1700 and 1900*. Amsterdam: John Benjamins. 15–34. https://doi.org/10.1075/z.173.02cre

Crespo, Begoña. 2016. "Genre categorisation in CEPhiT". In Moskowich, Isabel; Camiña, Gonzalo; Lareo, Inés and Crespo, Begoña (eds.), *'The Conditioned and the Unconditioned': Late Modern English texts on Philosophy*. Amsterdam: John Benjamins. 25–43. https://doi.org/10.1075/z.198.02cre

Davies, Mark. 2007. *TIME Magazine Corpus: 100 million words, 1920s-2000s*. Retrieved January 21, 2020, from http://corpus.buy.edu/time/

Dossena, Marina. 2016. "On the shoulders of giants: An overview of the discussion of science and philosophy in Late Modern times". In Moskowich, Isabel; Camiña, Gonzalo; Lareo, Inés and Crespo, Begoña (eds.), *'The Conditioned and the Unconditioned': Late Modern English texts on Philosophy*. Amsterdam: John Benjamins. 77–98. https://doi.org/10.1075/z.198.05dos

Ferguson, Charles. 1983. Sports announcer talk: Syntactic aspects of register variation. *Language in Society*, 12/2: 153–172. https://doi.org/10.1017/S0047404500009787

Gray, Bethany. 2011. Exploring academic writing through corpus linguistics: When discipline tells only part of the story. Unpublished PhD dissertation. Northern Arizona University.

Gray, Bethany and Biber, Douglas. 2012. "The emergence and evolution of the pattern N + PREP + V-*ing* in historical scientific texts". In Moskowich, Isabel and Crespo, Begoña (eds.), *Astronomy 'playne and simple': The writing of science between 1700 and 1900*. Amsterdam: John Benjamins. 181–198. https://doi.org/10.1075/z.173.09gray

Grieve, Jack. 2016. *Regional variation in written American English*. Cambridge: Cambridge University Press. https://doi.org/10.1017/CBO9781139506137

Hardie, Andrew. 2012. CQPweb–combining power, flexibility and usability in a corpus analysis tool. *International Journal of Corpus Linguistics*, 17/3: 380–409. https://doi.org/10.1075/ijcl.17.3.04har

Hardie, Andrew. 2016. "Infrastructure for analysis of the CEPhiT corpus: Implementation and applications of corpus annotation and indexing". In Moskowich, Isabel; Camiña, Gonzalo; Lareo, Inés and Crespo, Begoña (eds.), *'The Conditioned and the Unconditioned': Late Modern English texts on Philosophy*. Amsterdam: John Benjamins. 61–75. https://doi.org/10.1075/z.198.04har

Hymes, Dell. 1962. "The ethnography of speaking". In Gladwin, Thomas and Sturtevant, William (eds.), *Anthropology and human behavior*. Washington, DC: Anthropology Society of Washington. 13–53.

Hymes, Dell. 1984. Sociolinguistics: Stability and consolidation. *International Journal of the Sociology of Language*, 45: 39–45. https://doi.org/10.1515/ijsl.1984.45.39

Labov, William. 1963. The social motivation of a sound change. *Word*, 19: 273–303. https://doi.org/10.1080/00437956.1963.11659799

Labov, William. 1965. On the mechanism of linguistic change. *Georgetown Monographs on Language and Linguistics*, 18: 91–114.

Lareo, Inés. 2011. Colocacións con *make, take* e *do* + nome nun corpus do século XIX de textos ingleses científicos e literarios escritos por mulleres. *Cadernos de fraseoloxía galega*, 13: 191–214.
Lareo, Inés. 2012. "A corpus-driven approach to explore the use of complex predicates in 18th century English scientific writings". In Moskowich, Isabel and Crespo, Begoña (eds.), *Astronomy 'playne and simple': The writing of science between 1700 and 1900*. Amsterdam: John Benjamins. 155–180. https://doi.org/10.1075/z.173.08lar
Lareo, Inés; Monaco, Leida Maria; Esteve-Ramos, María José and Moskowich, Isabel (comps.). 2020. *The Corpus of English Life Sciences Texts (CELiST)*. A Coruña: Universidade da Coruña. https://doi.org/10.17979/spudc.9788497497848
Lareo, Inés and Esteve-Ramos, María José. 2008. 18th-century scientific writing: A study of *make* complex predicates in the Coruña Corpus. *ICAME*, 32: 69–96.
Lee, David. 1999. Modelling variation in spoken and written language: The multi-dimensional approach revisited. Unpublished PhD dissertation. Lancaster University.
Monaco, Leida Maria. 2017. A multidimensional analysis of Late Modern English scientific texts from the Coruña Corpus. Unpublished PhD dissertation. Universidade da Coruña.
Monaco, Leida Maria. 2019. "Exploring the narrative dimension in Late Modern English History texts". In Moskowich, Isabel; Crespo, Begoña; Puente-Castelo, Luis and Monaco, Leida Maria (eds.), *Writing history in Late Modern English: Explorations of the Coruña Corpus*. Amsterdam: John Benjamins. 84–101. 10.175/z.225. https://doi.org/10.1075/z.225.05mon
Moskowich, Isabel. 2021. "The making of CELiST, a bunch of disciplines". In Moskowich, Isabel; Lareo, Inés and Camiña, Gonzalo (eds.), *"All families and genera": Exploring the Corpus of English Life Sciences Texts*. Amsterdam/Philadelphia: John Benjamins. 1–19. https://doi.org/10.1075/z.237
Moskowich, Isabel. 2016. "Philosophers and scientists from the Modern Age: Compiling the Corpus of English Philosophy Texts". In Moskowich, Isabel; Camiña, Gonzalo; Lareo, Inés and Crespo, Begoña (eds.), *'The Conditioned and the Unconditioned': Late Modern English texts on Philosophy*. Amsterdam: John Benjamins. 1–24. https://doi.org/10.1075/z.198.01mos
Moskowich, Isabel and Crespo, Begoña. 2007. "Presenting the Coruña Corpus: A collection of samples for the historical study of English scientific writing". In Pérez-Guerra, Javier et al. (eds.), *Of varying language and opposing creed: New insights into Late Modern English*. Bern: Peter Lang. 341–357.
Moskowich, Isabel and Crespo, Begoña (eds.). 2012. *Astronomy 'playne and simple': The writing of science between 1700 and 1900*. Amsterdam: John Benjamins. https://doi.org/10.1075/z.173
Moskowich, Isabel and Monaco, Leida Maria. 2014. "Abstraction as a means of expressing reality: Women writing science in Late Modern English". In Gotti, Maurizio and Giannoni, Davide S. (eds.), *Corpus analysis for descriptive and pedagogical purposes*. Bern: Peter Lang. 203–224.
Moskowich, Isabel and Monaco, Leida Maria. 2016. Linking ideas in women's writing: Evidence from the Coruña Corpus. *Revista de Lingüística y Lenguas Aplicadas*, 11: 35–49. https://doi.org/10.4995/rlyla.2016.4660
Moskowich, Isabel and Parapar, Javier. 2008. "Writing science, compiling science: The Coruña Corpus of English Scientific Writing". In Lorenzo-Modia, María Jesús (ed.), *Proceedings from the 31st AEDEAN Conference*. A Coruña: Universidade da Coruña. 531–544.
Moskowich, Isabel; Camiña, Gonzalo; Lareo, Inés and Crespo, Begoña (eds.). 2016. *'The Conditioned and the Unconditioned': Late Modern English texts on Philosophy*. Amsterdam: John Benjamins. https://doi.org/10.1075/z.198

Nevalainen, Terttu and Raumolin-Brunberg, Helena. 1994. Sociolinguistics and language history: The Helsinki Corpus of Early English Correspondence. *Hermes, Journal of Linguistics*, 12: 135–143.

Nevalainen, Terttu and Raumolin-Brunberg, Helena. 1995. "Constraints in politeness: The pragmatics of address formulae in Early English correspondence". In Jucker, Andreas H. (ed.), *Historical pragmatics: Pragmatic developments in the history of English*. Amsterdam: John Benjamins. 541–601. https://doi.org/10.1075/pbns.35.28nev

Núñez-Pertejo, Paloma. 2013. From degree adverb to response token: *Absolutely* in Late Modern and contemporary British and American English. *Neuphilologische Mitteilungen*, 114: 245–273.

Pahta, Päivi. 2003. On structures of code-switching in medical texts from medieval England. *Neuphilologische Mitteilungen*, 104: 197–210.

Pahta, Päivi. 2004. "Code-switching in medieval medical writing". In Taavitsainen, Irma and Pahta, Päivi (eds.), *Medical and scientific writing in late Medieval English*. Cambridge: Cambridge University Press. 528–537.

Pérez-Guerra, Javier. 2012. "Discourse status and syntax in the history of English: Some explorations in topicalization, left-dislocation and *there*-constructions". In Meurman-Solin, Anneli; López-Couso, María José and Los, Bettelou (eds.), *Information structure and syntactic change in the history of English*. New York: Oxford University Press. 121–138. https://doi.org/10.1093/acprof:oso/9780199860210.003.0006

Puente-Castelo, Luis. 2016. "Explaining the use of *if… then…* structures in CEPhiT". In Moskowich, Isabel; Camiña, Gonzalo; Lareo, Inés and Crespo, Begoña (eds.), *'The Conditioned and the Unconditioned': Late Modern English texts on Philosophy*. Amsterdam: John Benjamins. 167–180. https://doi.org/10.1075/z.198.09pue

Puente-Castelo, Luis. 2017. On conditionality: A corpus-based study of conditional structures in Late Modern English scientific texts. Unpublished PhD dissertation. Universidade da Coruña.

Puente-Castelo, Luis. 2019. "'Were this estimation, however, to be depended on': Inversion conditionals as evidence of paradigmatic change in CHET". In Moskowich, Isabel; Crespo, Begoña; Puente-Castelo, Luis and Monaco, Leida Maria (eds.), *Writing history in Late Modern English: Explorations of the Coruña Corpus*. Amsterdam: John Benjamins. 130–147. https://doi.org/10.1075/z.225.07pue

Puente-Castelo, Luis. 2021. ""*If you will take the trouble to inquire into it rather closely, I think you will find that it is not worth very much*": Authorial presence through conditionals and citation sequences in Late Modern English life sciences texts". In Moskowich, Isabel; Lareo, Inés and Camiña, Gonzalo (eds.), *"All families and genera": Exploring the Corpus of English Life Sciences Texts*. Amsterdam/Philadelphia: John Benjamins. 189–208. https://doi.org/10.1075/z.237.10pue

Rissanen, Matti. 2011. "On the long history of English adverbial subordinators". In Meurman-Solin, Anneli and Lenker, Ursula (eds.), *Connectives in synchrony and diachrony in European languages*. Helsinki: VARIENG. /http://www.helsinki.fi/varieng/series/volumes/08/rissanen/

Rodríguez-Puente, Paula. 2013. "A turn of the screw in the semantics of phrasal verb: Phrasal verbs with *up* as a test case". In Hegedüs, Iren and Pödör, Dóra (eds.), *Periphrasis, replacement and renewal: Studies in English historical linguistics*. Newcastle-upon-Tyne: Cambridge Scholars Publishing. 243–265.

Rodríguez-Puente, Paula. 2016. Particle placement in Late Modern English and twentieth-century English: Morpho-syntactic variables. *Folia Linguistica Historica*, 37: 145–175. https://doi.org/10.1515/flih-2016-0005

Swales, John. 1990. *Genre analysis: English in academic and research settings*. Cambridge: Cambridge University Press.

Swales, John. 2004. *Research genres: Exploration and applications*. Cambridge: Cambridge University Press. https://doi.org/10.1017/CBO9781139524827

Taavitsainen, Irma and Pahta, Päivi. 2010. *Early Modern English Medical Texts: Corpus description and studies*. Amsterdam: John Benjamins. https://doi.org/10.1075/z.160

Taavitsainen, Irma; Pahta, Päivi; Hiltunen, Turo; Mäkinen, Martti; Marttila, Ville; Ratia, Maula; Suhr, Carla and Tyrkkö, Jukka (eds.). 2010. *Early Modern English Medical Texts*. CD-ROM. Amsterdam: John Benjamins. https://doi.org/10.1075/z.160

Weinreich, Uriel, William Labov and Marvin Herzog. 1968. Empirical foundations for a theory of language change. In Lehmann, Winfred and Malkiel, Yakov (eds.), *Directions for historical linguistics*. Austin: University of Texas Press. 95–195.

Zea, Sofía. 2015. A study on the distribution of the adjective in English scientific texts from the eighteenth and nineteenth centuries. Unpublished PhD dissertation. Universidade da Coruña.

Index